A TALE OF TWO GLOBAL CITIES

A Tale of Two Global Cities

Comparing the Territorialities of
Telecommunications Developments in Paris and London

JONATHAN RUTHERFORD
University of Newcastle, UK

ASHGATE

© Jonathan Rutherford 2004

Jonathan Rutherford has asserted his right under the Copyright, Designs and Patents Act, 1988, to be identified as the author of this work.

Published by
Ashgate Publishing Limited
Gower House
Croft Road
Aldershot
Hants GU11 3HR
England

Ashgate Publishing Company
Suite 420
101 Cherry Street
Burlington, VT 05401-4405
USA

Ashgate website: http://www.ashgate.com

British Library Cataloguing in Publication Data
Rutherford, Jonathan
 A tale of two global cities : comparing the
 territorialities of telecommunications developments in
 Paris and London
 1. Telecommunication - England - London 2. Telecommunication
 - France - Paris 3. Telecommunication policy
 I. Title
 384'.09421

Library of Congress Cataloging-in-Publication Data
Rutherford, Jonathan.
 A tale of two global cities : comparing the territorialities of telecommunications
 developments in Paris and London / Jonathan Rutherford.
  p. cm.
 Includes bibliographical references and index.
 ISBN 0-7546-3474-4
  1. Telecommunication--France--Paris. 2. Telecommunication--England--London. 3.
 Infrastructure (Economics)--France--Paris. 4. Infrastructure
 (Economics)--England--London. I. Title.

 HE8149.P37R88 2004
 384'.09421--dc22

2003060970

ISBN 0 7546 3474 4

Printed and bound in Great Britain by MPG Books Ltd, Bodmin, Cornwall

Contents

PART III: ANALYSIS AND CONCLUSIONS

List of Tables

Preface

The present study is the result of several years of research into the relationships between the development of telecommunications infrastructures and services in cities and a wide variety of territorial theories. The vast majority of the book is based on work carried out for my PhD degree between 1998 and 2002.

As such, the case study research at the heart of the study comparing telecommunications developments in Paris and London was mostly carried out during the period 2000-2001. This was perhaps the height of the 'boom' in the telecommunications and 'dot.com' sectors, when enormous capital was being invested by dozens of companies in building dense backbone infrastructure networks within and between the major cities of Europe. The analysis and comparison in this book, therefore, of strategies and developments in Paris and London by public and private actors is set within this global context of vast investment in infrastructure roll-out, which proved to be short-lived.

Indeed, in the period intervening between the carrying out and writing up of the research, and the completion of this study, a downturn in the telecommunications sector has substantially intensified on a global scale. Investment in advanced infrastructure deployment, even in the largest cities, has almost come to a standstill. Many companies have gone out of business or been taken over as a result of over-investment in the construction of networks within and between cities where infrastructure was already abundant. Others are in financial and / or legal difficulty for differing reasons. The case of WorldCom in 2002 was the most publicised of these, being largely the result of fraud and severe financial mismanagement. The brief analysis of their strategy in chapter 7 has, nevertheless, been left in the study, largely because WorldCom (or MCI as it has been renamed) was, and remains, one of the main competitive operators in both Paris and London, and it is still interesting to compare territorial aspects of their strategy as an American entrant to those of Colt as a key 'indigenous' European competitive operator.

As a result of these market shifts following the carrying out of much of the research, this book already has a certain historical element to it, and must be read bearing these shifts in mind. However, the main object of the study was never to provide a perfectly up-to-date account and analysis of the Paris and London telecommunications markets, but more to explore and compare the varying territorial contexts, specificities and implications of some of the telecommunications strategies and developments which have taken place in the two cities in recent years. As such, regardless of the above evolutions, the book still addresses its primary concern of highlighting the importance of territorial perspectives on telecommunications developments in global cities.

Many people have greatly assisted me during the research and the writing of this book. First and foremost, I would like to thank my supervisor, Stephen

Graham, for his help, support, encouragement and detailed comments from start to finish. It helps to be supervised by one of the leading experts in the field, but his confidence, enthusiasm and friendly criticism were equally vital in pushing me towards the finishing line. And all this, despite crucial footballing rivalries!

Many thanks also to Simon Marvin, now at Salford University, for supervisory assistance and encouragement during the first half of the PhD, which helped me to develop the study into its present form.

Thank you as well to other colleagues in the Centre for Urban Technology at Newcastle for a uniquely stimulating and friendly environment in which to work and have lunch. Special thanks go to Elizabeth Storey, without whose help and support I would have been at a loss, and to my office colleagues Tracey Crosbie and Graham Soult for their friendship, advice and excellent coffee-making skills. Latterly, the work has benefited from the perceptive comments of, and discussions with, Andy Gillespie, Ranald Richardson and Alan Southern. Thanks also to Jon Beaverstock and Richard Hanley for their suggestions.

An ESRC grant was invaluable in providing funding support. In addition, this research would have been made much more difficult without the assistance of numerous contacts in various organisations and companies in Paris and London. Thank you to all those who gave freely of their time for meetings, provided documents and information, and suggested further people to contact in each case. Special thanks go to Olivier Coutard and Marie-Claire Vinchon at LATTS in Paris and Mark Hepworth at the Local Futures Group in London who provided initial suggestions and names of relevant people and organisations.

Finally, absolutely nothing would have been possible without the support of my family and of Christelle Laumonier. They may not always have understood what I was doing or why I was doing it, but they deserve loving thanks for their patience and encouragement.

Institutional Glossary

AFOPT	Association Française des Opérateurs Privés en Télécommunications (the French association of private telecommunications operators)
ART	Autorité de Régulation des Télécommunications (the French telecommunications regulatory authority)
ARTESI	Agence Régionale des Technologies de l'Information et de l'Internet (the regional agency for information technologies and the Internet). Formerly known as both the Association Téléport Paris – Ile-de-France, and the Agence Régionale pour le développement du Téléport et de la Société de l'Information en Ile-de-France
AVICAM	Association des Villes pour le Câble et le Multimedia (the association of cities for cable and multimedia)
BT	British Telecom
CCIP	Chambre de Commerce et d'Industrie de Paris (the Paris chamber of commerce and industry)
CDC	Caisse des Dépôts et Consignations (the French state bank)
CRIF	Conseil Régional d'Ile-de-France (the regional council of Ile-de-France)
CROCIS	Centre régional d'observation du commerce et de l'industrie et des services (the regional observatory centre of commerce, industry and services)
DATAR	Délégation à l'Aménagement du Territoire et à l'Action Régionale (the government agency for territorial planning and regional action)
DEAL	Docklands East London – the development agency created after the winding up of the LDDC
DETR	Department of the Environment, Transport and the Regions
DGT	Direction Générale des Télécommunications (the general telecommunications directorate)
DiGITIP	Direction Générale de l'Industrie, des Technologies de l'Information et des Postes (the general industry, information technologies and postal service directorate)
DREIF	Direction Régionale d'Equipement en Ile-de-France (the regional infrastructure directorate of Ile-de-France)
DTI	Department of Trade and Industry
EPAD	Établissement Public d'Aménagement de la Défense (the public planning agency of La Défense)

FEDIA	Forum d'Études et de Développement des Infrastructures Alternatives (the forum for studies and the development of alternative infrastructures)
FOD	Fibres Optiques Défense
GLA	Greater London Authority
GLC	Greater London Council
GoL	Government Office for London
GPO	General Post Office
IAURIF	Institut d'Aménagement et d'Urbanisme de la Région Ile-de-France (the institute of planning and urbanism of the Ile-de-France region)
IAURP	Institut d'Aménagement et d'Urbanisme de la Région Parisienne (the institute of planning and urbanism of the Paris region, which preceded IAURIF)
LDA	London Development Agency
LDDC	London Docklands Development Corporation
LFC	London First Centre
ODEP	Observatoire du Développement Économique Parisien (the Paris economic development observatory)
Oftel	Office of Telecommunications
OTV	Observatoire des Télécommunications dans la Ville (the observatory of telecommunications in cities)
RATP	Régie Autonome des Transports Parisiens (the Paris city transport authority)
RDA	Regional Development Agency
SEM	Société d'Économie Mixte (a mixed public-private agency in economic development)
SIPPEREC	Syndicat Intercommunal de la Périphérie de Paris pour l'Électricité et les Réseaux de Communication (the intercommunal syndicate of the Paris periphery for electricity and communication networks)
SLIM	Society of London Information Technology Managers
SNCF	Société Nationale des Chemins de Fer (the French public railway organisation)

Technical Glossary

ADSL	Asynchronous Digital Subscriber Line – a means of offering fast Internet access by signal compression, through the use of existing copper analogue networks in the local loop, and thereby necessitating a modem for the user and a modem at the local exchange. A line can be simultaneously used for voice and data communications.
Aménagement du territoire	This particularly French notion and practice has broad connotations of an explicit form of equitable territorial planning and management, which has traditionally most concerned attempts at balancing Paris–provincial and urban–rural development.
ATM	Asynchronous Transfer Mode – the sending of identical 'packets' of data through a line for quick re-assembly at the destination, rather than the traditional method of dispatching differently-sized sets of data which then take longer to assemble after transmission.
Bandwidth	The quantity of spectrum needed for a particular purpose.
Broadband	Networks and/or services capable of handling large volumes of data.
CLEC	Competitive Local Exchange Carrier – a competitive, or new entrant, telecommunications operator, as opposed to an incumbent.
Coaxial cable	Cable used for the transmission of wideband information and data, although its capacity is less than fibre optic cable.
Dark fibre	Unactivated fibre optic cable.
EDI	Electronic Data Interchange – the industry term for direct computer to computer information exchange.
Fibre optic networks	Networks founded on the transmission of light-based signals (rather than electric ones), which are capable of providing very high communications capacities, but which need very great local investment for deployment.
Fixed wireless access	See Wireless local loop.
Frame relay	A means of broadband data transmission, which is an intermediary technique between packet transmission and ATM cells.
Gbps	Gigabits per second, roughly equating to 10^9 bits per second.

Gigabit Ethernet	A very high speed transmission service, which offers interconnection of backbone LANs.
Incumbent	The former monopoly operator in a particular country, eg BT in the UK or France Télécom in France.
IP	Internet Protocol – the definition of transmissions of data packets between network systems.
ISDN	Integrated Services Digital Network – one of the first attempts at offering simultaneous digital voice and data transmission through the conversion of a traditional analogue telephone line.
Kbps	Kilobits per second, roughly equating to 10^3 bits per second.
LAN	Local Area Network – a network focused on exchange and transmission within a small area, such as a building or cluster of buildings.
Leased line	A line dedicated solely to a private user. These are used particularly by large businesses and financial institutions for transmission of high volumes of data.
LLU	Local Loop Unbundling – the process which allows competitive operators to lease lines in the 'last mile' of the incumbent's network, and to have access to the latter's local exchanges for the deployment of their own equipment. The local loop is the last vestige of the PTT network monopoly era, and thus, its unbundling will theoretically create a fully open market in telecommunications networks.
Local loop	Alternatively the 'last mile' or 'last kilometre', or the copper pair, which is the part of the network where the operator directly provides a line from the local exchange to the final client. Until the process of unbundling, this was completely under the control of the incumbent operator.
MAN	Metropolitan Area Network.
Mbps	Megabits per second, roughly equating to 10^6 bits per second.
Multiplexing	The composite assembly of several signals from tributary routes for transmission on the same route at a higher speed.
MVNO	Mobile Virtual Network Operator.
PABX	Private Automated Branch Exchange – the equipment used to route calls over a private network, and to and from public networks.
Packet switching	A transmission method relying on the division of the data or information into specific sized packets.

PAGSI	Programme d'Action Gouvernementale pour la Société de l'Information (the French government programme for the information society).
POTS	Plain Old Telephone Service – the traditional basic voice telephony service.
PSTN	Public Switched Telephone Network.
PTO	Public Telecommunications Operator.
PTT	Postal Telegraph and Telephone (authorities), or (Ministère de) Postes, Télégraphes et Téléphones.
Renater	The French national infrastructure network connecting hundreds of research, technology and higher education institutes and sites.
RER	Réseau Express Régional (the regional public train system in Ile-de-France).
SDH	Synchronous Digital Hierarchy – a digital transmission channel, based on synchronous multiplexing, which can maintain high data speeds and can carry ATM and frame relay.
Shorthaul	Point to point fibre links over a short distance.
3G	Third generation mobile networks.
Transpac	A subsidiary of France Télécom providing the physical infrastructure for the data transmission networks of the latter.
UMTS	Universal Mobile Telecommunication Standard – the industry basis for third generation mobile networks.
VANS	Value Added Network Services – services which supplement that of basic telephony, such as call handling, voice mail, data services, itemised billing, and specialised corporate services like virtual private networks.
WAN	Wide Area Network – a network focused on exchange and transmission between locations some distance apart.
Wireless local loop	Recent broadband communications technology which uses very high frequency Hertzian wavelengths to provide fast Internet access through radio links instead of through a telephone line.

Chapter 1

Introduction

It is not a question of denying existing powers and types of legitimate areal territoriality, but of recognising the existence of other powers and of reticulate territorialities. For that, we have to equip ourselves with the means of thinking about and analysing networks in and for a new urbanism. We need to promote the tools which allow efficient action on what has become urban.

To restore the great reticulate utopias; to bring reticulate urban thinking out of its marginality; to better base theoretically the territorial notion of network; to favour in modern urban debate a contradictory discussion on territoriality; to present, explain and develop the tools allowing the 'retistic' to be taken into account in the 'urbanistic': an ambitious, but necessary, programme... (Dupuy, 1991, p.13, author's translation).

Large urban projects have, at the end of the day, less impact than the networks which irrigate the territorial city. It is in the construction of bypasses and motorways, and in the procedures of cabling and interconnection that the urban future is played out, more than in the organisation of urban sequences and buildings, however spectacular they may be (Picon, 1998, p.78, author's translation).

Context

In the last ten or fifteen years, there has been increasing interest in such disciplines as geography, planning, sociology, and urban studies in the relationships between cities and information technologies (IT) and telecommunications (see Dutton, Blumler and Kraemer, 1987; Moss, 1987; Castells, 1989; Batty, 1990; Hepworth, 1990; Brotchie et al, 1991; Gillespie, 1992; Bakis, Abler and Roche, 1993; Graham, 1994; Ascher, 1995; Mitchell, 1995; Musso and Rallet, 1995; Castells, 1996; Graham and Marvin, 1996; Wheeler, Aoyama and Warf, 2000). This is in recognition of the increasingly important part these information and communications technologies (ICTs) are playing in the social, economic, political and cultural processes and transformations which are occurring in and between urban places across the globe, underlying the well-documented 'rise of the network society' (Castells, 1996).

Although society, and its urban concentrations in particular, has always been organised in, and bound up with, technological networks in their broadest sense, it is posited that never before have we seen such a presence of, and demands placed upon, nodes, hubs, and flows as in the contemporary world, and one of the reasons held out for this is a proliferating development and use of telecommunications and IT. As Schiller observes:

As business users' dependence on network systems grew more concerted, more multifaceted, and more extensive, an unparalleled telecommunications boom was triggered. Capital investment surged forward; the number of worldwide installed main telephone lines has grown *eightfold* since 1960, and increased by nearly 60 percent just between 1990 and 1997 – from 520 to 800 million. Cellular phone systems added hundreds of millions of additional units to the worldwide base (Schiller, 1999, p.37).

This telecommunications boom is bound up with a dual regulatory transformation, which has seen, on the one hand, the liberalisation of many local markets, and on the other hand, a substantial move by the largest telecommunications companies around the world into each other's national markets. For some, these transformations illustrate nothing less than the diminishing geographical organisation of the telecommunications industry – indeed, 'the end of territoriality in communications' (Noam and Wolfson, 1997). For others, the increasing concentrations of telecommunications infrastructures in large cities and the continuing importance of national regulatory practices, amongst other processes, problematise or dispel such theories. For Guy, Graham and Marvin, thus, the emerging parallel logics of globalisation and localisation in utility provision:

> undermine old assumptions that the logic of supply of infrastructure networks is to fill territories with standardised, expansionary, homogeneous services; that these can then be largely taken for granted; and that the whole process of utility development is an unproblematic, technical and rather dull exercise (Guy, Graham and Marvin, 1997, p.197).

These processes are most evident in relation to the apparently increasingly globalised economy. The financial sector and the key urban financial centres across the world now rely firmly on telecommunications technologies, infrastructures and services, as their principal means of information, exchange and transaction (Sassen, 1991; 2000). In increasingly competitive twenty four hour global markets, the need for instantaneous access to the relevant information and the ability to be able to act on this is absolutely crucial.

The importance of telecommunications and IT has not gone unnoticed by authorities and policy-makers at the local, urban, regional and national levels, in their search for 'innovative' responses to territorial processes and requirements (Briole and Lauraire, 1991; Briole et al, 1993; Offner, 2000b). This is a major challenge, given the traditional dominance of telecommunications operators in the development and deployment of networks, 'whose investment and pricing decisions are normally subject to national regulatory agencies but not to municipal authorities' (Hepworth, 1990, p.545). Gillespie and Williams captured the gap between public and private strategies, when they wrote that:

> there is a predictable tension between the development of public telecommunications networks, which have historically been nurtured under the auspices of regulatory regimes designed to benefit the polity as a whole, and private (corporate) networks, which are being developed to capture competitive advantage for specific players over others (Gillespie and Williams, 1988, p.1315).

Nevertheless, a veritable profusion of different types of urban strategies, initiatives and projects has developed across the world, based on telecommunications and IT. Following the steady liberalisation of telecommunications markets in the last two decades, these strategies sometimes involve gearing or shaping the relevant territorial market to increase competition amongst operators, and therefore, choice and quality of infrastructures and services for business and residential consumers. Beyond this regulatory role, however, some authorities and policy-makers have taken a more interventionist role and have actually tried to implement their own strategies and projects in the telecommunications domain, for economic and / or social development purposes. Graham and Marvin (1998) identify three types of, what they call, 'relational planning': integrated transport and telecommunications strategies, city-level new media and IT strategies, and information districts and urban televillages. Indeed, these projects are starting to be viewed by some strategic authorities and development agencies as essential components of wider economic development strategies in cities, and as playing particularly innovative roles in the strategic and competitive positioning of these cities in the global inter-urban competition environment. They illustrate how urban places and information and communications technologies can be shaped and configured in parallel, as:

> Material space and electronic space are increasingly being produced together. The power to function economically and link socially increasingly relies on constructed, place-based, material spaces intimately woven into complex telematics infrastructures linking them to other places and spaces (Graham, 1998, p.174).[1]

This strategic role in urban competitive advantage is just one illustration of how telecommunications and IT strategies in cities are always shaped by, and respond to, varying cultures, territorialities and place specificities. Yet, this key aspect of such developments is frequently either neglected, ignored, or negated by confused, simplistic and determinist (yet surprisingly influential) hyperbole, which tends to take one of two perspectives. Firstly, technological determinist views have been criticised for seeing the 'impacts' of telecommunications on urban places as being relatively straightforward and happening independently of social and political processes (see, for example, Pascal, 1987):

> In the technological determinist view, technology is conceived as a separate entity that follows a linear path. Technology is like a train, with a track that is fixed, although not

---

[1] All this builds on William Mitchell's notion of 'recombinant architecture', the idea that key urban institutions now have parallel physical and virtual characteristics, and that the latter 'is increasingly taking over' from the former (Mitchell, 1995, p.49). However, rather than suggesting this transcendence of material space, the key must surely be to emphasise the *parallel* production of urban place and electronic space, and their 'recursive interaction' (Graham, 1998, p.174), hence the important notion of 'relational planning' (Graham and Marvin, 1998; see also Graham and Healey, 1999). In this way, Mitchell's arguments about decreased need for built space and physical accessibility can be viewed as a form of utopian hyperbole.

known in detail. One cannot hope to change the train's direction, only to check its speed and the safety of the crossing (Bijker, 1993, p.129).

Secondly, utopianist–futurist views have been criticised for seeing telecommunications as solving many widespread urban problems as well as leading to anything being able to be done anywhere and at anytime. As Gillespie observes, 'in all such utopian visions, the decentralising impacts of communications technology are regarded as unproblematic and self evident' (Gillespie, 1992, p.69). Telecommunications are thus alleged to be rendering cities redundant, and making distance irrelevant (Negroponte, 1995; O'Brien, 1992).[2] These hypotheses tend to derive from theorists such as Marshall McLuhan and his 'global village' suppositions (McLuhan, 1964), Melvin Webber (Webber, 1964)[3] and Alvin Toffler with his idealistic 'third wave' society (Toffler, 1981). The information and communications technologies held to be responsible for all these assertions are themselves held to be 'new', allowing society to produce information and use communication in ways we have never seen before. As Thrift puts it:

> To expound any other point of view is to be open to the criticism of a 'retro-orientation' which denies the existence of a whole new world of 'warp-speed accelerations', 'telecommercial hypermanic cultures', 'hungrily expanding spatialities' and 'virtual human obsolescence' (Thrift, 1996b, p.1463-1464; citing Land, 1995).

These points of view on the relations between ICTs, cities and geography have been heavily criticised in some quarters as unhelpful to say the least:

> The strong leaning of contemporary technological discourse towards substitution and transcendence perspectives, I would argue, tends to perpetuate little but dangerous myth and fallacy. In proffering new technologies as some complete and simple *substitutes* for the material body, the social world, and for space and place, its proponents do little to advance understanding of the complex *co-evolutionary* processes linking new information technologies and space, place and human territoriality (Graham, 1998, p.171).

---

[2] Related to this are the alleged possibilities for telecommunications technologies to substitute for transport and the need to travel, particularly through the concept of teleworking. However, teleworking has been argued either to be an unconvincing and somewhat inflated utopian ideal, unlikely to extend beyond a handful of exponents (see Ascher, 1995, p.55-60), or to be linked possibly to increased use of transport and longer journeys, thus dispelling any 'myths of transcendence' (Gillespie and Richardson, 2000).

[3] In 1968, for example, Webber wrote about how:

> For the first time in history, it might be possible to locate on a mountain top and to maintain intimate, real-time, and realistic contact with business or other associates. All persons tapped into the global communications net would have ties approximating those used today in a given metropolitan region (Webber, 1968; quoted in Moss, 1987, p.535).

Instead, in contrast to these approaches, a socio-technical perspective allows telecommunications to be seen as being shaped within society, and not independently of it. As Mattelart argues: 'The contrast is striking between utopian discourses on promises for a better world due to technology, and the reality of struggles for control of communication devices and hegemony over norms and systems' (Mattelart, 2000, p.21).[4]

Fortunately, the earlier purely theoretical arguments are now being surpassed by actual empirical research, which is leading to the realisation that cities and territories do still matter, and that they may actually be increasing in importance in the light of the ways in which they interact with telecommunications and IT. In this way, Ascher (1995, p.63) argues that the paradox of telecommunications is that 'they enhance the value of everything which is not telecommunicable'. As Lynch suggests:

> A city's attractiveness as a telecom hub is a major determinant of whether it will be a prosperous place in the 21st century. In the same way that access to maritime and overland trading routes determined prosperity in the past, access to affordable and bountiful bandwidth services will be the marker of the future (Lynch, 2000).

Nevertheless, the complex interactions between cities and telecommunications are still not really fully understood. Graham (1996) criticised the generally descriptive and rather unsophisticated nature of the analysis of urban telecommunications strategies. There appears to be some difficulty in grasping the implications of the juxtaposition between the place-based city and space-transcending telecommunications, which could very well be a question of the scalar complexity that this juxtaposition suggests, and the ways in which these are mediated by different local, regional and national territorialities and political and policy specificities.

In terms of urban telecommunications strategies, this impinging scalar complexity is composed of both the forms and processes which necessarily influence the formulation and implementation of these strategies from above the urban scale (global, national, regional) and below it (local), and the subsequent post-implementation interactions of these strategies with practices above and below the urban scale, which in turn re-influence the forms and processes that influenced the strategies to start with. It is therefore a two-way incremental interaction with logics and policies from a combination of scales both influencing and being influenced by the telecommunications strategies.

The ways in which logics from a number of different scales seem to be bound up in urban telecommunications strategies is another illustration of how the 'urban question' is becoming a 'scale question' (Brenner, 2000). In particular, it is becoming increasingly difficult to justify the conceptualisation of geographical scales as a nested, or overlapping, hierarchy. It might instead require the notion of

---

[4] Kevin Robins goes even further in criticising nearly the entire agenda of drawing out the relations between cities and ICTs, arguing instead in favour of a disordered Byzantine city over the reconfigured rational order of the real-time city (Robins, 1999).

a 'telescoping' of scales: 'the idea of a multi-layered territory, of a superposition over a single space, of territories functioning according to different matrixes and on different levels, and more or less well interconnected' (Offner, 2000a, p.173).

This scalar complexity is tied up with much wider processes and logics of 'globalisation', and the narrower regulatory practices of deregulation and liberalisation of telecommunications markets. Some writers have associated all these forces within a restructuring shift which is concentrating political economic power both at the supranational and sub-national levels, and therefore leading to the 'hollowing out' of the traditional nation state (Jessop, 1994). On a local level too, the negotiation and implementation of policy and decision-making is suggested to have devolved from being firmly the domain of local *government* to involving many wider actors and groups in a process of *governance* (Stoker, 1995).

These shifts would suggest more of a role for urban authorities and development agencies in designing and implementing their own economic development strategies, such as telecommunications initiatives. Global or world cities, in particular, as command centres of the world economy (Sassen, 1991; 2000), or key nodes in the global network society (Castells, 1996), appear to be in extremely favourable positions to shape their relational and interconnection networks in order to be able, for example, to articulate in parallel their electronic spaces and urban places (Graham and Marvin, 1996; Kitchin, 1998; Sassen, 2002), or their local – global juxtapositions (Cohen et al, 1996; Borja and Castells, 1997).

On the other hand, it does not appear that we can simply dismiss the role of the nation state altogether (Anderson, 1996; Hirst and Thompson, 1996). It is still the state which plays the major regulatory role in national planning and economic development practices (Newman and Thornley, 1996), and in shaping national telecommunications policies (Thatcher, 1999).

Having provided some broad initial context, the rest of this introductory chapter outlines some of the traditional areas of neglect and limitation in urban telecommunications research. The objectives of the study are then discussed, and specifically how it is hoped that the study can fill in some of the gaps evident in previous research on cities and telecommunications developments. The theoretical basis of the thematic strands of the study as a whole is also explained in more detail, before the structure of the study is outlined.

Areas of Neglect and Limitation in Previous Urban Telecommunications Research

Despite the increasing numbers of studies of, and researchers becoming interested in, the relationships between cities and telecommunications, it can be countered that this area of research is still neither achieving the attention it deserves, nor delivering the quality of empirical research it demands.

Both cities and telecommunications have been important parts of societal organisation in the developed world for a long time, yet developing understandings of their interactions has rarely been high on the research agenda in urban studies, geography or planning. This is perhaps especially surprising with regard to geography, given the traditional focus of the discipline on relations between people, and between people and place (Hillis, 1998). Yet, Barney Warf suggests that: 'Telecommunications is one of the few topics in geography that richly illustrates the plasticity of space, the ways in which it can be stretched, deformed, or compressed according to changing economic and political imperatives' (Warf, 1998, p.255). Such neglect has been put down in some quarters to the invisibility and intangibility of communications technologies in the city (Moss, 1987; Graham and Marvin, 1996; Kaika and Swyngedouw, 2000, p121).[5] It might also be a question of a traditional connotation of communications technologies with the technical or the 'inhuman' elements of societal mobility (Thrift, 1996a). Furthermore, even when researchers have ventured into the communications field, their analysis and understanding has often been too simplified and narrow: 'Either communications is subsumed and made to operate within an inherently metaphysical and disembodied matrix of economic progress, or it is the adjunct of social relations. It is rarely perceived as 'the language' that links both' (Hillis, 1998, p.555). Urban telecommunications research, then, clearly needs to emphasise more the interrelations between social and economic processes in city initiatives.

Whilst there has been widespread recognition of the role of telecommunications in opening up the 'space of flows' and facilitating the development of a new intensity of flows and networks of social relations and economic linkages across this global space, this has, in turn, led to a profound neglect in understanding how telecommunications are equally tied up in the individual attributes of particular cities in the 'space of places' (Castells, 1989; 1996). The main reason for this neglect relates quite simply to the lack of actual *empirical* research of telecommunications developments in cities. This is especially the case from a European perspective, as there has been slightly more progress to producing such research in a US context (see Schmandt et al, 1990; Longcore and Rees, 1996; Wheeler, Aoyama and Warf, 2000, for an example of the predominance of empirical research on telecommunications in North American cities over European cities).

On a wider level, a further area of neglect within world city research is beginning to be highlighted, namely the 'critical empirical deficit within the world-city literature on intercity relations... we know about the nodes but not the links in this new metageography' (Beaverstock, Smith and Taylor, 2000, p.124).[6] Telecommunications infrastructures are obviously one of the key providers of these links, but it could be argued, certainly from the point of view of telecommunications developments, and probably from a wider point of view as well, that we still do not know enough about the nodes in global networks – the greatly varying territorialities of world cities. This seems to be brushed aside a little too quickly because it is held that the crucial and intrinsic element of 'world

---

[5] This is part of the broader way in which many key socio-technical processes and practices in the city have become implicitly or explicitly invisible (see Calvino, 1979; Boyer, 1995, for her notion of the 'disfigured city'; Latour and Hermant, 1998).

[6] This lack of relational data has been termed as 'the dirty little secret of world cities research' (Short et al, 1996).

cities' is their relational links, as if these links existed either prior to or outside of the territorial specificities of each city. This study focuses instead on telecommunications developments *in* the 'nodes' of Paris and London, rather than on the telecommunications-based links or 'intercity relations' *between* these two cities, because we know very little about how the differing territorial contexts and specificities of Paris and London, from parallel scalar levels, influence and are influenced by telecommunications developments. Only when we have investigated this territorial basis to telecommunications networks within world cities, can we begin to investigate the telecommunications connections between these world cities. In any case, it is possible that, in terms of the relationship between telecommunications and global cities, we know more about how global cities are linked by telecommunications networks than we do about how these networks help to shape and configure individual urban places or nodes. It is much easier, for example, to procure maps of the global intercity networks of telecommunications operators than maps of their intra-city networks. Within the broader basis of world city research as a whole, then, there is still more work to be done on world city 'nodes' than has been suggested, beyond the traditional quantitative-based comparisons of graphs and tables of economic activities. The agenda of world cities research has, thus far, offered both insufficient consideration of the territorial specificities and contexts of a city which are inherently bound up in its 'world city' status, and insufficient comparative, empirical work on how these territorial specificities and contexts shape and interrelate to actual territorial developments in world cities in different ways (although see Savitch, 1988).[7] An understanding of how telecommunications developments relate to these world city 'nodes' is an increasingly important part of this.

In terms of urban telecommunications developments, the tension in placing and defining the nation state within global-local processes and practices seems to open the door to empirical research of a comparative and cross-national nature, which focuses on how these wider relational processes and practices are influencing the negotiation, formulation and implementation of telecommunications developments in key (global) cities. Very few people, however, have so far taken the step through this open door. Indeed, actual comparative research on urban telecommunications developments is far from common, never mind actual comparative, cross-national research. With social science debate taking very seriously the complex and often multi-faceted interactions between cities and territories, the role of the nation state, and the nature of economic development in the light of globalising, or indeed 'glocalising', political economic logics, the distinct lack of comparative, cross-national research on urban telecommunications developments means that the latter cannot be convincingly incorporated into such debates, despite the ways in which they appear to offer a new and innovative dimension to economic development policy and practice. As Hantrais and Mangen suggest: 'Comparisons can lead to fresh,

---

[7] Sassen though argues the contrary in asserting that "international studies of cities tend to be comparative. What is lacking is a transnational perspective on the subject..." (Sassen, 2000, p.8).

exciting insights and a deeper understanding of issues that are of central concern in different countries' (Hantrais and Mangen, 1996, p.3). Furthermore, the lack of *cross-national* comparative empirical research of telecommunications developments in cities[8] also makes it extremely difficult to relate the domain to broader structural debates, because if we are unsure of the similarities and differences between telecommunications developments in cities in different nations, we cannot be sure of whether, and in what ways, they are bound up in global – local processes. Hantrais and Mangen have summarised the advantages of cross-national studies as:

> a means of confronting findings in an attempt to identify and illuminate similarities and differences, not only in the observed characteristics of particular institutions, systems or practices, but also in the search for possible explanations in terms of national likeness and unlikeness (Hantrais and Mangen, 1996, p.3).

Comparative, cross-national research is critically needed then on two levels – firstly, to engage with these important academic debates, and secondly, following on from this, to develop an understanding of how telecommunications developments in cities are negotiated, formulated and implemented in different ways, in order that urban telecommunications policy becomes richer and more mature, and can be subsequently grasped by city authorities and all those with a vested interest in the shaping of such policies at the urban level.

The other key aspect of the quality of urban telecommunications research relates to the level of analytical approach taken to exploring the relations between cities and telecommunications. As we saw earlier, reliance on simplistic and conjectural ideas of technological determinism or utopianism – futurism will never produce the necessary complex understandings of these relations. Empirical studies of the mutual territorial influences on and implications of the relations between cities and telecommunications are, then, desperately required. The present study aims to contribute to this requirement.

The Objectives and Focus of the Study

In response to all this, the main objective of this study is to produce an explanatory, comparative, cross-national analysis of urban telecommunications developments, founded on two empirical case studies of developments in Paris and London, which are widely regarded as the two pre-eminent 'global' cities in Europe (Llewelyn-Davies et al, 1996; Savitch, 1988; Beaverstock, Smith and

---

[8] The survey of urban and regional 'technopoles' by Castells and Hall (1994) constitutes a rare exception here, but their broad, worldwide approach inevitably lessens both the detail into which they delve on individual strategies, and the subsequent comparative element which is based on the technopole concept as a whole, rather than two or three detailed strategies from different countries, from which we can clearly see the similarities and differences in national approaches.

Taylor, 2000). It is perhaps especially surprising, given what was said earlier about the increasingly important role of ICTs in global cities, that there has been so little empirical research focusing on this role (although see Moss, 1987; Hepworth, 1992; Longcore and Rees, 1996; Graham and Guy, 2002), and that, in turn, there has been absolutely no comparative research in this domain actually comparing developments in global cities. There would appear to be a certain tendency to think of global cities as somewhat homogeneous and territorially similar entities (Taylor, 2000), rather than bound up with a ''multiplexing' of diverse economic, social, cultural and institutional assets' (Amin and Graham, 1997, p.412; see also Graham and Healey, 1999), or characterised by 'a set of spaces where diverse ranges of relational webs coalesce, interconnect and fragment' (Amin and Graham, 1997, p.418). A key objective for this study then is to explore the level of similarity and difference between the territorialities of telecommunications developments in Paris and London. It would seem to be especially important to investigate whether or not certain dynamics and processes in global cities remain tied to, and do not exist totally outside the realm of, nation states.

The research aims to provide a complex and detailed exploration of:

- the territorial contexts, logics and processes from differing, but parallel scalar levels, which are both influencing and are influenced by the construction of telecommunications developments and strategies within the two cities;
- the subsequent processes and practices of negotiation, formulation and implementation of these telecommunications developments and strategies;
- the theoretical and policy insights and implications that can be gained from a comparative, cross-national analysis of urban telecommunications developments.

The principal approach taken to cut across this overall exploration is the definition of three key thematic strands within the research, developed from prior consideration of the main policy and general theoretical dynamics relating to the study. The theoretical, empirical case study, and analytical parts of the book are then underpropped by, and structured according to, these three strands. The aim in this way is to facilitate the construction and flow of the arguments and narrative, while emphasising the recursive interaction between theory, empirical contribution and analysis. Rather than the traditional structure of separate chapters for each case study, followed by a comparative analysis, the empirical contribution of this study, then, is founded on three main thematic case study chapters, followed by a broader comparative analysis. The aim here is to reflect the complex and generally innovative nature of urban telecommunications developments in the structure of this study, and more importantly, to keep the comparative and cross-national aspect of the study very much to the fore, rather than 'relegating' it or synthesising it into just one analytical chapter after the individual case studies.

Methodological Matters

This study is based on a comparison of two detailed case studies of contemporary telecommunications developments in the metropolitan regions of Paris and London. They are intended to illustrate how these developments relate to broader strategic practices and perspectives, how they have been initiated and shaped by various groups and processes at various scales, and their implications for these groups and processes, and the metropolitan region as a whole. The comparison of the two cases will, in this way, illustrate the similarities and differences in the complex and interacting territorialities of telecommunications developments in the two cities. The case study approach is thus appropriate here because we are interested in 'how' and 'why' questions about a contemporary set of events over which we have little or no control (Yin, 1994, p.9). Furthermore, cross-national research in the fields of geography, planning and urban studies has frequently employed this approach (see, for example, Savitch, 1988; Fainstein, Gordon and Harloe, 1992; Booth and Green, 1993; Keating and Midwinter, 1994). In turn, cross-national, comparative case studies have also been designed to study processes and practices in the field of the political economy of telecommunications (see, for example, Morgan, 1989; Palmer and Tunstall, 1990; Hulsink, 1998; Thatcher, 1999), and, most importantly for this study, in previous research on urban telecommunications developments (see, for example, Schmandt et al, 1990; Noam, 1992; Briole et al, 1993; Castells and Hall, 1994; Graham, 1996). Given the complex and detailed background to telecommunications developments in large global cities such as Paris and London, it is unlikely that any other research strategy apart from case studies would be able to achieve the sufficient depth required for a descriptive, explanatory and analytical study of the initiation, shaping and growth of telecommunications developments in these cities. The case study approach is ideal for delving purposefully into the policy, institutional, and territorial influences and practices behind these developments.

The choice to investigate telecommunications developments in Paris and London reflected the economic and territorial importance of these two European 'global' cities, but also their divergent historical contexts, and national systems of planning, governance and regulation. The mixture of contextual similarities and differences provides a complex and interesting background to an exploration of the shaping of telecommunications developments in the two cities. Indeed, specific comparative case studies of Paris and London are not unusual in the planning and urban studies field (see, for example, Savitch, 1988; Newman and Thornley, 1994; Llewelyn-Davies et al, 1996; Chevrant-Breton, 1997; Nelson, 2001).

Definition of Key Terms

Core geographical concepts such as scale, place and territory are at the heart of our investigations. We have already discussed above how a reconfigured notion of scale seems to be increasingly pertinent here, which emphasises flexible, parallel and intertwined scalar levels rather than a fixed hierarchy. This will be expanded on in our theoretical chapters. Place and territory are often used interchangeably, and seem to connote similar notions and ideas, yet the latter appears to have an extra dimension to the former. Castells offers a definition of place as 'a locale whose form, function and meaning are self-contained within the boundaries of physical contiguity' (Castells, 1996, p.423). Territory, though, is not necessarily a locale, and it can be viewed as far less 'bounded' than place, as its form, function and meaning may extend well beyond a single contiguous place. In this study, then, it appears important to develop a notion of territory in relation to our reconfigured notion of scale. In this way, the form, function and meaning of a territory are intertwined with contexts, processes, mechanisms and specificities from multiple parallel scalar levels. However, the recursive interaction between territoriality and multiscalar contexts, processes, mechanisms and specificities means that the latter are continually territorialised, deterritorialised, and reterritorialised in different ways. As we shall see, then, the focus of this study on comparing the territorialities of telecommunications developments in Paris and London means investigating and analysing the ways in which contexts, processes, mechanisms and specificities from multiple parallel scalar levels shape and are shaped by their territorialisation as bound up in these telecommunications developments.

This notion of territoriality is therefore especially important within our study. Previous research on cross-national differences in national telecommunications policies and regulation is not short on the ground (Thatcher, 1999; Hulsink, 1998; Curwen, 1997; Eliassen and Sjøvaag, 1999), but this research falls short of focusing on the territorial dynamics of policy and regulatory differences. As we have already noted, empirical work which has been based on the territorial development of telecommunications networks has rarely, if ever, had a cross-national, comparative element to it.

There also seems to be considerable room within urban telecommunications research for including and drawing on broader relational urban theories concerned with, for example, governance, scalar complexity and urban cohesion and fragmentation. Graham and Marvin (2001) have begun to address this situation in their work on infrastructure networks, but a greater and more powerful complexity to research on cities and telecommunications would be achieved by incorporation of, and interaction with, these influential theoretical debates.

The objectives of this study derive, then, from gaps in previous research in which a focus on a cross-national comparison of the territorialities of telecommunications developments in global cities can be successfully and productively inserted. Excellent theoretical and, in some cases, empirical research has been done in each of the areas discussed above, but we now need firmly multiscalar, empirical studies which draw on an incremental combination of all these areas to highlight the strengths and weaknesses of these individual theoretical debates. The complex, cross-cutting, interdisciplinary and multiscalar nature of urban telecommunications research has the potential to successfully draw on and add to these theories as well as any other area of research in geography, urban studies or planning. As a result, one of the main contributions of this study as a whole can be to critically combine these theoretical debates and influences, and within this, in particular, to offer a fresh multiscalar perspective on telecommunications developments in global cities.

The Approach and Research Questions of the Study

Analytical, comparative and cross-national research into urban telecommunications developments, as in this study, evidently requires a solid framework on which this research can be built. The theoretical, empirical and analytical parts of the study need to cover similar ground in their different ways, and also follow on very closely from each other, so that the contribution of the study is constructed steadily and becomes increasingly clearer.

The complexity of building a rigorous framework for studying urban telecommunications developments should not be underestimated. As has become clear already in this introduction, these developments influence and are influenced by many differing areas of theoretical debate and public policy. They span theories and policies related to both cities and technologies, and crucially bring together the traditionally bounded and place-based nature of cities, and the traditionally unbounded and space-annihilating nature of telecommunications. The most important implication of this juxtaposed tension is the subsequent need in this study to consider and analyse urban telecommunications developments within a multiscalar approach, from global reach to local detail.

As has already been mentioned, the choice has been made in this study to achieve this solid framework through the development of three tightly-focused, but cross-cutting thematic strands. These strands are then able to structure the theoretical, empirical and analytical parts of the study in an innovative, yet clear and symmetrical fashion. Whilst it is obviously very important not to simplify the complexity of the subject of the study, it is equally critical to present such a complex subject in a way which makes it easier to follow the arguments of the study.

The three thematic strands making up the theoretical – empirical – analytical framework to this study have been chosen following careful and detailed consideration of the main policy and general theoretical dynamics relating to the study. This prior choice might be viewed as problematic, as the strands structuring the research are apparently not clearly being chosen as a result of theoretical discussion within the study. However, as has already been made clear, the aim is to symmetrise theory and empirical research to create a clearer, more tightly-focused analytical framework, and this necessitated such a decision. It is hoped that the juxtaposition and interaction of the following thematic research strands creates a framework for successfully exploring empirical urban telecommunications developments.

The first thematic strand relates to influences on telecommunications developments in global cities from the *national* level, and the subsequent cross-national differences in the nature of these influences. Specifically, given the coming together in urban telecommunications developments of theories and policies of both cities and technologies, this theme needs to closely draw on the policy theory in the fields of national telecommunications policies and regulation, and national intergovernmental systems, and urban and economic development policies, which focuses particularly on differences between France and the UK (chapter 2). This dual set of policy theories is then used in the analysis of national influences on these developments in Paris and London (chapter 6), by asking firstly how telecommunications developments in the two cities are being influenced by the telecommunications policies and regulatory practices of France and the UK, and secondly how they are being influenced by the intergovernmental systems, and urban and economic development policies of France and the UK.

The second thematic strand addresses the *urban regional* level within telecommunications developments in global cities. This strand draws on recent theoretical work within urban studies and geography which has begun to theorise the relationships between global cities and telecommunications / IT (chapter 3), and goes on to relate this work to the differing roles of telecommunications developments within the broader competitive global city economic development 'package' in Paris and London (chapter 7). The main research question for this strand examines, then, how telecommunications developments are being materially and discursively constructed and packaged to meet and respond to the requirements of Paris and London as global cities.

The third thematic strand is concerned with the *local or intra-urban* level within telecommunications developments in global cities. Here, we need to draw on broader relational theories from the fields of political economy and planning (chapter 4), in order to explore how these developments both relate to the different spaces within global cities, and subsequently increase or decrease tensions in the domain of overall territorial cohesion in Paris and London (chapter 8). We investigate here the ways in which the roll-out and territorial fixing of local telecommunications developments vary within Paris and London, reflecting increasingly complex and mutually constitutive processes and practices of local governance and multiply scaled economic development, plus the implications of telecommunications developments for Paris and London in terms of the possibilities for reconciling overall economic development and territorial cohesion objectives.

The Structure of the Study

Having set the contextual scene, and outlined the objectives and the approach of the study as a whole, we can move into the first part, in which in chapters 2-4, the aim is to build the necessary theoretical framework, from which we can approach and underpin the empirical contribution. In chapter 2, this constitutes focusing on critically reviewing the policy theory of cross-national differences between France and the UK both in telecommunications policies and regulation, and in intergovernmental structures, urban policies and economic development practices. In chapter 3, the focus shifts to the theorising of relationships between global cities and telecommunications / IT, while in chapter 4, we concentrate on wider relational urban theories, especially from the fields of political economy and planning.

Part Two of the study focuses on the empirical contribution. Chapter 5 moves in to the domain of the case studies proper, with a French and UK historical perspective on the areas of communications technology, urbanism and cultures of territoriality, which are at the heart of this study. In chapters 6-8, the two case studies of telecommunications developments in Paris and London are comparatively explored and analysed according to each of our three respective thematic research strands. Chapter 6 focuses, therefore, on the national influences on telecommunications developments in Paris and London. Chapter 7 is concerned with how these developments fit within the competitive global city economic development 'package', while chapter 8 looks at how telecommunications developments in Paris and London relate to the different urban spaces of each city, both in the particular ways in which new networked spaces are being constructed, and in the tensions in territorial cohesion through the propinquity of connected and disconnected spaces and groups.

The third and final part draws the different elements of the study together. In chapter 9, an analytical conclusion relates the findings of the three case study chapters back to the theoretical framework, and consequently illustrates in detail the main overall contributions of the study to the field of urban telecommunications research. Finally, chapter 10 considers the research and policy implications of the findings of the study, and suggests fruitful areas for future research to explore.

PART I
GLOBAL CITIES, NATION STATES AND TELECOMMUNICATIONS

Introduction

Having positioned the study and specified its objectives in the introduction, we need, in the next three chapters, to consider, discuss, and critically analyse the relevant theoretical literatures that impinge on the subject of this study, and that will underprop the development and exploration of the research questions and actual empirical research in the following chapters. As Yin notes, the discussion of previous research is a means 'to develop sharper and more insightful *questions* about the topic' (Yin, 1994, p.9). This analysis necessarily encompasses a wide range of interrelated literatures that touch on urban studies, geography, planning, communications studies, and the social sciences in general, because a detailed analysis of urban telecommunications strategies demands primarily an understanding of the complex intertwining of social, economic, political and cultural contexts and processes bound up in the strategies (Graham and Marvin, 1996).

These theoretical discussions and analyses have been organised into three chapters, according to the three main thematic strands of the research as a whole, in order to privilege simplicity and continuity of argument throughout the study. Such a structure is not, however, meant to draw artificial theoretical divisions between discussions and contexts which clearly interconnect, particularly given that our chosen approach has been developed in order to highlight and analyse the ways in which the development of urban telecommunications strategies are bound up with parallel and intertwined multiscalar processes and practices. Thus, chapter 2 develops an understanding of cross-national urban and telecommunications policy contexts, with particular reference to France and the UK. In chapter 3, we consider some of the theories which explore the territorialities of global cities and telecommunications, while chapter 4 discusses political economies of urban governance, rescaling and economic development.

Chapter 2

Understanding Cross-National Urban and Telecommunications Policy Contexts

Introduction

This chapter considers and analyses from a theoretical perspective the first thematic strand of the study as a whole, namely the cross-national differences in the influence of national contexts and policies in urban telecommunications developments. This is the first layer of our theoretical analysis, which, in parallel with the following two chapters, will provide a solid base for our empirical investigations later in the study. The discussions in this chapter will be particularly relevant in relation to chapter 6, which examines the same thematic strand from an empirical perspective, yet this is not to deny the intertwined nature of our theoretical – empirical structure, within which the three thematic strands cannot be completely separated, but must be considered incrementally and in parallel to reflect the multiscalar and concurrent territorial contexts and specificities of telecommunications developments in global cities.

This focus on the national (and cross-national) level is prompted by various recent suggestions and arguments that the national level is no longer quite the determining force it was in terms of shaping policy contexts and practices in states around the world (see, for example, Jessop, 1994; Thrift and Leyshon, 1994; Veltz, 1996; Taylor, 2000). This strand aims to examine the extent to which this may or may not be the case in relation to urban telecommunications developments. This will illustrate whether transnational and global processes and dynamics 'displace' or 'overtake' processes and dynamics at the national level, or whether they simply overlap or intertwine with them. We can, however, argue already that comparing cross-national responses to all these processes and dynamics is therefore still very worthwhile.

In examining and analysing urban telecommunications developments, we start from a proposition that the national policy context might still matter in two main domains:

- Telecommunications policies and regulatory practices.
- Political cultures, intergovernmental systems, and urban and economic development policies and practices.

These will be looked at in turn, with a particular focus on previous cross-national studies in these two areas, as an illustration of the continuing benefits of a cross-national approach.

National Telecommunications Policies and Regulation

Telecommunications is a particularly interesting sector to study in cross-national policy research. While, on the one hand, it is held up to be illustrative of space-transcending, globalising processes and practices, on the other hand, it is also subject to sometimes great variation between different nation states. The former view is understandable when we consider only the realms of networks and communications flows which stream across borders and around the world, all the time propulsed by multinational companies and channelled through the infrastructure of international telecommunications operators. Placing this within a context of differing types of national telecommunications policies and differing national levels of regulation, however, renders rather simplistic and exaggerated this globalising view, and complicates significantly the actual situation.

An approach to telecommunications developments in global cities based in part on the political economy of telecommunications emphasises, then, the diversity of, and interaction between, relevant political and economic processes and practices in individual nation states and systems of governance. These processes and practices, of which types and levels of liberalisation and competition are likely to be among the most important, filter down to shape the deployment of telecommunications infrastructures and services in particular cities and regions: 'The regulatory environment for telecommunications determines not only the nature, availability, cost, and supply of services to any locality and/or customer but also the *incremental* development of the telecommunications network between localities' (Gillespie and Williams, 1988, p.1315). There are many factors in telecommunications developments which need to be considered by public sector actors in cities and regions. As Graham suggests: 'In entering into the telecommunications arena, municipal policy makers will need to confront many new and challenging learning curves in finding out about, first, the *existing* telecommunications situation within their city, and second, the *scope* for urban telecommunications initiatives' (Graham, 1996, p.77).

Cross-National Differences of French and British Telecommunications Policies and Regulation

National telecommunications policy in France and the UK in the last two decades has been dominated by a general shift in the nature of infrastructure networks from being bundled to the implementation of unbundling processes, and different responses to the decline of Keynesian 'natural monopolies', and its 'modern infrastructural ideal' of telecommunications as 'public goods' (Graham and Marvin, 2001).

The liberalisation or demonopolisation of the telecommunications sector is often pictured as a simple, uniform switch from monopoly supply to competition. The actual process of 'unbundling' telecommunications monopoly networks has been considerably more complex in most western countries, and has tended to involve differing levels or types of liberalisation or reconfigurations of network packages according to the different telecommunications markets. For Graham and Marvin (2001), the unbundling of infrastructure networks needs to be seen as concerning processes of 'network segmentation', which:

> involves detaching activities and functions that were previously integrated within monopolies and opening them to different forms of competition. [...] The process of unbundling, therefore, attempts to separate the natural monopoly segments of a network and then promote new entrants and competition in segments that are potentially competitive (Graham and Marvin, 2001, p.141).

They discuss three forms of this segmentation (Graham and Marvin, 2001, p.141-143):

- Vertical segmentation divides the core of networks from other elements further along on the supply chain, closer to the user.
- Horizontal segmentation divides network activities according to geographical or sectoral markets.
- Virtual segmentation concerns the emergence of competitive services over continuing monopoly networks, via the management of information technology systems.

In terms of telecommunications, it is not unusual to see all three forms of segmentation present in parallel in contemporary markets (Graham and Marvin, 2001). Vertical segmentation has created a divide between the new network services of competitive operators and the continuing control of the 'last mile' of local, user contiguous networks of incumbent operators. This is closely tied to virtual segmentation, where, prior to the local loop unbundling process at least, new entrants have been providing their competitive services over the local monopoly networks of incumbents, monitored by IT systems. The development of mobile networks in parallel with traditional fixed telephony networks is an example of horizontal segmentation.[9] In addition, the processes of licence attribution for new telecommunications networks such as the wireless local loop

---

[9] The mobile and fixed telephony telecommunications markets differ greatly, however, in competitive terms. As Gadault and Saget observe:

> In the first case, the operators have all, more or less, had to start from zero, without a single client, and build a whole network. Hence the similar cost constraints for the three or four companies licensed in each country, and, finally, a healthy competition in the cost and quality of the service. In the second case...it is impossible for new entrants to overtake completely the long amortised infrastructure of the incumbent operator, while also hoping to make a profit (Gadault and Saget, 2001, p.56-57, author's translation).

have been characterised frequently by geographically-based horizontal segmentation.

This complex situation is likely to further problematise the relations between telecommunications and urban development, given the already general neglect of the former in the policies of local and regional governments. As Graham and Marvin question: 'With this wide range of options how do policy makers identify elements of a network that are provided as a monopoly and those that are contestable?' (Graham and Marvin, 2001, p.144).

Partly in relation to these common emerging market unbundling dynamics, France and the UK have approached and handled telecommunications in very different ways in the last thirty years or so (Palmer and Tunstall, 1990; Hulsink, 1998; Eliassen and Sjøvaag, 1999; Thatcher, 1999). As Thatcher summarises:

> Institutional reforms in Britain led to increasing divergence between the two countries during the 1980s. The French reforms of 1990 reduced differences somewhat, but it was only after further reforms in 1996-7 that significant convergence occurred and, even so, important institutional contrasts with Britain remained (Thatcher, 1999, p.143).

Even in the light of common supranational EU regulation dictating the liberalisation of national voice markets from 1998,[10] there are still variations in institutional and policy organisation which reflect the differing political, technological and historical cultures of the two countries. It is against these deviating national backdrops in telecommunications policies and regulation that telecommunications developments are shaped and implemented in Paris and London.

The essential differences between French and UK telecommunications policies and regulation derive from the political-economic stances taken by central government in both cases. While the UK government was an early adopter of an overwhelmingly neo-liberal, market-led approach, not just for telecommunications, but right across the policy board,[11] the French government approach to the telecommunications sector was more autonomous and characterised by continuing *dirigiste* state intervention (Morgan, 1989; Hulsink, 1998). In the UK therefore, privileging the market meant opening up the telecommunications sector very early to competition and privatising the former state monopoly British Telecom (BT).[12] As Schiller suggests:

> These measures permitted the United Kingdom to offer itself as a hospitable site for the information system operations of major US firms needing access to European markets. By 1997, no less than 120 rival companies competed in all segments of the British

---

[10] For a review of EU telecommunications policy, which we do not have the space to discuss in detail here, see Eliassen, Mason and Sjøvaag, 1999; Thatcher, 1999, chapter 4.
[11] See Guy, Graham and Marvin, 1997.
[12] This followed the US lead with its divestiture of AT&T in 1982, creating a long distance company and five smaller Baby Bells.

telecommunications market, which in turn had become the main hub for Continental Europe. Pressure correspondingly ratcheted up on adjacent countries to liberalise their own policies (Schiller, 1999, p.43).

However, this newly deregulated, liberalised and globalising context[13] in telecommunications is increasingly proving to exacerbate socio-spatial polarisation in the provision of telecommunications services, with companies competing for the custom of lucrative users, whilst leaving more marginal groups and areas of the city (and the nation as a whole) often completely disconnected (Graham and Marvin, 1996; Sussman, 1997).

In France, in complete contrast to UK policy, the emphasis in the 1980s was on the state keeping control of the sector and protecting the state monopoly of the *Direction Générale des Télécommunications* (DGT) / France Télécom (Morgan, 1989). The national *Plan Câble* in the 1980s was, for example, notable for the relative absence of private sector involvement (Morgan, 1989, p.41-42). As Noam writes:

> Under Mitterand,[14] telecommunications were not a goal in and of themselves, but one tool for a more general modernisation of society and the economy. As a result, telecommunications policy became more political and less technocratic; engineering bureaucrats yielded the centre stage to empire builders straddling the public and private sectors (Noam, 1992, p.146).

France was therefore probably the European nation which was the least enamoured with the EU directive on liberalisation by 1998, and unsurprisingly, they left it until the last possible minute before finally complying and opening their markets.

The main reason for the reluctance of the French government in the 1980s to create more competition in their telecommunications market seems to have been international pressures: 'the French Socialists had aimed to strengthen France's competitive position in international markets by a policy of building up national champions even (indeed deliberately) at the cost of pluralism in French markets' (Humphreys, 1990, p.210). Indeed, as Humphreys continues: 'French policy-makers were becoming increasingly aware that the neo-liberal tide of international telecommunications developments was being propelled, to an important extent, by the demands of international markets beyond their direct control' (Humphreys, 1990, p.214). In addition, there was not the same pressure for liberalised markets in France from business that there was in Britain, from the City of London in particular (Morgan, 1989). As a result, some commentators a decade ago were already questioning how the 'interventionist and mercantilist French state tradition' would face up to and adapt to moves towards deregulation (Humphreys, 1990).

---

[13] The terms and practices of 'liberalisation' and 'deregulation' should not be seen as synonymous. A competitive telecommunications market still requires a regulatory framework in order to control and configure the roles of dominant operators and new entrants to produce fair and efficient competition, as has been evident in the UK over the last twenty years.

[14] The French president from 1981 to 1995.

Nevertheless, moves in France towards a more liberal and deregulatory environment in the telecommunications sector date back further than the only recent opening up of the French market might suggest. Firstly, the markets of the DGT for value-added services were opened up to private sector competition in 1986. Secondly, a law was drafted during the mid 1980s that notably changed the status of the DGT from being an element of the state to being a state-controlled private company with more market independence, under the name of France Télécom from 1988 onwards (Humphreys, 1990). It was finally transformed into a public trading organisation or corporation in 1990-91 (Thatcher, 1992). Morgan (1989, p.43) has called these developments in the mid to late 1980s 'pragmatic liberalism'. Liberalisation was therefore steadily impending and inescapable. As Thatcher observes:

> Competition in network operation became accepted as inevitable (both because of technological factors and then EC law), especially after the Right won the 1993 legislative elections. Nevertheless, France Télécom's monopoly over fixed-line public voice telephony and network infrastructure continued until 1 January 1998, the last date stipulated by EC law. The main obstacle to increased competition was the desire of policy-makers to protect France Télécom: the maximum time was sought for France Télécom to adapt to new market conditions, through institutional reform but also by tariff rebalancing and internal reorganisation (Thatcher, 1999, pp.216-217).

This protection of France Télécom put paid, for example, to any thoughts of the creation of a new national broadband fibre optic network, called for by a report for the government, which would have cost France Télécom more than 150 billion francs (Théry, 1994; Thatcher, 1999, p.244).

There were several other developments in the early 1990s which foresaw the complete opening up of the French telecommunications market. More advanced services in addition to voice telephony were permitted via private teleports and then cable television networks. When the public *Caisse des dépôts et consignations* sold its cable infrastructure in 1994, half went to France Télécom and half to the private utility giant La Lyonnaise des Eaux. Both private and public utility companies with their own networks were called upon to offer competition to France Télécom for after 1998 (Thatcher, 1999, p.217).

Global Logics? The Growth of Transnational Telecommunications

The telecommunications sector was traditionally dominated by national monopolies, but since the early 1980s, there has been a gradual shift towards liberalisation which culminated in 1998 with the end of national monopolies in basic telecommunications services in EU countries. At the same time however, the original operating companies have been able to expand their interests into the global market (see Mouline, 1996). This is an equally important process in national and cross-national restructuring. Schiller (1999) views these developments as part of a 'neo-liberal project in transnational telecommunications'.

At least until a few years ago, deregulation was not complete, with global competition still restricted in extent by the nation state. Wells and Cooke observed the changing spatial organisation of telecommunications firms at around this time: 'These regulatory conditions, combined with the rapid rate of technological change in the computer and communications industries, and burgeoning demand for global, customised telecommunications services, provide the structural conditions for strategic alliances' (Wells and Cooke, 1991, p.91). Cross-border mergers and acquisitions in the telecommunications sector were worth $17 billion in 1997, with much of this taken up by European operators in the pre-liberalisation environment (Schiller, 1999, p.64; see also Elixmann and Hermann, 1996; Curwen, 1999). For Curwen, 'alliance-forming can be seen either as an offensive or, more realistically, as a defensive strategy, or indeed as both at the same time' (Curwen, 1997, p.39). In the former case, the aim is usually to create a viable competitive alternative to a powerful incumbent in a new market, while in the latter case, the objective is more to do with reducing the level of 'cherry-picking' of key customers or sectors by new entrants (Curwen, 1997). The key global alliances of the 1990s were Concert (BT and MCI), Global One (Deutsche Telekom, France Télécom and Sprint) and AT&T-Unisource, but these have since suffered from the downturn in the telecommunications sector and, in the cases of Concert and AT&T-Unisource, broken up, following WorldCom's successful takeover of MCI and AT&T's decision to abandon its alliance respectively.

With liberalisation of European telecommunications markets in 1998 came the opportunity for operators – both new and relatively small entrants as well as the large incumbents – to build infrastructure on a pan-European level in order to deploy their own services or rent out fibre on the wholesale market. The major cities of Europe are now linked by numerous such networks, although, as we shall see, this has in turn intensified spatial and territorial differentiations both at the national level between these major city nodes and smaller provincial urban and rural areas which are not on these networks, and at the urban regional level between the key business sectors of major cities, which are the most heavily networked, and other areas, often residential, which are also disconnected. The telecommunications sector, therefore, is very much dominated by global(ising) logics, as well as political, technological and historical cultures at the national level.

The parallel, and intertwined, result of all this is, on the one hand, increasing demand for 'international systems solutions' or one-stop shops from heavy business users, and, on the other hand, an alleged downplaying of traditional universal service policy by western nation states (Gillespie and Robins, 1991; Gille, 1995). This shift from supply-led universal service towards demand-led market segmentation focused on key customers clearly has serious socio-economic repercussions for telecommunications access, for the more disadvantaged groups in these countries, and for the telecommunications sector as a whole (Gillespie and Robins, 1991).

As we shall see in the following chapters, these 'global' or 'transnational' processes are, nevertheless, more complex and less uniformly bound up in specific territorial contexts than has been portrayed here thus far. However, highlighting the incremental nature of these processes poses the question of whether we are seeing an increasing level of convergence in the relations between telecommunications and territorial logics, whether bound up in the national, regional, and / or local levels (Négrier, 1990; Offner and Pumain, 1996). The example of the Internet, the primary global communications network, illustrates the question: 'Never before has a functioning medium made such a hash of geopolitical boundaries. By annihilating traditional territorially anchored controls, the Internet has been configured to comprise what one legal analyst terms a 'universal jurisdiction' that endangers every lesser sovereignty' (Schiller, 1999, p.72). Armand Mattelart goes even further in summarising what is at stake with the increasingly global logics of telecommunications:

> With privatisation and deregulation, first in broadcasting and later in the break-up of public telecommunications monopolies, contradictions in the constant call by those in power for social benefit through technological revolution were quickly apparent. Welfare systems nationally and internationally have been subordinated to technical and financial logics. Geo-economics protagonists have superseded geopolitical actors, until recently sovereign in the enunciation of macro-strategies for network expansion. The former basically await the latters' removal of the final legal barriers to the construction of a totally fluid global market (Mattelart, 1999, p.188).

A more historical perspective on the development of telecommunications in France and the UK will be given in chapter 5, but here, we have already begun to see the broad types of differences present in French and UK telecommunications policy, which, despite a parallel set of more global dynamics in the telecommunications sector, suggest that it is worthwhile taking a cross-national policy perspective within our study.

Political Cultures, Intergovernmental Systems, and Urban and Economic Development Policies

In the same way as the previous policy strand was based on the assumption that national telecommunications policies and regulation strongly influence telecommunications developments in global cities, the second theme of this chapter highlights another set of national policy contexts, in the shape of the intergovernmental systems and the formulation of urban and economic development policies, which are also important backdrops to these developments, albeit in a slightly more indirect way than national telecommunications policies and regulation.

From a cross-national perspective, intergovernmental systems and urban policies are likely to vary even more between states than telecommunications policies and regulation: 'There are important differences between countries in local economic interventionism, depending on national institutional structures and national and local political orientations' (Pickvance and Preteceille, 1991b, p.218).

Equally though, there are more global processes within which national processes need to be set:

> A major element in any explanation of trends in the institutional and financial character of the territorial organisation of the state must clearly be the broader economic and political context in which these trends have taken place. The process of economic restructuring since the economic crisis of the mid-1970s and the rise of neo-liberalism are the most important aspects of this context (Pickvance and Preteceille, 1991b, p.214).

The implementation, shaping and growth of telecommunications developments in Paris and London takes place, then, against a backdrop of very differing political and historical cultures in France and the UK, which has meant that urban and economic development policies (including interventions relating to telecommunications) have traditionally been formulated and enacted in different ways.

Political Cultures and Intergovernmental Systems in France and the UK

Whilst both France and the UK are unitary states within the European Union with similar population sizes, the former covers a territory twice as large as the latter, and their administrative structures vary quite greatly. France has a four-tier system, in which the region and the department play key roles between the highest level, the state, and the lowest level, the commune. The French system is quite fragmented with 36,000 communes, 96 departments, and 22 regions – a sub-national government arrangement which has stayed more or less the same as in 1789 (Preteceille, 1991).[15] In fact, 'France has as many local councils as the rest of the European Union countries put together, before the last enlargement' (Sallez, 1998, p.107). In keeping with the French Jacobin heritage,[16] this system has traditionally been relatively centralised, and local government has had a quite limited role in delivering services (Keating and Midwinter, 1994).

Until quite recently, the UK was said to have a two-tier structure with 'no formal intermediary between the nation state and local authorities' (Parkinson, 1998, p.413), although the (non-elected) county level does have limited administrative responsibilities. The new Regional Development Agencies (RDAs) in England have been given responsibility for economic development, but they remain appointed by government rather than directly elected (Parkinson, 1998, p.413). The British structure is relatively rationalised, compared to France, with 400 units of local government. Local government has traditionally been more

---

[15] French regions only became an elected level of government in 1984, following the initiation of the decentralisation reforms of the Socialist government. Prior to this, they lacked specific executive functions, with the *département* being the key tier of government below the state (Budd, 1997).
[16] As Keating has noted: 'Centralisation is buttressed by the Jacobin ideology, stressing equality, uniformity, nationalism, and the obligations of citizenship' (Keating, 1991b, p.444).

autonomous than in France, with a larger responsibility for service delivery (Keating and Midwinter, 1994).

Intergovernmental relations are stronger and less top-down in France than the UK. A process of decentralisation was set in motion by the French government in 1982, which 'has progressively reoriented urban policies and development, until then directed by central government, towards economic contracts signed by the state, the region and the urban community' (Sallez, 1998, p.129).[17] This decentralisation did not though lead to complete administrative and financial independence at the commune level. However, it did lead to regions being allowed to draw up their own budgets (Sallez, 1998), although the negotiation of *contrats de plan* every five years officialises financial relations between the regions and the state. Keating summarises the shifts in French central-local politics:

> The development function has progressively been taken out of the hands of the Parisian technical state elite and put into those of the territorial political elite, which themselves have been changing. New relationships have been established among the central state, local governments, and the business sector. For the last several years, central governments have sought to rationalise these local efforts and harness local development initiatives to the needs of national competitiveness (Keating, 1991b, p.455).

In complete contrast to French decentralisation, the UK government in the 1980s and the first half of the 1990s pursued a policy of 're-centralisation', taking back power and responsibility from local authorities and restraining their financial budgets and expenditure (Parkinson, 1998). Pickvance discussed a total of eight measures which central government took with regard to local government, with varying success: controls on spending; controls on income; the abolition of the Greater London Council (GLC) and other metropolitan county councils; housing privatisation; contracting out some council services; the creation of bodies which 'bypass' local government, particularly the Enterprise Zones and Urban Development Corporations in the planning domain; reform of the rating system; the possibility of non-council control of schools and housing (Pickvance, 1991, pp.68-77). The subsequent Labour government has, to some extent, reduced slightly the highly centralised control of local authorities, but the state remains the dominant force (see Wollmann, 2000, pp.38-39), and there is little question of contractual arrangements being forged in key economic areas as in France. Weak intergovernmental relations in the UK are seen to have significant consequences:

> At present, the links between national government, regional organisations and local authorities remain uncertain. In many commentators' eyes, this also affects the capacity of larger cities to respond on a strategic level to economic, social, cultural and

---

[17] Wollmann attributes the decentralisation measures to 'the perceived need to remould and modernise France's politico-administrative system vis-à-vis the growing challenges from the international political and economic environment, not least from European integration and the new role of regions in it' (Wollmann, 2000, p.42).

technological changes which cross narrowly defined local authority administrative boundaries (Parkinson, 1998, pp.413-414).

Pickvance and Preteceille have related the general trends in nation state restructuring to key global political-economic processes. For them, the centralising policy of the UK (in terms of control, more than functions) can be seen as emblematic of 'a response to local opposition to neo-liberalism in countries with strong central states seeking to promote neo-liberalism', while French decentralisation (in terms of control and functions) is 'a means of promoting neo-liberalism by appealing to local support, in countries with a weak central state' (Pickvance and Preteceille, 1991b, pp.222-223). Despite the differing policies of France and the UK then, both countries share broader capitalist economic objectives.

The implications of intergovernmental relations for urban policy in the two countries are clear: 'Decentralisation gave French cities greater room to manoeuvre than their British counterparts during the 1980s where Conservative governments centralised power and decision-making' (Parkinson and Le Galès, 1995, p.32). Through case studies of local technopole developments, Eberlein (1996) was able to illustrate change in French territorial regulation following the decentralisation policies of the 1980s, with an increasing emphasis on 'intergovernmental conflict and cooperation', which led him to characterise this regulation as 'quasi-federalism in unitary disguise':

> Subnational governments in post-decentralist France are capable of autonomous policy initiatives. Progressively, local authorities have acquired management responsibilities and resources. A new group of local politicians launched ambitious public policies in a territorial market characterised by competition and cooperation between different political entrepreneurs and levels of government. Though the cities are at the forefront of these developments, *départements* and regions, too, participate in the emergence of a truly 'governmental' logic of centre-periphery relations (Eberlein, 1996, p.370).

French politicians and elites have, then, also had more room to manoeuvre than those in Britain. The clearest illustration of the overlapping nature of the French administrative structure is the way in which politicians can occupy several powerful positions at all levels, the so-called 'cumulation of mandates'. In the mid 1980s, for example, Jacques Chirac was both Mayor of Paris and French Prime Minister at the same time. In the UK, central government removed some of the powers and responsibilities of local politicians, although this situation is beginning to change as the Labour government partly reduces the grip of the state on local government, and introduces more of a regional administrative level. Political leaders in cities, though, have a slightly contradictory responsibility to reflect urban territorial cohesion, while equally representing the diversity of a city and addressing tensions of urban fragmentation (John and Cole, 1999). Nevertheless, as John and Cole summarise:

> The French pattern of leadership, operating in its political culture and state tradition, seems to offer more possibilities for creativity by giving mayors extensive resources

and a source of legitimacy from the electorate. The interchange between central and local politics by joint office holding reflects the intense localism of French society embedded in the centralised state, and such a culture exists far less in Britain (John and Cole, 1999, p.114).

Given these tensions and oppositions in central-local government relations and political leadership in both France and the UK, it is not surprising to find an increasing focus on the notion of partnership within urban political and economic agendas (Heinz, 1994; Newman and Thornley, 1996; Newman and Verpraet, 1999). This reflects the increasing difficulty for single 'urban' authorities in configuring a cohesive form of government that has the resources and the capacity to act in the interests of all city groups, communities and spaces. This difficulty is clearly further increased in the global city context. Many types of partnership have subsequently seen the light of day as a means of addressing this difficulty, varying quite substantially both within individual cities according to the differing projects being initiated (and sometimes within individual projects at different stages), and cross-nationally according to the differing political and cultural contexts within which the partnerships are being created. In France, for example, inter-communal cooperation has become more and more widespread, with communes belonging to an average of five cooperation organisations, which become 'a way of making the management of some services (water, waste management, etc) more coherent, and if possible, a way of creating some financial solidarity between communes' (Sallez, 1998, p.109). Another example of the different types and natures of partnerships in France and the UK has been provided by Nelson in her analysis of urban renewal projects in Paris and London (Nelson, 2001). She concluded:

> In the French case study the long history of collaboration between public and private sectors had created an institutional structure and a culture which was more conducive to cross-sectoral working. However, this more collaborative approach was not at the expense of public control over development. In the British case there was greater conflict between different levels of government and local government for a long period had little involvement in the redevelopment process. The initial emphasis on private sector initiative did not result in organisations working in partnership. However, to complete the process of urban renewal and secure the development of appropriate transport and social infrastructure a more collaborative approach was needed (Nelson, 2001, p.500).

This discussion of the notion of partnership within its differing cross-national public political contexts should not lead to the neglect of the inherent connotative breadth of this notion. A partnership does not have to be only between public sector actors, or initiated by public sector actors. As we shall see later in the study, private sector involvement in urban development partnerships is growing (Lorrain, 1991; Heinz, 1994; Newman and Verpraet, 1999), for example, with the emergence of *Sociétés d'Économie Mixte* (SEM) in France. As Nelson describes:

> Rather than passively awaiting invitations to submit tenders for the provision of particular services, the private sector is now anticipating the requirements of

municipalities and offering new products. Initially private sector firms became involved in providing services, but they are now extending their role and becoming involved in the conception of new projects (Nelson, 2001, p.485).

Urban and Economic Development Policies in France and the UK

Following the abolition of the level of metropolitan government in 1986, the urban and economic development of London (which therefore lost the GLC) has been the responsibility of the state and the one remaining elected tier of local government, the boroughs, but this was a lopsided, top-down style of power relationship, in which the emphasis has been on private sector-led urban entrepreneurialism. Physical capital and wealth creation overtook social capital and welfare delivery in policy importance (Parkinson, 1998, p.416). The government did create a number of separate organisations to deal with certain aspects of urban policy, such as the Urban Development Corporations (Pickvance, 1991), but in other cases, the availability of central government funding for local urban development or regeneration has been dependent on a competitive bidding process, as in City Challenge, which created at least as many losers as winners (Booth and Green, 1993).

Meanwhile, in contrast, two planning instruments in France have placed emphasis on intergovernmental relations. The *contrat de ville* has represented 'an effort to create links between the state and local authority or inter-communal groupings, within the framework of a contract negotiated between the two sides' (Parkinson and LeGalès, 1995, p.34). Booth and Green state how most people have seen the *contrat de ville* as a government attempt 'to produce a global policy that coordinated ministerial intervention in an effective manner and brought together all actors at the local level' (Booth and Green, 1993, p.387). This would include central and local government, and the voluntary and business sectors.

In a comparative study of urban policy in the UK and France, Booth and Green (1993) were able to identify some of the similarities and differences between the City Challenge and *contrat de ville* initiatives in particular. Notably, there was their parallel focus on partnerships involving the private sector, but with a substantial role for local authorities. However, on the other hand, the national specificities of political cultures and administrative arrangements still provided an important diverging overall context for the policies. There is, in addition, the French regional *schéma directeur*, prepared by the *Préfecture* and the *Direction Régionale d'Equipement* (both regional arms of the state), with the participation of the elected *Conseil Régional* (Llewelyn-Davies et al, 1996, p.34), which needs to consider the agendas and orientations of the state, the region and local authorities in its territorial planning negotiations.

One restricting factor in France is that the central government agency, the *Délégation à l'Aménagement du Territoire et à l'Action Régionale* (DATAR – the agency for territorial planning and regional action), is still the main body in the domain of economic development. With regard to Paris, this represents a further disadvantage as DATAR tries to discourage the location of major economic activity in the capital in favour of other French regions (Chevrant-Breton, 1997, p.152). Nevertheless, overall assessment of the urban policies of France and the UK seems to nearly always come out in favour of the former:

> The integration of levels through the accumulation of political offices gave French urban planning a degree of coordination lacking in the conflictual central-local relationships in Britain... As a result of conflict, London lacked the institutional framework around which a coherent pro-growth strategy could develop (Newman and Thornley, 1996, p.247).

One of the major plus points for the implications of the UK political culture on urban policy has been that urban policy in the UK has allowed public-private sector relations to flourish, whereas in France, it has been the state which has taken the lead, and the private sector has been relatively less involved in initiatives. Developments in London have been seen as part of the trend towards increased permeability of public and private-sector boundaries in urban governance (Newman and Thornley, 1997).

French urban policy has, nevertheless, been characterised by a search for better sub-national structures, and the use of SEMs for urban development. The decentralisation measures of the 1980s gave full responsibility for urban planning to municipalities (Preteceille, 1991). In Britain meanwhile, we have seen the use of quangos in urban policy, as well as new agencies at the subregional and city levels. Central government also encouraged economic objectives, while restricting social programmes. In France, it has been cities which have pursued economic goals, and central government has introduced urban social policy initiatives.

In France, both the national and local public sector orchestrated new city strategies, while in Britain, in London for example, private interests supported by central government have continually dominated the construction of city-wide visions. The private sector though has had difficulty in contributing to urban policy, as 'partnership' still implies an arrangement within the public sector. French urban planning, however, now also deals with major urban projects involving both the public and private sectors and local communities (Parkinson and LeGalès, 1995).

The ways in which these cross-national similarities and differences have been territorialised over the years in the urban and economic development of Paris and London will be discussed in a general manner in chapter 5. In summary, though, we need to examine the extent to which, and the ways in which, these differing intergovernmental systems, and urban and economic development policies of France and the UK influence the implementation, shaping, and growth of telecommunications developments in Paris and London.

The Roles of Differing Levels of Government in Telecommunications in France and the UK

Having discussed the cross-national differences in both telecommunications policies and regulation, and intergovernmental systems and urban policy, we can begin to put the two sets of policy contexts together and lay out some assumptions about the roles of differing levels of government in telecommunications in France and the UK.

As we have seen, in both telecommunications policies and political culture, the UK has recently very much been characterised by a neo-liberal stance on the part of the state. This led to early liberalisation of telecommunications markets on the one hand, and a recentralisation of administrative power and responsibility on the other hand. A market-led approach to the development of telecommunications in the UK, then, would appear to have designated a very limited role for local government. Indeed, the priority given to 'private enterprise and economic factors' (de Gournay, 1988, p.337) over the state diminishes any kind of territorial perspective or objective for telecommunications at all. As Graham writes: 'In fact, it could be argued that the combination of extreme political centralisation with the radical liberalisation of telecommunications markets makes British local authorities the most powerless over telecommunications in the western world' (Graham, 1996, p.78).

Telecommunications policies in France have had more of a *dirigiste* state interventionist nature to them. It was the state which instigated the intensive *Plan de Rattrapage Téléphonique* during the 1970s to bring the French national system up to speed with their neighbours. Up to the 1980s, then, French telecommunications policy largely failed to include much of a role for local government, as in the UK. In France, however, the reason had more to do with an actual conscious central government decision to prevent local government intervention, than its incompatability with an overall neo-liberal stance. As de Gournay observed:

> More than in other nations, the concept of local territorial jurisdiction underlies political cohesion, setting up tangible limits to the spatial extension of networks. It is therefore no surprise that the administrative reform carried out under Gérard Théry during the telephone boom of the 1970s was above all an attempt to impose a telecommunication management structure that did not correspond geographically to local administrations. The Directions Régionales des Télécommunications, or DRT (Regional Telecommunications Management Offices) and the Directions Opérationnelles des Télécommunications, or DOT (Operational Telecommunications Management Offices) were laid out in zones that purposely did not coincide with local administrative divisions. It was claimed that this was for technical reasons, but in fact its effect was to prevent local administrations from controlling network management. Through its 'extraterritoriality' the telecommunication management thought it would be able to escape the local political control that had traditionally blocked 'transnational' circulation of information (de Gournay, 1988, p.337; see also Briole et al, 1993, pp.74-75).

Decentralisation of some political responsibilities in the 1980s and the subsequent reaffirmation of an important role for local government in French politics has given *collectivités locales*[18] more input into urban and economic development policy. This suggests a far greater role for local government in telecommunications developments in France than in the UK (see Briole and Lauraire, 1991; Négrier, 1990; Vinchon, 1998b; Offner, 2000b), which, together with the continuing overall direction of the telecommunications sector by the state, seems to crucially recognise the inherent territorial nature of telecommunications (see Briole and Lauraire, 1991; Musso and Rallet, 1995; Lefebvre and Tremblay, 1998).

National telecommunications policies, intergovernmental systems, and urban and economic development policy are, then, part of the ways 'through which nation states can use their regulatory powers to ease or block movements of capital, labour, information, and commodities', thereby illustrating how 'the fading away of the nation state is a fallacy' (Castells, 1997, p.307; see also Hirst and Thompson, 1996). The cross-national policy context perspective of this chapter appears, therefore, to be a first crucial foundational element to our theoretical – empirical explorations of the differing, parallel territorial contexts and specificities of telecommunications developments in Paris and London.

---

[18] *Collectivité locale* usually refers to a local authority or commune, but, frequently, the term is also used to designate authorities at departmental or regional levels, which are actually *collectivités territoriales*.

Chapter 3

The Territorialities of Global Cities and Telecommunications

Introduction

This chapter considers and analyses from a theoretical perspective the second thematic strand of the study as a whole, namely the relationships between telecommunications networks and global cities. This is the second layer of our theoretical analysis, which, in parallel with the last chapter and chapter 4, will provide a solid base for our empirical investigations later in the study. The discussions in this chapter will be particularly relevant in relation to chapter 7, which examines the same thematic strand from an empirical perspective, yet, as mentioned earlier, this is not to deny the intertwined nature of our theoretical – empirical structure as a whole.

Whereas the first strand in this study concerns itself with the territorial contexts and specificities of the national level, the focus in this second strand is more on those of the urban regional level. We are interested here in how telecommunications infrastructures and services are being constructed and packaged to meet the requirements of global cities. Within this, there are at least three points of interest which need to be considered: firstly, the conceptualisation of global cities as all-powerful centralised nodes of the world economy; secondly, the broader forces underpinning the intensive equipping of global cities with telecommunications infrastructures; and thirdly, the steady emergence of a number of theories and conceptual ideas which have attempted to make sense of the increasingly complex relations between telecommunications and cities.

The Territorial Sites of Global Processes and Practices: Conceptualising Global Cities

'World cities' are not just national capitals and centres of national economies, but also concentrate a number of key functions in the global economy as hubs on global networks of exchange and transaction (Hall, 1977; Friedmann and Wolff, 1982; Friedmann, 1995; Knox, 1995). The literature on world cities has been concerned, to a large extent, with constructing urban hierarchies based primarily on the location of multinational companies in the key producer service industries (such as law, banking and insurance).

The more recent proliferation of interest in and studies of the 'global city', particularly from an economic viewpoint, owes much to the work of Saskia Sassen (1991; 2000). Here, the major cities of the world become command and control centres of the global economy through the ways in which they concentrate key activities and practices (including labour, skills and ICT infrastructure). Through her analysis of New York, London and Tokyo, she was able to suggest that global cities have four major (new) functions:

> first, as highly concentrated command points in the organisation of the world economy; second, as locations for finance and for specialised service firms, which have replaced manufacturing as the leading economic sector; third, as sites of production, including production of innovations, in these leading industries; and fourth, as markets for the products and innovations produced (Sassen, 1991, pp.3-4).

We do not have the space here, however, to discuss in detail the complex changes in the nature of global finance that are inherently bound up in the emergence of these command points of the world economy (for a summary, see Fainstein, 1994, pp.27-33; Sassen, 1991, chapter 4). More relevant from the point of view of this study is the relationship, or apparent lack of relationship, between global cities and their territorial contexts and specificities.

Despite asserting that 'they are specific places whose spaces, internal dynamics, and social structure matter; indeed, we may be able to understand the global order only by analysing why key structures of the world economy are *necessarily* situated in cities' (Sassen, 1991, p.4), Sassen, like many writers on global or world cities, does not devote much attention to the specific territorialities of these cities. There is only limited focus, for example, on local and urban governance, the continuing national regulation of space, and their inherent historical contexts and positionalities. The dominant view, in most of the global or world city literature, seems to be of cities 'whose form depends more on the world economy than on local conditions' (Longcore and Rees, 1996, p.355), or cities which are merely 'a function of a network' (Sassen, 2000, xiii). In this way, it has actually been suggested that there are no global cities, but only one networked global city. Castells, for example, argues that:

> [t]he global city is not a place, but a process. A process by which centres of production and consumption of advanced services, and their ancillary local societies, are connected in a global network, while simultaneously downplaying the linkages with their hinterlands, on the basis of information flows (Castells, 1996, p.386).

While Sassen emphasises, then, the cities (but not so much their territorial specificities) and their agglomerative processes, Castells sees the connections to networks, and the characteristics of these networks, as the primary factor in the power of certain cities. As John Allen suggests: 'In other words, it comes back to the question of whether the networks themselves 'generate' cities as sites of power through their interconnections *or* whether cities 'run' the networks through their concentration of resources and expertise' (Allen, 1999, p.202). However, while we will come back to the work of Castells in more detail later in the chapter, we should note here that something seems to be missing from both these accounts, namely territoriality. Sassen may focus more on the power of cities than the power of networks, but her view of how the former derives from concentrations of particular command and support activities in New York, London and Tokyo seems to neglect looking inward to the inherent territorial specificities of these cities in favour of looking outward to see how their activities relate to and shape global networks. It seems difficult in some ways, then, to concur with the emphases of either Sassen or Castells on process and flow over place and territoriality. New York, London and Tokyo remain unique urban places with their own historical conditions and social and cultural practices. Their function as command positions in the world economy takes full account of, and is bound up with, these conditions and practices, instead of ignoring them. Therefore, the form of global or world cities emanates from the dialectical relations between the specific local conditions and their command functions in the world economy – that is, the local conditions can be shaped by the command function, but probably only after the command function has been significantly shaped by the local conditions. This study will go on to explore how the parallel shaping of the articulations between telecommunications/IT and the global cities of Paris and London certainly does not lead to the latter becoming somehow more aterritorial in nature. Indeed, such cities are actually in the midst of subsequent reterritorialising processes and practices.

Global Cities as Financial Centres

Processes of globalisation have been said to be menacing the future importance of cities (O'Brien, 1992). It is something of a paradox, then, that it is in global cities, and in particular in those global cities which are the major financial centres of the world, where these processes can be most clearly seen. Leslie Budd argues that:

> If one claims that globalisation is a 'radicalising of modernity', as Giddens does, then globalisation cannot be divorced from the development of the metropolis. The basis of the metropolis is territoriality and the basis of the comparative advantage of financial centres is based upon agglomeration economies (Budd, 1995, p.359).

He goes on to suggest that 'even where electronic trading systems appear to overcome the constraint of locality, one finds that their development and that of global alliances between exchanges is informed by notions of territory and territoriality' (Budd, 1995, p.359). These territorialities within and between financial centres have many endogenous 'social and cultural determinants' (Thrift, 1994), which are likely to ensure the continuing specificity of financial centres, despite their threatened domination by electronic spaces of flows:[19] 'Indeed, the volume and speed of such flows may make it even more imperative to construct

---

[19] See Pryke (1991) for an example of how the social and territorial traditions and contexts of the City of London have interrelated with changes in the flows, networks and markets of international finance, leading to the formation of a 'new spatial matrix'.

places that act as centres of comprehension' (Thrift, 1994, p.337). In particular, the continuing, and even increased, need in financial centres for face-to-face communication to exchange ideas and information has been well documented (Pryke, 1991; Thrift, 1994; 1996b; Thrift and Leyshon, 1994), and is one of the major undiminished elements in 'the compulsion of proximity' (Boden and Molotch, 1993). The territorially localised production complexes which continue thus to 'dominate' global capitalism, such as Santa Croce in Tuscany and the City of London, have been viewed as 'neo-Marshallian nodes in global networks', requiring territorial specificities such as know-how, skills and institutional and technological infrastructures in order to retain their competitive advantage (Amin and Thrift, 1992). Thrift and Leyshon reflect further on the continuing territorial requirements of the international financial system by projecting that:

> the interdependent connectedness of disembedded electronic networks promotes dependence on just a few places like London, New York and Tokyo where representations can be mutually constructed, negotiated, accepted and acted upon. In effect, these are the places that make the non-place electronic realm conceivable (Thrift and Leyshon, 1994, p.312).

Martin also supports such a view by stating that 'while the speed of information communication has annihilated *space* it has by no means undermined the significance of location, of *place*' (Martin, 1994, p.263). For example, in a study of the financing of the restructuring of office provision in the City of London in the mid to late 1980s, Pryke showed the pertinence of both traditional and reconfigured sociospatial relations and practices:

> The task of physically extending the geography of the City's property markets as it turned global was not only about rethinking the financing of an element of the City's structure of office provision, it also involved bringing new territories within the social boundary of the old City *and* building around a mix of sociofinancial practices, that brought together the latest in high-technology dealing floors and the most traditional of banking and broking practices, previously rooted in the old spatial matrix (Pryke, 1994, p.245).

The inherent geographical nature of financial production seems, then, to be becoming increasingly recognisable (Lee and Schmidt-Marwede, 1993; Pryke and Lee, 1995).

However, following on from all this, the reciprocal relationship between globalisation processes and global financial centres is further complicated and altered by the changing and seemingly ever more contradictory role of the nation state. If the continuing importance of the city in the light of globalisation has been illustrated by the various 'territorial fixes' of the international financial system, then it has not been as much the case for the nation state. The importance of London, New York and Tokyo as financial centres is not so closely tied to national economic regulatory practices as in the past. A kind of 'cities versus states' theory has begun to take shape revolving around 'the release of cities towards a more global destiny' (Taylor, 2000; see also Sassen, 2002). Thrift and Leyshon also talk instead of the emergence of a new 'phantom state':

> both constituting and representing money power, that is based on the communicative power of electronic networks and a few, selected (g)localities. It consists of actor-networks which increasingly rely on money power and communicative power without having to call on the degree of bureaucratic administrative power usually associated with the state form. [...] To an extent at least, the old nation-state form has been outfoxed by a combination of money power and communicative power. In one sense, the power of the new 'phantom state' is still based in institutions, but in another sense it is based in the flow of communication itself. [...] More and more, we might argue that, in the modern world, money power and communicative power have been able to replace state authority based on administrative power with a discursive authority which is based in electronic networks and particular 'world cities'. This discursive authority is the stuff of a phantom state whose resonances are increasingly felt by all (Thrift and Leyshon, 1994, pp.323-324).

In the intense competition between global financial centres such as London, New York and Tokyo, advantages are narrowing. London is facing increasing competition from New York in particular, and, against the backdrop of a growing knowledge-based economy, developments in ICTs and electronic financial flows (Local Futures Group, 1999b; Power, 2000a), will, at some point, have to find ways to hold its current position. Thrift (2000) has suggested that innovation will be crucial to the City of London, as will the role of ICTs in enhancing the quality of financial information, so vital in the past. As Power observes then: 'The example of recent moves to create a pan-European stock market demonstrates that increasingly rivalries between financial centres are being 'fought' and 'forged' through ICT issues' (Power, 2000b, p.5).

Global Cities and Telecommunications

The focus, then, for much work in geography and urban studies on global cities has been the continuing territorialities of the international financial system.[20] This work has become a key element in explorations of the ways in which cities in general develop through juxtapositions or tensions between processes and practices of fixity and mobility. Even more specifically, however, a body of work has emerged in recent years to construct theoretical perspectives on one aspect of this urban fixity–mobility duality, namely the relations between global cities and telecommunications infrastructures. It is to these perspectives that we now turn in this section.

As we have touched upon, one of the major forces in the broad development of global cities, and in the restructuring of their financial and business sectors, has

---

[20] Although social and cultural processes and practices in global cities have certainly not been ignored (see Eade, 1997).

been information and communications technologies. Power summarises why these technologies are so crucial to these sectors in particular:

> More recent times have seen a rapid increase in the level of dependence upon and use of high technology and telecommunications in the financial services sector. This technologisation of the sector has been mainly due to the computerisation of trading operations, increases in cross border trading activities, the increasing complexity of markets and instruments, and associated needs for high speed state of the art communications and computer networks. Many of the major banking and finance activities that City firms are now engaged in are either entirely electronic or entirely reliant upon telecommunications for their functioning and settlement. Under such conditions the speed, quality, and reliability of telecommunications services as well as product availability and cost are of paramount importance to firm's access to markets, and competitive advantage within; if a firm can transfer information from New York to London a second or two faster than their rivals then they will be the one to make the profit (Power, 2000b, pp.7-8).

Whilst the importance of these technologies for global cities has been generally recognised, relatively few empirical studies in geography and urban studies have investigated in any great detail their appropriation and implications. Given that this is a key objective for us in this study, we must consider and analyse the emergence of a body of work which is now attempting to theorise the complex relations between (global) cities and information and communications technologies.

The Rise of Informationalism

The starting point for theories of cities and ICTs has been the increasing importance of notions of information and knowledge within western economies, societies and cultures. A new social structure, 'associated with the emergence of a new mode of development, informationalism, historically shaped by the restructuring of the capitalist mode of production towards the end of the twentieth century' (Castells, 1996, p.14), has evolved in recent years. Information and knowledge have become the key assets of an emerging informational economy. High technology industries are involved in the production of information processing.

The informational economy may be developing on a global scale across national boundaries, but it remains bound up with a number of political processes emanating from the state. The global informational economy is, therefore, characterised by the interaction between an increasing use of information and communications technologies, the continuing requirements of mobile capital for territorial fixity in certain places around the world, and the continuing importance of local, regional and national specificities and contexts to the dual dynamics of this territorial fixity of both capital itself, and the telecommunications (and other) infrastructures which serve its production and reproduction.[21] Territorial fixity in

---

[21] Not the least of which are the employees and labour processes and practices of the 'information society' (see Sussman and Lent, 1998).

the global informational economy has become concentrated in, amongst others, offshore banking centres, global back offices and global cities (Warf, 1995). This process has also become tied up with a reproduction or renewal of social and spatial inequalities, as:

> Telecommunications simultaneously reflect and transform the topologies of capitalism, creating and rapidly recreating nested hierarchies of spaces technically articulated in the architecture of computer networks. Indeed, far from eliminating variations among places, such systems permit the exploitation of differences between areas with renewed ferocity (Warf, 1995, pp.375-376).

Cities, Telecommunications, and the Circulation of Capital

With so much being discussed and written about in recent times, relating to the pace and intensity of political, cultural, and in particular, socio-economic change (see Harvey, 1989b; Castells, 1996; 1997; 1998, for analyses of how all these interrelate), it is necessary to recall briefly the underlying currents which can have a significant bearing on the development of urban telecommunications strategies. Equally however, it is important to emphasise the ways in which telecommunications technologies can have a significant bearing on the nature of these underlying currents. The expansion of telecommunications, for example, is widely recognised to be inherently bound up in the restructuring of the capitalist mode of production. For David Harvey (1989b), the means by which information technologies promote both the overcoming of the distance between two places and the more rapid movement of flows of capital around the globe signifies the new concept of the process of 'time-space compression'. However, globalisation and time-space compression processes have been around since the beginnings of road and rail communications, for at least two hundred years (Mattelart, 2000).

Nevertheless, this would appear to tell only half the story of the interrelations between telecommunications and the circuits of global capital. As Michael Storper makes clear, 'global capitalism is being constructed through interactions between flow economies and territorial economies' (Storper, 1997, p.181). In order to expound the relationship between telecommunications and global capital further, we need to draw a parallel between 'the implications of the contradiction between *fixity* and *motion* in the circulation of capital − between capital's necessary dependence on territory or place and its space-annihilating tendencies − for the changing scalar organisation of capitalism' (Brenner, 1998b, p.461) and the implications of the same contradiction in the expansion of telecommunications − between the physical, place-based infrastructures that are the foundation of telecommunications networks and their promotion of connections to the intense flows in global space − for the differing scalar organisation of urban telecommunications strategies. In other words, as Graham elaborates:

> relatively immobile and embedded fixed transport and telecommunications infrastructures must be produced, linking production sites, distribution facilities and consumption spaces that are tied together across space with the transport and

communications infrastructure necessary to ensure that a spatial 'fix' exists that will maintain and support profitability (Graham, 1998, p.176).

The notions of 'fixity' and 'motion' here though seem to offer a further example of what Latour (1993) calls 'a perverse taste for the margins', in so far as defining what is fixed necessitates contrasting it with what is dynamic, and vice versa. Just as 'the words 'local' and 'global' offer points of view on networks that are by nature neither local nor global, but are more or less long and more or less connected' (Latour, 1993, p.122), so 'fixity' and 'motion', which closely relate to 'local' and 'global' must offer similar points of view on networks, such as those of capital and telecommunications, as cited above. Nevertheless, there seems to be an implication that 'fixed' cities are waiting for 'mobile' capital. Such a view does not take into account the potential autonomy of cities as a whole, and in particular the ways in which city authorities and other actors can shape global circuits of capital (see Hubbard and Hall, 1998, p.16).

Neither does it take into account the supply-side logics of the relationships between the 'network firms' providing the capital and the transformations of global – local spatial organisation (Rimmer, 1997). In an analysis of the level of globalisation of Japanese firms in the container shipping, telecommunications and airline industries, Rimmer was able to conclude that 'all three activities are apparently part of a wider corporate restructuring process which uses strategic alliance activity to address changing market forms' (Rimmer, 1997, p.110). Such a focus on the interaction between global – local restructuring and important national firms is rare in the literatures of geography, planning, urban studies and sociology, and suggests a useful avenue of research for the future.

The Space of Flows Meets the Space of Places

Perhaps the key theoretical idea in recent years relating to the interactions between cities and information and communications technologies has been Manuel Castells' deliberations on the nature of our 'network society', and specifically, the juxtaposition of a space of flows and a space of places (Castells, 1989; 1996). In the first volume of his colossal 'Information Age' trilogy, 'The Rise Of The Network Society', Castells argues that the principal spatial form of the global informational society and economy is the space of flows: 'the material organisation of time-sharing social practices that work through flows' (Castells, 1996, p.412). This is not to suggest that we have only recently begun to experience the structuring practices of networks. As Armand Mattelart has convincingly shown, the networking of the world began in the late eighteenth century: 'The widespread interconnection of economies and societies is the ultimate goal of global integration initiated at the turn of the nineteenth century [...] ...networks have never ceased to be at the centre of struggles for control of the world' (Mattelart, 2000, vii-viii). Nevertheless, society has never been more networked than it is today, so it is worth spending some time here analysing how the notion of the 'space of flows' fits in with this study, because, as Castells himself makes clear, it is a little more complex than it might first appear.

Indeed, the space of flows is actually made up of three layers of (mutually constituting) physical supports: a circuit of electronic impulses such as telecommunications; a network of place-based nodes and hubs; and the spatial manifestations of dominant groups and interests (Castells, 1996, pp.412-418). In this way, the technological infrastructure of the first layer provides a foundation for contemporary socio-economic processes, which take place both in and between the key cities of the second layer, while the powerful and wealthy elites control, configure and articulate their own spatial requirements in the third layer, frequently at the expense of the disconnected masses.

This complex vision of the space of flows negates therefore both any idea that it is purely about global telecommunications connections, and the suggestion that it is strictly placeless. Indeed, Castells (1997) goes on to illustrate through numerous examples the different ways in which communities and nations are trying to shape and configure their places in relation to the logic of the space of flows. This is a process quite frequently characterised by struggle and resistance. It is argued that, given the general urban polarisation tendencies of many IT and telecommunications developments, such local, place-based, bottom-up strategies, shaping territories and telecommunications in parallel, are becoming more and more important (Graham, 2000c). Alan Southern, for example, offers a good example of how the space of flows is necessarily bound up with the space of places through an investigation of the way the local governance of differing ICT projects in Sunderland became an effort at territorialising the former, through both the development of the Doxford International Business Park and the recycling of computers for the benefit of local communities (Southern, 2000).

While Castells argues that the space of flows is the dominant spatial form of the network society, 'because it is the spatial logic of the dominant interests / functions in our society... [and] societies are asymmetrically organised around the dominant interests specific to each social structure' (Castells, 1996, p.415), he does not necessarily mean that the space of flows is completely superseding the space of places. The space of places remains important because it is the spatial perception and experience of the numerical majority, even if power and function exist in the space of flows and influence the dynamics of places (see also Allen, 1999).

So how does the space of flows and its relationship to the space of places manifest itself in this study? We have to be concerned with all three layers of the space of flows – how the technological infrastructure of the first layer in the form of telecommunications developments is being territorialised in the global city nodes of the second layer in the form of Paris and London, to support these production sites of the global economy. It is, however, the third layer, the actual spatial manifestations of dominant groups and interests, which would appear to hold the key to how telecommunications infrastructures are being shaped and territorialised to bolster Paris and London as important global cities. In essence then, we are investigating two dialectical interplays in parallel:

- The interplay, in terms of telecommunications developments, between Paris and London as nodes in the global economy and the space of flows, and Paris and London as places, where national and urban processes and practices are territorially specific, rather than necessarily supportive of, or responsive to, some global logic.

*   The interplay, in terms of Paris and London, between telecommunications developments as the supportive infrastructure of the global economy and the space of flows, and telecommunications developments as place-based infrastructures, which are influenced by territorially specific national and urban processes and practices.

The one question which seems most pertinent in relation to the theory of the space of flows, especially in the light of our discussions in Chapter 2, is 'where is the national level in the space of flows?' Does the space of flows not give limited consideration to still very relevant national processes and practices, and to inter-urban and intra-urban competition, because its second layer, a network of place-based nodes and hubs, suggests the existence of a rather homogeneous 'network' of rather homogeneous (and non-competitive) 'nodes'? These are questions which we will explore throughout this study, particularly in our empirical investigations into telecommunications developments in Paris and London in Part Two.

Cities as Nodes on 'Global Grids of Glass'

Cities are, then, on one level, the nodes and hubs of the space of flows. It is within them that telecommunications networks, the physical infrastructure of the space of flows, concentrate. Finnie, for example, observed how:

> when London-based Cable and Wireless plc and WorldCom laid the Gemini transatlantic cable – which came into service in March 1998 – they ran the cable directly into London and New York, implicitly taking into account the fact that a high proportion of international traffic originates in cities. All previous cables terminated at the shoreline (Finnie, 1998, p.20).

Yet, however developed their urban infrastructures are, a single city remains a single node on a global network of connections:

> Dedicated optic fibre grids *within* the business cores of global cities are of little use without interconnections that allow seamless corporate and financial networks to piece together to match directly the 'hub' and 'spoke' geographies of international urban systems themselves (Graham, 1999, p.945).[22]

The study of telecommunications networks would therefore seem to demand a more relational perspective:

> by attempting to address intra-urban, inter-urban and transplanetary optic fibre connections (and disconnections) in parallel... [for] discussions of restructuring *within*

---

[22] The recent work of Edward Malecki offers interesting analysis into these inter-urban fibre backbone networks from an American perspective (see, for example, Gorman and Malecki, 2000; Malecki, 2003).

cities increasingly must address the changing relations *between* them, whilst also being cognisant of the importance of these changing relations within broader dynamics of geopolitics and geoeconomics (Graham, 1999, p.931).

Such a relational perspective on cities and telecommunications concerns a reconfiguring of our ideas of geographical scale, and also a recognition of the analytical importance of either what lies beyond these networks or their wider socio-spatial implications, in other words the creation of disconnections as well as connections (Castells, 1996, p.404; Amin and Graham, 1999). As Graham suggests: 'Such infrastructures are carefully localised to include physically only the users and territory necessary to drive profits and connect together global corporate, financial and media clusters' (Graham, 1999, p.947). These are the 'premium network spaces' of global cities (Graham, 2000b),[23] which seem to be an example of how global processes of spatial and economic restructuring become territorially fixed: 'Being terminal nodes of networks through which direct connections with distant locations are maintained without impediments, and being points of reference for 'communities without propinquity', these clusters frequently become the places where globalisation is condensed' (Ezechieli, 1998, pp.25-26). On another level, even specific buildings within global cities have become the focus of these networks, and of the wider global economy. Pawley (1998) describes this feature as 'terminal architecture', because, within a context of information technology and dispersed urban nodes, these buildings (banking and financial service premises, computer centres, warehouses and the like) are more important for being practical 'terminals for information' than for their aesthetical appearance.

The development of these premium network spaces in cities is a prime example of the ways in which the proliferation of contemporary telecommunications networks seems to be bound up in parallel processes of uneven socio-spatial development (see Graham, 2000c). This is a mutually constitutive form of uneven development as well. Massey reflects on how:

> it does seem that mobility and control over mobility both reflect and reinforce power. It is not simply a question of unequal distribution, that some people move more than others, some have more control than others. It is that the mobility and control of some groups can actively weaken other people. Differential mobility can weaken the leverage of the already weak. The time-space compression of some groups can undermine the power of others (Massey, 1993, p.62).

By this token, the time-space compression of premium network spaces in global cities can weaken the level of connectivity of juxtaposed local spaces and communities. A high level of investment in infrastructure deployment in a concentrated business quarter is likely to diminish in some way the possible level of investment in nearby residential areas. In this way, Graham and Marvin (2001) have produced a significant typology of three interacting forms of 'infrastructural

---

[23] See Coutard (2002) and Graham (2002) for further discussion of the concept of 'premium network spaces'.

bypass' to develop an understanding of these complex relations between urban restructuring and 'unbundled infrastructure':

- 'Local bypass' is the development of new or segmented parallel infrastructure networks, which offer an alternative to more traditional networks for favoured groups and spaces, whilst neglecting less favoured groups and spaces.
- 'Glocal bypass' is the development of new infrastructure networks, which connect the favoured groups and spaces of a city to other favoured groups and spaces in cities across the globe. As in 'local bypass', these networks leave many urban groups and spaces disconnected.
- 'Virtual network bypass' is the development of information technology systems to allow the privileging of a proliferation of 'seamless' competitive services over single infrastructure networks, aimed at specific types of users. This process promotes the distinction of favoured and less favoured groups and spaces, both in itself and as a support to 'local bypass' and 'glocal bypass', thereby suggesting an intensification of the socio-spatial implications of these latter forms.

In terms of telecommunications, the 'premium network spaces' of global cities are most commonly associated, then, with the 'glocal bypass' networks of private fibre optic infrastructures, specifically targetted at key business and financial areas. Related to questions of profitability, there are also technical reasons why fibre networks tend to be concentrated within and between these premium network spaces. As Longcore and Rees state:

> Fibre-optic wire, although exceeding traditional twisted copper lines and coaxial cable in carrying capacity, speed, security, and signal strength, is not as easily spliced and hence its use favours high-volume, point-to-point communications. Fibre-optic systems, therefore, are used to link major hubs and thus reinforce the existing urban hierarchy (Longcore and Rees, 1996, p.356).

In this way, the parallel configurations and articulations of place and telecommunications infrastructures – of the space of places and the space of flows – seem to be increasingly bound up with socio-territorial disjunction and uneven development. We will return in more detail to this point in chapter 4.

Urban Competition and Place Marketing

An important element in the relationships between global cities and telecommunications is the way in which these are configured in parallel as part of wider competitive objectives. Global cities have to be inherently 'competitive' in any case, as the roles they principally undertake within a world economy are derived from their advantages or assets in relation to other cities. In this context, the place marketing efforts and actions of urban economic development agencies are seen to be crucial (see Kearns and Philo, 1993; Gold and Ward, 1994).

However, the world economy is not characterised by numerous command centres, which all fulfill differing and complementary roles in the whole. There is significant overlap in the roles of global cities, which means that each city is looking to concentrate or agglomerate in itself a majority of capital, inward investment and multinational firms, at the expense of other cities, or in other words, 'securing economic growth by finding the most suitable forms of insertion of local economic (and often, indeed, micro-economic) spaces into the broader spatial division of labour' (Jessop, 1998, p.80).

This process is viewed as having increased as the importance of national frontiers has decreased (Lever, 1993; Jensen-Butler, 1997), so that global cities now appear to have more in common with, and more connections between, each other, than with their smaller national counterparts (Sassen, 2000). This latter observation may be true in certain respects, but the overall impression of much work on inter-urban competition is of vast, sweeping statements in which the 'impact' of globalisation is necessarily to reduce the influence of the nation state and to create supranational urban systems in which all cities are competing for every bit of economic activity. These assertions are often not backed up by empirical evidence. Thus, Lever suggests that: 'Cities now increasingly perform in ways which are transnational. They compete for mobile capital, employment, institutions and events. [...] there has been throughout the 1980s an increasing tendency for cities to break out of their national identities...' (Lever, 1993, p.947). In a similar vein, Jensen-Butler argues:

> Competence is gradually being transferred from states to regions and cities, which has in itself increased differentiation in the spheres of public service provision and regulation of economic activity and has thus intensified interurban competition. The local state becomes more differentiated and these differences are important for investors (Jensen-Butler, 1997, p.6).

The most common source of evidence for such assertions is not detailed and comparative empirical investigation of individual cities, but generalised league tables of cities, containing broadly defined data on employment, Gross Domestic Product (GDP) per capita, inward investment, multinational firms, and the like. It is not difficult, as a result, to share the opinion of Leslie Budd, when he argues that the notion of territorial competition is as much of an abstraction as the notion of globalisation (Budd, 1998).

Customising infrastructure networks for urban competition   The ways in which local and regional authorities are privileging the importance of infrastructure networks such as telecommunications within their territorial competition strategies are becoming increasingly diverse. Clarke and Gaile call this 'the fourth wave of local policy initiatives', whereby there is 'greater attention to the integration of human capital and economic development concerns and to the trade links and information infrastructure necessary to link local economies with the global web' (Clarke and Gaile, 1998, p.183). From a North East England perspective, Southern (1997) has drawn up a typology of local authorities according to how favourably they view the potential of IT in economic development, thereby demonstrating that the instigation of telecommunications strategies depends, to a large extent, on how 'computopian' a local authority is. Meanwhile, within a more general infrastructural context, Peck also utilises a North East perspective to show how:

> once all other criteria have been satisfied, success in levering-in new investment can still depend critically upon the ability of public authorities, regional organisations, and development agencies to produce and reproduce an infrastructure which is customised to the requirements of incoming firms (Peck, 1996, p.328).

In territorial competition, then, there seems to be an increasing focus on possible competitive advantages to be gained from being able to provide a high quality infrastructure package tailored specifically to the demands of either individual firms or a small group of firms:

> In some cases, public investment in infrastructure may create the 'collective' and 'integrative' basis of economic activity but some forms of expenditure can become 'individualised' and 'exclusive' to a very narrow range of users. Inward investors may be interested not only in the general modernity of the infrastructure of a region, but also in the degree to which they can exercise control over its present and future development (Peck, 1996, p.337).

In this way, then, we are beginning to touch on the complex sets of public – private sector relations and tensions bound up in the construction and packaging of infrastructure networks such as telecommunications in key cities. It is perhaps unsurprising that these relations, and the subsequent deployment of infrastructures, are, in turn, inherently bound up with issues and notions of power and control.

Globalisation and the 'Overexposed City'?

More than three decades ago, Marshall McLuhan predicted the emergence of a 'global village' which would mean that the city 'as a form of major dimensions must inevitably dissolve like a fading shot in a movie' (McLuhan, 1964; quoted in Amin and Graham, 1997, p.412). Perhaps for some, the globalising forces of the last two decades or so mean that we are well on the way to seeing a 'global village' eventually, but very few would argue that the city is somehow disappearing, or even diminishing in importance as a result. In fact, far more would argue that globalisation is augmenting the importance of urban places. It is clear though, that the question of the extent to which relations and processes at the urban scale are influenced by forces at the global scale is becoming more and more critical (Beauregard, 1995). As we have seen, some authors such as Castells are intent on arguing and demonstrating how the global city is more of a process than a place.

In this one respect, his emphasis on processes / flows / networks, Castells is perhaps not so theoretically distant from Paul Virilio and the latter's philosophies of speed, technology and virtuality. As Luke and Ó Tuathail have suggested: 'Typically, Virilio's argument is a more extreme technological vision of the literature identifying the emergence of an interlinked system of global cities' (Luke and Ó Tuathail, 2000, p.377). On the other hand, however, we can clearly distinguish between the reterritorialisation arguments of Castells and the deterritorialisation theses of Virilio. For the former, the increasing location of power and control in the space of flows does not lead to the disappearance or the diminishing importance of the space of places, but in many cases, the former actually rejuvenates and refocuses the influence of the latter and its unique set of territorialities (Castells, 1996). By contrast, the latter has envisioned a new era in which, with cities becoming increasingly 'overexposed' and 'disorganised' through their relations with ICTs (Virilio, 1987), the urbanisation of real time has become predominant over the urbanisation of physical space (Virilio, 1997). For Virilio, 'Territory has lost its significance in favour of the projectile. *In fact, the strategic value of the non-place of speed has definitely supplanted that of place*, and the question of possession of Time has revived that of territorial appropriation' (Virilio, 1986, original emphasis; quoted in Luke and Ó Tuathail, 2000, p.370). His hyperconcentrated omnipolitan metropolisation thesis (Virilio, 1997) has a familiar connection – disconnection implication: 'Politics is eclipsed by technology as citizens separate out into either caches of netizens networking in the fast lanes of the global economy or the trashbins of lumpen techno-proletarians stuck at the dead ends of networks' (Luke and Ó Tuathail, 2000, p.377). Yet, in contrast to Castells and other writers, Virilio seems to deny the intrinsic territorialities of technology and of networks. The key difference, then, is an understanding of the continuing spatial variations and territorial specificities of technological networks for one, and a distinctly unrelational reading of these technological networks as omnipotent and homogeneous for the other.

Many writers have, however, begun to question the overpowering influence of interrelated technological and globalising processes and their supposed responsibility for 'the end of geography' or 'the death of distance', observing instead that 'globalisation seems to demand localisation in the favoured places of the emerging planetary urban system' (Graham and Healey, 1999, p.632). For Swyngedouw, 'every social activity is *inscribed in space* and *takes place*' (Swyngedouw, 1993, p.306). Perhaps one of the key hypotheses of this study, therefore, is that the space of flows is not increasingly displacing the space of places, but that connections and mobilities in the space of flows depend intrinsically on telecommunications and IT being integrated into localised development in the space of places. As Amin and Thrift suggest: 'globalisation need not necessarily imply a sacrifice of the local… [as] global processes can be 'pinned down' in *some* places, to become the basis for self-sustaining growth at the local level' (Amin and Thrift, 1994, p.11).

However, accepting that urban telecommunications initiatives have to be territorially fixed brings up a new set of prejudices to take into account, as these infrastructures are usually explicitly deployed to serve key, profitable sectors and users. This deliberate concentration of connections in a small number of cities around the globe, or in certain parts and among certain groups in those cities, is symbolic of 'a new geography of the centre' (Sassen, 2000). In addition, this relationship between the infrastructure and the region into which it has been implanted is necessarily complex, as '[the] uneven penetration [of global forces] is not simply a matter of *which* institutions, industries, people, and places are affected but also *how* they are affected. They might be overwhelmed, exploited, or enhanced' (Beauregard, 1995, p.235). Overall however, in the words of Amin and Thrift, '[i]ncreasingly, the pressure posed by globalisation is to divide and fragment cities and regions, to turn them into arenas of disconnected economic and social processes and groupings' (Amin and Thrift, 1994, p.10). Such fragmentation has led to the increasing relevance of the concept of the 'dual city' (Castells, 1989).

Networks, Territoriality and 'Network Urbanism'

For Manuel Castells, then, networks have become 'the new social morphology of our societies' (Castells, 1996, p.469). Yet, as has already been emphasised, the presence of networks in cities is not a recent phenomenon, nor has it only recently started to be remarked upon. We should talk about a more intense network society, instead of 'the rise of the network society' (Castells, 1996). By the same token, 'network urbanism' cannot be seen as a novel concept for cities – indeed, Dupuy (1991) himself argues that the theoretical basis of 'network urbanism' is as old as urbanism itself – but has become more widespread and, in some cases, has taken on a grander scale than before, with the proliferation of IT and telecommunications infrastructures being deployed in urban regions.

The development of theoretical approaches for understanding the interaction between networks and urban places, or territoriality, has been a particular feature of French literatures, in which this discussion has been arguably most extensive and revealing (see, for example, Dupuy, 1991; Cassé, 1995; Offner and Pumain, 1996). The technical networks underpinning the city appear to have been understood to be a crucial component within urban planning theory and practice in France more universally than in many other countries.

The history of the notion of 'network' shows that it has become a progressively more complex notion (Offner, 1999). Offner charts the chronological development of the meaning of the term from the time of Saint-Simon around the beginning of the nineteenth century through to our age of 'interactive computer modes'. According to his narrative, components of networks that we would think to be common today such as connectivity, speed, and relational perspectives were not among the key characteristics of earlier networks (Offner, 1999).

The ubiquity of the notions of 'network' and '*réseau*' needs some explanation. It would appear that this can be at least partly derived from the differing broadness of their definitions and connotations. For both Dupuy (1991) and Offner (1999), the French word '*réseau*' is more abstract and 'polysemous' than the word 'network'. Dupuy, indeed, finds 'network' to imply in particular the cabled networks of infrastructure such as telecommunications.

However, 'network' and '*réseau*' do arguably share major implications: 'The network ['*réseau*'] opens things up; the network ['*réseau*'] closes things in. The whole symbolic power ('*imaginaire*') of the notion lies in this paradox' (Mercier, 1988; quoted in Offner, 1999, p.218). We could suggest here that it is not actually the network which opens things up and closes things in – it is the various social actors who develop the network, and the (power) relationships between them. To see it in any other way is to adopt the rather simplistic, determinist view of networks themselves resulting in connections and disconnections. However, ignoring this point and the linguistic differences, if a network is simultaneously an open and closed system, then one of the main questions in any study of networks needs to be 'to whom is it open, and to whom is it closed?' In other words, which groups and spaces does the network connect, and which groups and spaces does it disconnect?[24]

In many ways, it has also been French writers who have devoted most time to investigating the nature of the links between networks and territory. For example, Jean-Marc Offner summarises these links:

> To understand the true nature of the territorial efficacy of networks, it is enough to accept the following two propositions: networks make possible the creation or the strengthening of interdependencies between places. These places may then be considered as belonging to one territory. In other words, it is through the networks that territories form a system (Offner, 2000a, p.170; see also Offner and Pumain, 1996).

Here then, Offner stresses the importance of seeing network connections as the means by which places belong to the same territorial system. From this, we can imply that a disconnected place cannot be viewed as part of this territorial system, even if it is geographically proximate to a connected neighbouring place. Linked in to this, for Alain Lefebvre (1998, p.22), there seem to be three ways of considering the relationship between networks and territory:

- *inscription territoriale des réseaux* / territorial enrolment of networks;
- *territoires en réseaux* / networked territories;
- *territoire des réseaux* / territory of networks.

From one angle at least, these three perspectives do not seem to be distinctive so much as successive, as the initial development of networks is influenced by the territorial specificities of particular places, leading to the creation of a number of specific networked territories. Following on from this, building on the assertions of Sassen and Castells from earlier in the chapter, and Offner's place interdependency through network connections, a single territorial system may eventually emerge in which previously separately connected territories are themselves connected globally. However, while Sassen and Castells emphasise the homogeneous functions of global cities within a territory of networks, we can perhaps instead view this tripartite network – territory perspective as parallel and incremental in nature rather than sequential. In this way, a global city such as Paris or London is not just part of a 'global' territory of networks, but also at once both a specific networked territory, and a place where the development and deployment of networks must intrinsically take account of important territorial contexts,

---

[24] See Castells (1996, p.470) for his conceptualisation of networks in which connection between nodes in the same network is essential for frequent interaction.

specificities and practices. This latter point is not therefore somehow bypassed or overtaken because Paris and London are 'nodes' on global networks. Indeed, we shall see throughout the study the ways in which the status or function of Paris and London as 'global cities' actually reinforces the importance of their territorial contexts and specificities in the construction and implementation of telecommunications developments.

Chapter 4

Political Economies of Urban Governance, Rescaling and Economic Development

Introduction

This chapter considers and analyses from a theoretical perspective the third thematic strand of the study as a whole, namely the relations between global economic restructuring, the rescaling of political and economic governance and territorial fragmentation in global cities, and urban telecommunications developments. This is the third layer of our theoretical analysis, which, in parallel with the previous two chapters, will provide a solid base for our empirical investigations later in the study. The discussions in this chapter will be particularly relevant in relation to chapter 8, which examines the same thematic strand from an empirical perspective, yet, as mentioned earlier, this is not to deny the intertwined nature of our theoretical – empirical structure as a whole.

From the national and urban regional levels of the two other strands in this study, we move here to look at the local or intra-urban level. The hypothesis here is that increasingly complex processes and practices of local political and economic governance are necessarily bound up with the ways in which local telecommunications developments within global cities may reinforce or combat uneven territorial development, by either involving intensive infrastructure deployment in key networked spaces, or promoting a socio-territorial cohesion component to the ICT domain. By focusing in this chapter on territorial cohesion, we can bring the discussions in all three theoretical chapters together, in order to think about and highlight how the interactions between global cities and telecommunications developments are characterised by mutually constitutive multiscalar territorialities. This means that telecommunications developments in global cities might be viewed as both influencing and being influenced by national territorial cohesion, urban regional cohesion, and local territorial cohesion in parallel.

Global Economic Restructuring and Scaling Governance

Global Economic Restructuring and the 'Crisis' of the Nation State?

Contemporary political and economic processes and practices of restructuring are substantially altering both the ways in which cities work and develop, and the ways in which the differing elements of urban development are being theorised and analysed. These restructuring processes and practices are generally associated with the wider mechanisms of political and economic 'globalisation', which we touched upon in the previous chapter in the discussion on global cities as global financial centres.

The political and the economic aspects and features of globalisation have sometimes been discussed and analysed separately. Whilst the inherent complexities and interactions of this area of debate make it very difficult to analyse in a coherent manner, many writers have attempted to draw a link between the crisis of the Fordist social mode of economic regulation and the apparent shift to a post-Fordist, or more flexible, mode of capitalist accumulation, as well as the emergence of new types of economic and political governance in the light of local state restructuring (Harvey, 1989a; Amin, 1994; Mayer, 1994; Jessop, 1995).

This has often been done using the regulation approach, which it is argued offers a highly significant way of understanding these shifts and of formulating their possible implications (Lipietz, 1994; Peck and Tickell, 1994; Jessop, 1995; Painter, 1995; Collinge, 1999). Some commentators have, however, questioned the ability of the dominant macro-level focus of regulation theory to explain shifts and transformations at the local level (for example, Jones, 1997).

The Fordist to post-Fordist shift is also linked to the development of new technologies, internationalisation and regionalisation (Jessop, 1995). In the first case, the intensive development of telecommunications infrastructures in key cities and regions has been viewed as an important part of attempted resolutions of the 'crisis of Fordism', especially in terms of the reconfigured and more flexible nature of the organisation of production and market control (Gillespie, 1991).

Jessop has illustrated the theoretical parallels between the regulation approach and theories of governance 'in terms of their common interest in the path-dependent, constitutive relationship between modes of governance / regulation and objects of governance / regulation' (Jessop, 1995, p.326). On one level, there are political processes at work which are supposed to be eroding away the power of the nation state, both from above (the growth of supranational institutions such as the European Union), and from below (the greater administrative roles and responsibilities handed down to local government, and indeed to non-governmental organisations). This 'hollowing out' (Jessop, 1994) of the role of the nation state, which Jessop sees as an important element of the 'Schumpeterian workfare state', has been captured on one level in the notion of a shift from govern*ment* to govern*ance*. The full extent of these transformations has, however, been contested and argued to be less than portrayed in some accounts. Rather than a complete displacement or overpowering of the nation state, it is suggested that 'a mixture of old, new, and hybrid forms and processes are now operating at and between different levels, above, below, and including the nation state' (Anderson, 1996, p.150). Anderson, in turn, draws links between global processes, reconfigured territorialities, and the medieval and/or postmodern conditions that recent political economic restructuring has been held to partly illustrate:

> The condition of postmodernity involves the partial and selective unbundling of territoriality, especially in the sphere of economic production. The unbundling is both a consequence and a cause of globalisation and Europeanisation, and its partiality reflects their uneven impacts. Territory is losing some of its importance as a basis of sovereignty and political rule, but states and territorialities are being qualitatively transformed, rather than states 'declining' or 'territorially based sovereignty' ending (Anderson, 1996, p.150).

These political processes have been happening in parallel with the many well-discussed processes and consequences of economic globalisation.[25]

What we seem to have now, then, are a set of complex and interacting political and economic processes and practices at different geographical scales. In an urban context, just as it becomes difficult to discuss the economic development of cities without taking into account the influence of a globalising economy, it is also becoming increasingly difficult to talk about urban governance without admitting that the many different aspects of economic and political regulation in cities are subject to the influence of actors and institutions at all geographical scales. We must therefore discuss and analyse the development of forms of 'multiply scaled territorial governance', and the ways in which they are perhaps differentially territorialised in (global) cities. Jessop (1997b, p.60; quoted in Keil, 1998, p.635) and Keil (1998, p.617) argue that this is the challenge of a 'governance of complexity'. As Neil Brenner concludes:

> problems of urban governance can no longer be confronted merely on an urban scale, as dilemmas of municipal or even regional regulation, but must be analysed as well on the national, supranational and global scales of state territorial power – for it is

---

[25] Harding and Le Galès (1997, p.184; see also Amin and Thrift, 1994) list the following key features of this:

- The growing spatial 'reach' and centrality of the international finance system and its ability to organise complex and rapid transactions on a global scale.
- The growth and expanding influence of transnational corporations, able to maximise potential returns by switching investments between a vast range of localities.
- The growing importance of less traditional economic factors of production, not least the productive value of the storage, generation and retrieval of knowledge and information.
- The capacity, afforded by innovations in information technology, to organise production on a global scale and to transmit information instantaneously across the globe.
- The constant expansion of international, compared to domestic, trade in goods, services and labour.

ultimately on these supra-urban scales that the intensely contradictory political geography of neoliberalism is configured (Brenner, 1999, p.447).

The global-local or 'glocal' notion captures this increasing scalar complexity, as 'the increasingly dense superimposition and interpenetration of global political-economic forces and local-regional responses within the parameters of a single, re-scaled framework of state territorial organisation... a 'glocal' re-scaling of state territorial organisation' (Brenner, 1998a, p.16). The political and economic restructuring processes and practices underway can thus perhaps be best understood as relating very closely to the ways in which spatial and territorial organisation and practices are bound up with a set of dynamics and mechanisms of rescaling, which acts at once as a presupposition, medium and outcome of this restructuring (Brenner, 1998b; 1999; 2000).

Processes and Practices of Urban Governance

Recent political economic restructuring at multiple scales has led to the reconfiguring of political power in many cities, one of the main elements of which has been frequently captured, as we have seen, in notions of a shift from government to governance. However, whilst emerging forms of urban governance are including a broader range of territorial groups and interests, which were often left out of previous official political negotiations, the main bodies of metropolitan governments have become increasingly concerned with situating their cities favourably with regard to global economic competition and the attraction of inward investment (Jouve and Lefèvre, 1999).

The notion of 'governance' in this regard is necessarily a complex one, as it must incorporate both:

> The capacity to integrate and give form to local interests, organisations and social groups and, on the other hand, the capacity to represent them outside, to develop more or less unified strategies towards the market, the state, other cities and other levels of government (Le Galès, 1995; quoted in Le Galès, 1998, p.496).

Le Galès also offers a slightly simpler perspective: 'From a fairly traditional definition of governance as coordination, we have arrived at a definition as 'linkage between different regulations in certain territories'' (Le Galès, 1998, p.502).

One of the key issues in considering governance and regulation is that we must be careful not to confuse the object of regulation (that which is regulated) with the regulatory process itself (that which does the regulating). In other words, in this study, the telecommunications market with local and regional authorities and the state. For Goodwin: 'We should thus be more concerned with looking at the role of local governance in helping to promote and sustain a broader mode of regulation, than with the changing structures and practices of local governance *per se*' (Goodwin, 1996, p.1401). In other words, a study of the role of local governance in telecommunications developments should take into account how it is inherently part of an overall mode of regulation. Within regulation theory, Goodwin (1996) also shows that there have been two main approaches: analysing the notion of post-Fordism in local governance, and analysing the role of local governance in post-Fordism. Here, the former is most important for its contribution to the latter:

> We should not merely be identifying the changing practices of local governance in the global city in order to chart a new mode of regulation. We are instead trying to trace the connections and dynamics that operate (in both directions) between social and economic restructuring and changes in local governance (Goodwin, 1996, p.1402).

In any case, it is far from clear that a new mode of regulation is emerging in global cities such as London. Writing in 1996, Goodwin suggested that global city processes and functions in London might be hindering the development of local regulatory capacity. In this way, globalisation and regulation are connected through their impairment, rather than their privileging, of any possible new mode of regulation. However, following the creation of the new London government, we might be able to chart more closely the connections and dynamics operating in both directions between this new regulatory capacity and global city processes and functions.

Governance regimes and entrepreneurial governance The regulationist approach is just one theoretical approach to developing an understanding of the complexities and implications of transformations in the global political economic climate. In terms of understanding specifically the negotiation and shaping of local responses in cities to these transformations, geographers and planners have advocated, in turn, the merits of the not unrelated notions of urban regimes, which focuses on political power, and entrepreneurial governance, which focuses on constructing economic development benefits.

Globalisation and global – local ideas are at the heart of moves towards entrepreneurial cities. As Harvey describes:

> The new entrepreneurialism has as its centrepiece the notion of public – private partnership in which a traditional local boosterism is integrated with the use of local governmental powers to try and attract external sources of funding, new direct investments or new employment sources (Harvey, 1989a, p.7).

However, whilst acknowledging the increasing attempts made by entrepreneurial city leaders and urban governments in the realm of urban competitiveness, Jouve and Lefèvre have found limits to the extent of such strategies:

> European cities have not yet become the place of expression and of conquest of the power of a hegemonic political elite, autonomous with regard to other local and national elites. They still remain crossed by tensions between social groups, between systems of representation and long-lasting diverse interests, which constitute powerful and efficient opposite powers, which in turn prevent all possibility of seeing the emergence of a model of unitary urban government. Local societies probably, but

certainly not communities, nor the only spaces of aggregation and mobilisation which would rise from the ashes of states incarnating an outmoded political order (Jouve and Lefèvre, 1999, p.44, author's translation).

Developing, in turn, a new and more complex understanding of the role of power in city politics is helpful. Stoker explains how this has been done in urban regime theory:

> To understand the politics of a complex urban system it is necessary to move beyond a notion of power as the ability to get another actor to do something they would not otherwise do. Politics is not restricted to acts of domination by the elite and consent or resistance from the ruled... In a complex society the crucial act of power is the capacity to provide leadership and a mode of operation that enables significant tasks to be done. This is the power of social production. Regime theory suggests that this form of power involves actors and institutions gaining and fusing a capacity to act by blending their resources, skills and purposes into a long-term coalition: a regime (Stoker, 1995, p.69).

It has been suggested that there is a close correlation between the British focus on urban governance and the more US focus on urban regimes: 'The commonality refers to the mixing of public and private power to deal with increasingly complex urban problems and the emergence of new networks aiming to provide urban leadership' (Newman and Thornley, 1997, p.968). However, Newman and Thornley (1997, p.969) go on to identify how relating the regime approach to global cities may be problematic because of the probable absence of a single coherent regime, and because of the increased involvement of central government. Regime theory has also been criticised on other grounds, especially lack of focus on the state, and too much focus on the urban at the expense of other scales (see Le Galès, 1998, p.497; Wood, 1998, p.277).

In turn, MacLeod and Goodwin (1999) argue that the recent proliferation of theoretical approaches to changing urban processes and governance is marred by limited consideration of the role of the state and limited problematisation of scale. This theme in our research aims to counter these problems by foregrounding both the role of the state in terms of planning, governance and telecommunications policy processes (as already discussed in chapter 2), and the multiply scaled parallel territorial contexts within which cities are increasingly situated, and which necessitate the intervention of authorities and agencies in order to reconfigure urban places to their best advantage, for example, through the shaping of infrastructure networks (as we have seen in the last chapter and this chapter). There is also a conscious attempt to look at the implications of the three interrelated processes, discussed by MacLeod and Goodwin, of the 'denationalisation of the state' (cities and regions linking to other cities and regions in other states), the 'destatisation of the political system' (government to governance), and the 'internationalisation of policy regimes' (strategic importance of the global context), rather than just one or other of these processes in isolation (MacLeod and Goodwin, 1999; see also Jessop, 1997a). Nevertheless, what this theme is stressing is the continuing importance of the national level for particular cities. The case studies will demonstrate that urban telecommunications initiatives are one of the clearest illustrations of the ways in which policy and regulation at the national level still has strong territorial implications and influence at the urban regional level.

Problematising Scale and Territoriality

Thinking about the notions and connotations of territoriality and scale goes very much hand in hand, because territoriality can cover a variety of different scales, and scale enlarges the notion of territory. A brief discussion of an interesting recent paper by Joe Painter on cities, territoriality and scale (Painter, 1999) can illustrate some of these points. Painter attempts to argue in favour of a reconceptualisation of the city which is not based on territoriality, but which instead adopts a more relational approach which recognises the heterogeneity of cities. He suggests that the 'aterritorial city' is a notion which aims 'to emphasise that the diverse elements that make up urban life are parts of networks of social relations that are not confined spatially within the city and to contest the assumption that cities have any territorial integrity' (Painter, 1999, p.12; see also Doel and Hubbard, 2002, on urban competitiveness). Thus, 'places are no longer seen as 'entities' but as the unstable effects of multiple webs of relations, which sometimes intersect and sometimes remain autonomous' (Painter, 1999, p.24). We might appear to be in the domain of Castells here, but Painter also dismisses the concept of 'global-local relations' as implying a reified view of both place – 'in part because it is insufficiently attentive to heterogeneity and difference' (Painter, 1999, p.14) – and scale, which 'involves representing (maybe unintentionally) regions and cities as differentiated, bounded (perhaps even organic) wholes' (Painter, 1999, p.13). It is not difficult to agree with his assertion that we need a reconceptualisation of scale, but his wider argument rejecting the territorialities of the city appears to be somewhat problematic. Firstly, there seems to be a reliance throughout on an implication that territoriality equates merely to singular ideas of coherence, boundedness, reification, and non-relational entities, and, because there is a need to 'break decisively with territoriality', that the very idea of territoriality has perhaps become staid and outmoded. Secondly, opposing this, he sets up the notion of aterritoriality as constituted in and through a series of relations and flows, part of the implication here being that these relations and flows must be fairly novel to the city. In discussing the 'territorial city', there is an assumption that the 'territorial' must inherently and exclusively relate to the 'urban'. This view of the notion of 'territoriality' is itself unjustly bounded. In the light of all this, we can question why territoriality necessarily needs to be opposed to relations, flows, connections, and networks. Surely this can be seen as another example of forced duality in urban research – the setting up of two conflictual sides to an argument, and implying that there is nothing in between. We have already seen how Castells overcame this problem by asserting that the increasing influence of the space of flows does not necessarily negate the influence of the space of places in some quarters. Thus, Castells is not denying the presence of the 'territorial integrity of cities' in the same way as Painter. It must be possible instead to reassert the crucial nature of the territorial integrity of cities. We need to emphasise how cities are bound up with processes and practices from many different scalar levels, in other words, with different national, urban, local, and sub-local level territorialities in parallel. It is therefore not just that the territorial city does not deny 'the diverse elements that make up urban life as parts of networks of social relations that are not confined spatially within the city'. It can be argued that it is these diverse elements which the *territorial* city actually thrives on. Discussing the 'territorial city' is a recognition of how 'the city' and its 'territorialities' are mutually constituted – territorial processes, practices and mechanisms from many different mutually constituted scalar levels can shape the development of the city, which in turn can shape the nature of the territorial processes, practices and mechanisms. The aterritorial city must therefore be a contradiction in terms, and the suggestion that a reconceptualisation of geographical scale can only be achieved by denying the 'territorial integrity' of the city, or even the status of the city as an 'entity', is surely wide of the mark. Economic activity which is territorialised is suggested to be 'dependent on territorially specific resources' and characterised by an 'economic viability [which] is rooted in assets (including practices and relations) that are not available in many other places and cannot easily or rapidly be created or imitated in places that lack them' (Storper, 1997, p.170).

Until recently, however, the notion of geographical scale had been viewed as relatively unproblematic. The local, the urban, the regional, the national, and the global have been fixed entities: 'relatively stable, nested geographical arenas inside of which the production of space occurred...' (Brenner, 1998b, p.460). This traditional view is exemplified by Massey's description of the spatial as:

> constructed out of the multiplicity of social relations across all spatial scales, from the global reach of finance and telecommunications, through the geography of the tentacles of national political power, to the social relations within the town, the settlement, the household and the workplace (Massey, 1994, p.4).

This meant that capitalist territorial organisation was underpinned by what Harvey terms a 'spatial fix': 'a tendency towards... a structured coherence to production and consumption within a given space' (Harvey, 1985, p.146). The work of people such as Erik Swyngedouw (1997; 2000), Neil Smith (1993; 1995) and Neil Brenner (1998b; 1999; 2000), however, has begun to complicate the question of spatial scales, with subsequent implications for the analysis of processes of capitalist territorial organisation. In particular, scales now have to be seen as important components, rather than containers, of the process of the production of space, especially given the globalising nature of the economy. This does not mean, however, that the local (or any other geographical level) is subsequently surpassed:

> Adapting to changes of scale does not mean that one no longer privileges the observation of small units, but that one takes into consideration the worlds that cross through them and extend beyond them and, in so doing, constantly constitute and reconstitute them (Augé, 1999; quoted in Mattelart, 2000, p.108).

Equally, for Beauregard, 'unless wholly isolated and insulated, the local is the mediated outcome of forces operating at all spatial scales' (Beauregard, 1995, p.238).

Neil Smith argues that scale can be viewed as 'the geographical resolution of contradictory processes of competition and co-operation' (Smith, 1993, p.99). Following this, the traditional urban scale can be conceived as a 'territorial compromise' between the need for competitive production in cities and the need for a set of harmonised regulatory conditions to promote such production. The urban scale has, then, been the focus for the telecommunications industry because the intertwined relations between accumulative competitive markets and co-operation among producers and between producers and the state over the setting of market conditions have traditionally been most intense and favourable there. The global city, however, would appear to further problematise the question of scale, because of the difficulties of achieving a 'territorial compromise' between 'contradictory processes of competition and co-operation', when these processes are increasingly mobilised at and between differing scales at the same time. This would appear to suggest a deterritorialisation of capital accumulation, as telecommunications companies, for example, are now concerned with competitive markets at multiple scales in parallel. They are, however, constantly looking to reterritorialise activities in places which especially offer market conditions which have evolved to take account of the influence of multiply scaled processes and practices. These places are frequently global cities.

It follows from all this that we need to expound the notion of a 'scalar fix' (Smith, 1995) as having superseded that of the 'spatial fix', as 'spatial scales constitute a hierarchical scaffolding of territorial organisation upon, within, and through which the capital circulation process is successively territorialised, deterritorialised, and reterritorialised' (Brenner, 1998b, p.464). The displacement of the national level by the increasing economic and political importance of both supranational organisations and the cities and regions of the nation state, coupled with the growing pertinence of strategies of 'global localisation' (Swyngedouw, 1997; Cooke, 1992), in which place-specific developments are proliferating in the search for competitive advantages in the global marketplace, mean that Brenner's (1998b) notion of the 'glocal scalar fix' becomes crucial to understanding the spatial implications of the process of the circulation of capital, especially the contradiction between fixity and motion in its 'rising power to overcome space and the immobile spatial structures required for such a purpose' (Harvey, 1985, p.150). For Brenner (1998b, p.462), 'the current round of globalisation can be interpreted as a multidimensional process of re-scaling in which the scalar organisation of both cities and states is being reterritorialised in the conflictual search for 'glocal' scalar fixes'. It can be argued, therefore, that the (re)configuration of place and IT / telecommunications in juxtaposition that constitutes the basis of urban telecommunications strategies is one of the clearest examples of this quest for a 'glocal scalar fix'.

IT and telecommunications are frequently held up to be the principal means by which capital has been able to transcend traditional spatial barriers and become truly global. Equally however, much less attention has been paid to the more local, place-based developments which these technological networks require as a foundation. It is these developments which embody the reterritorialised scalar organisation of cities and states, whereby 'contemporary urban regions must be conceived as pre-eminently 'glocal' spaces in which multiple geographical scales intersect in potentially highly conflictual ways' (Brenner, 1999, p.438). The 'fixing' in place of a telecommunications infrastructure allows a city, or a part of a city, to be configured in important ways at and beyond the local, urban scale, thereby becoming a multiscalar form of territorial development. In this particular type of 'glocal scalar fix', however, it is not always the case that the territorial state takes the key role, as Brenner (1998b) has argued. As we shall see in subsequent chapters, urban telecommunications strategies frequently involve and actually evolve from the diversity of interactions between different actors at varying levels. It can be suggested that we are not only seeing re-scaling or glocalisation at the national state level, but that critical analysis of urban telecommunications initiatives demonstrates that institutions and organisations at the subnational, urban level are equally tied up in processes of re-scaling. In addition, just as Brenner (1998b, p.476) observes how 'a key result of these processes of state re-scaling has been to *intensify* capital's uneven geographical development through an internal redifferentiation and redefinition of national social space', the ways in which local authorities and organisations are involved in the reconfiguration of urban social space in their attempts to promote multiscalar connections in urban telecommunications strategies in their cities is equally likely to lead to exacerbated spatial inequalities within urban regions. In this way, the notion of 'glocal scalar fixes' becomes pertinent to the analysis of the intrinsic connections and disconnections of urban telecommunications initiatives that is an important element of this study.

Globalisation has been seen as a process of reterritorialisation, in which the re-scaling of cities is crucial (Brenner, 1999). The global-local connections of urban telecommunications strategies might be one of the principal examples or components of this re-scaling. Brenner argues that '[t]he territorial organisation of contemporary urban spaces and state institutions must be viewed at once as a presupposition, a medium and an outcome of th[e] highly conflictual dynamic of global spatial restructuring' (Brenner, 1999, p.432). Taking one form of this territorial organisation, namely urban telecommunications strategies, we can suggest that the planning authorities and organisations involved in the development of these strategies are trying to anticipate how best to configure their urban spaces to connect up with and take advantage of potential global-local processes. Equally, these strategies become a medium between the local and the global, the means through which urban spaces are linked to global-local processes. In addition, they are the actual corollary of these processes, as a diverse range of enterprises and organisations from the local to the multinational level attempt to locate themselves where they can prosper from a global competitive advantage. Here, again, Brenner argues that:

> Contemporary reconfigurations of both urbanisation patterns and forms of state territorial organisation are best conceived as contradictory, contested strategies of

reterritorialisation through which the place-based and territorial preconditions for accelerated global capital circulation are being constructed on multiple spatial scales (Brenner, 1998a, p.3).

If this is the case, then urban telecommunications strategies, in creating these multiscalar 'place-based and territorial preconditions', would appear to typify recently reconfigured capitalist territorial development.

Cities are, then, increasingly becoming a series of spaces where a variety of specific social, cultural and economic assets converge (Amin and Graham, 1997). Graham and Healey argue, therefore, that 'we need new conceptions of place and the city based fundamentally on relational views of time and space and the notion of 'multiplex', socially constructed time-space experiences within urban life' (Graham and Healey, 1999, p.629). It is these ideas of the multiple spaces of cities that we discuss in the next section, with particular reference to urban planning literatures.

Urban Cohesion and Fragmentation – Connections and Disconnections

With the establishment of a new set of urban limits, introduced by the citadels, cities are not going to disappear but rather tend to become less and less understandable as a whole system. The city of the future is likely to become an archipelago of clusters: sorts of islands floating in a fluid system of connections (Ezechieli, 1998, p.24).

Much of what has been discussed so far, in terms of processes and practices of restructuring at varying scalar levels, has been suggested as leading to an intensification of the fragmentation of cities. Indeed, the attempts of states, city governments and authorities, and the private sector (sometimes individually and sometimes in varying types of partnership) to shape city-level responses to these processes has led to an identification of implicit and explicit types of practices of 'splintering urbanism' (Graham and Marvin, 2001). It is clear, therefore, that multiply scaled processes of political, economic and technological restructuring have major implications for the social and territorial cohesion of contemporary cities, and for urban planning in general.

Planning theory has already begun to reflect these changes by moving away from a view of planning as 'all activities of the state which are aimed at influencing and directing the development of land and buildings' (Brindley, Rydin and Stoker, 1996, p.2), towards, for example, attempts at formulating a more communicative and collaborative planning ideal, which emphasises 'place' instead of 'state', and specifically 'represents a continual effort to interrelate conceptions of the qualities and social dynamics of places with notions of the social processes of 'shaping places' through the articulation and implementation of policies' (Healey, 1997, pp.7-8).

In their typology of forms into which planning 'fragmented' in the 1980s, Brindley, Rydin and Stoker (1996) show that 'styles' of planning were essentially distinguished by their varying institutional arrangements. Regulative, trend, popular, leverage, public-investment, and private-management all suggest the way in which they were managed and implemented. This fragmentation, together with the related shift from the nation state to the municipality, has led more than one writer to suggest that contemporary urban planning is suffering something of a crisis (Dupuy, 1991; Sandercock, 1998).

The 'Multiplex' City: Reconfiguring Cores and Peripheries?

Cities have long been conceptualised in terms of their different spaces or areal variations. We only have to think back to the Chicago School's model of concentric urbanisation of the 1920s, or the subsequent sectoral and multiple nuclei models (see Dear and Flusty, 1998). The core and periphery model, for example, although not specific to urban form or spatial organisation, remains influential in urban studies and geography. The dominance of a core area has traditionally been contrasted with the subordination of peripheral areas, thus offering a compelling, if oversimplified, description of the spatial and territorial unevenness of socio-economic power. Within a US context however, Atkinson (1997) argues that we now need to talk about metropolitan-wide rather than just core economies, as industry is no longer so concentrated in the core areas of cities, but has become more footloose and spatially dispersed partly as a result of technological change. He suggests that many outer suburbs of cities have the potential to be the strongest parts of metropolitan economies. Similarly, albeit in a more philosophical manner, Antoine Picon writes about the increasing ambiguity of the core and periphery question:

> Time is otherwise taking on a more and more strategic character in the urban economy, an economy in the process of globalisation of which the spatial constraints perceive themselves from now on in terms of accessibility more than distance. The substitution of the notion of accessibility for that of distance puts in crisis crucial distinctions, such as those of centre and periphery. Insufficiently served, some districts of old centres are finding themselves in a more peripheral situation than airport or industrial zones where motorways and rail lines interconnect (Picon, 1998, pp.22-23, author's translation).

The increasing emergence of a variety of types of territorial enclaves in cities suggests a revival of core – periphery concerns, albeit based around the overwhelming dominance of the former. For Ezechieli, this dominance is reinforced by the erection of specific boundaries: 'The establishment of large urban enclaves, whose evident purpose is protecting a specific territorial circle, reveals the rise of the new paradigms of occupation and control of space of the 'network society'' (Ezechieli, 1998, p.7). These enclaves and their boundaries can be explicitly defined, as with the rise of gated communities in North America and the Ring of Steel security cordon at the limits of the City of London (see Amin and Graham, 1999; Power, 2000a; Graham and Marvin, 2001, pp.272-273, p.320), but they can also be more implicit, as in the case of downtown San Francisco where 'dot-com' entrepreneurs and industries now dominate the urban landscape, despite the resistance and protestations of numerous social movements (Graham and Guy,

2002). We shall also see this in relation to the key networked urban spaces of Paris and London, where the prevalence and significance of 'global connection' juxtaposes and reinforces spaces and communities characterised by 'local disconnection' (Amin and Graham, 1999). Ezechieli here distinguishes between 'global citadels' and 'local citadels' (Ezechieli, 1998, pp.11-14). The territorial fragmentation of the city is, then, quite evident, and seems to be bound up in, and manifest itself in, parallel multiscalar processes. As Picon again suggests:

> The globalisation of the economy has come to accentuate this fragmentary character by leading spectacular disparities in development to increasingly reduced scales. Linked to the rest of the planet by high-performance information networks, a business centre or an industrial zone can prosper amidst suburbs with problems. The importance taken by the notion of accessibility reinforces this process...Such spatial fragmentation has something paradoxical about it at a time when behaviours and lifestyles are tending to show uniformity. It also constitutes a handicap with regard to the necessity for cities to acquire a bright image in order to attract capital and businesses within a context of widespread economic competition. Never has the global economy been as urban; never has the notion of the city showed itself to be as blurred (Picon, 1998, p.24, author's translation).

Nonetheless, urban spatial restructuring is not always a case of movements or exchanges between core and periphery or the other way round, in terms of a dominant core and a dependent periphery. Longcore and Rees (1996) have offered a case study of the spatial restructuring of financial institutions in Manhattan in relation to changing technological requirements. Here, the increasing complexity of IT in banking and insurance forced several major firms to relocate from the traditional financial core around Wall Street to more peripheral midtown sites where the buildings had the structures to support the technology and the larger floorspace for the traders. In this way, the core and periphery switched round, with Wall Street becoming a 'subdistrict' rather than a core (Longcore and Rees, 1996, p.366). The only process suggesting the dependency of the periphery was the continuing need for face-to-face communication which remained strongest in the traditional financial core. However, Longcore and Rees were still able to conclude that the urban spaces of global finance are beginning to undergo a form of spatial restructuring:

> As highly competitive major financial firms retreat to secretive, security-conscious structures and a building technology that stresses large horizontal over vertical spaces, the traditional tightly focused financial district land market may finally demonstrate geographical flexibility. Monitoring the future extent of this morphological disaggregation by the one district that had for so long maintained its coherence and stability will tell us much about the value of urban propinquity (Longcore and Rees, 1996, p.368).

This restructuring of the traditional core and periphery development has been highlighted as a major component of an emerging Southern Californian form of postmodern urbanism, which is henceforth partly characterised by 'a reterritorialisation of the urban process in which hinterland organises the centre' (Dear and Flusty, 1998).[26]

In much the same way as Longcore and Rees, Daniels and Bobe (1993) questioned whether the development of Canary Wharf in London's Docklands as a new large-scale office space could be seen as an extension to the traditional boundary of the City of London, because, at least at first, it was being marketed as a response to the shortage of office space in the City.[27] The subsequent relationship between the traditional City 'core' and the new regenerated Docklands 'periphery' became more complicated:

> Nevertheless, diversification rather than specialisation of client functional types suggests the Canary Wharf development was seen (at least by Olympia & York) as essentially a 'City centre' type of development. To this end the functional dynamic is that of a development that is part of the City of London and yet in a peripheral location. This interpretation lends credence to the idea that Canary Wharf should become the focus of London's third business district, after the City and the West End. [...] In one sense Canary Wharf can be seen as being in direct competition with the City for office tenants. Conversely, it can be viewed as complimentary to the City, and this is probably the most valid interpretation in that success for Canary Wharf relies in essence upon the continued success of the City of London as an international financial and business centre (Daniels and Bobe, 1993, p.548).

In this way, the Docklands can be seen as an example of the 'large-scale polynuclear and extended metropolitan corridors' of Gottmann's concept of the 'megalopolis'. This is also reinforced by more recent redevelopments just to the east of the City of London, as at Cutlers Gardens and Spitalfields Market (see Power, 2000a, p.8). Power sees these redevelopments as emblematic of an attempt at reaffirming the identity and distinctiveness of the Square Mile as a centre of global finance:

> By playing a major role in the re-walling of the City and in the 'policing' of its territory, the technology installed by the state and businesses has a central role in underpinning and reinforcing the City as a cohesive centre and territory (Power, 2000a, p.9).

Core areas thus become a key element in state territorial development strategies, which Jones has suggested are becoming focused around a practice of 'spatial selectivity', whereby certain places are promoted, materially and / or symbolically, as strategic poles for development over others (Jones, 1997). The role of telecommunications infrastructures in the development of such spaces is

---

[26] The wider arguments and implications of Dear and Flusty's 'postmodern urbanism' research agenda were much discussed and, in some cases, criticised in a special issue of *Urban Geography* (1999).

[27] See Pryke (1991; 1994) for a detailed analysis of the spatial changes in office provision in the City of London during the 1980s, associated with transformations in broader sociofinancial processes and practices, which led notably to some banks relocating beyond the traditional core, thereby contesting the boundary of the City.

becoming increasingly important here, as Longcore and Rees (1996) also showed, and we will consider the implications of the possible structuring of these infrastructures in polynuclear and extended corridors in Paris and London on urban form and territorial cohesion in chapter 8.

In these case studies then, the description of the urban spaces as core and peripheral seems interchangeable with new and old, or traditional and contemporary spaces. These latter dual notions imply similar oppositions, and can still be shaped through processes of 'exchange, production or evaluation' between the two, but they remove some of the dominant – dependent tensions present in the notion of core and periphery, even if they do not go as far as implying the inverted urban restructuring processes of Dear and Flusty (1998). The interaction between telecommunications / IT and urban spatial restructuring between new and old, or traditional and contemporary spaces, will be an important empirical element of the case studies of Paris and London, particularly in chapter 8.

The Implications of 'Spatial Selectivity' Within Urban Regions: Connection and Disconnection?

> Connections are what make successful cities. Unsuccessful cities are unconnected (Cowan, 1997, p.3).

The importance of a socially inclusive aspect to the development of new technologies has been widely recognised. Sussman, for example, argues for more inclusive social access to means of communication: 'A democratic and egalitarian communication system would be one that promotes community initiative, equitable distribution of wealth, and public and private policy based on equal opportunity' (Sussman, 1997, p.275).

In the same way, the uneven nature of geographical development has traditionally meant that, for all the benefits that urban planning strategies are able to bring to some groups and areas of the city, there will always be others which lose out and are forced to remain marginalised. This is even more the case in the context of 'spatially selective' state territorial strategies (Jones, 1997). As Jones argues: 'The impetus for spatial selectivity is ultimately driven by political ideology, a need by the state to maintain hegemony, suppressing counterhegemonic interests and in the process attempting to gain, through pursuing a particular accumulation strategy, international competitiveness' (Jones, 1997, p.849).

Until recently, these marginalised groups have very rarely been the main focus of study in planning or other urban-related disciplines, although popular planning in the 1980s invoked community involvement and planning 'from below'. However, the work of people like Leonie Sandercock (1998) has begun to demonstrate the richness and the importance of the 'insurgent' viewpoints and stories of previously neglected social groups. For Sandercock, the traditionally-recounted history of planning has been too selective and state-oriented, in which the main actors are always white, middle-class men. Planning historians have committed many 'sins of omission [which] are the noir side of planning' (Sandercock, 1998, p.37). This can be related in turn to the recent identification of some writers of the importance of individual action within planning (Healey, 1997).

Sandercock's 'noir side of planning' can also be seen to have left its mark on the urban landscape in the form of a kind of noir side to the city. The city cannot possibly be embraced or understood 'with one glance' (Latour and Hermant, 1998). Christine Boyer (1995), for example, has identified a neglected, invisible 'disfigured city' which is overlaid by a more imageable 'figured city'. The poverty of IT / telecommunications infrastructure in the former means that excluded urban spaces can be characterised as 'lag-time places' (Boyer, 1996). Such places and their inhabitants are clearly the losers in the information society so far. While others benefit, it is they who are on the receiving end of what Doreen Massey calls 'the power-geometry of time-space compression':

> For different social groups and different individuals are placed in very distinct ways in relation to these flows and interconnections. This point concerns not merely the issue of who moves and who doesn't, although that is an important element of it; it is also about power in relation *to* the flows and the movement. Different social groups have distinct relationships to this anyway-differentiated mobility: some are more in charge of it than others; some initiate flows and movement, others don't; some are more on the receiving end of it than others; some are effectively imprisoned by it (Massey, 1993, p.61).

Few have argued about the detrimental social consequences of new forms of communication more forcefully than Erik Swyngedouw, for whom 'the increased liberation and freedom from place as a result of new mobility modes for some may lead to the disempowerment and relative exclusion of others' (Swyngedouw, 1993, p.322). This is all linked in to how 'cities hold a *diversity* of power arrangements, with different parts of a city locked into different sets of rhythms and connections, with different degrees of influence and authority' (Allen, 1999, p.186). However, Swyngedouw points out that communication networks still require a certain level of fixed territorial organisation which may permit an increasing number of connecting flows, but will still inhibit others. For Swyngedouw, therefore, freedom from the binding effects of place for some is liable to contribute to others being more firmly tied to their native neglected places with little prospect for reversing this trend because they lack the economic and social power that comes from an ability to command place (Swyngedouw, 1993). As he argues:

> The two-speed and three-speed Europe is not one linked to a geographical core and periphery in terms of their determination to accelerate integration, but is rather an internal differentiation between those who revel in and benefit from greater command over space and those who remained trapped in the doldrums of persistent marginalisation and exclusion (Swyngedouw, 2000, p.73).

The problems posed to planning and urban development by a need for social and territorial cohesion in the parallel shaping of places and IT / telecommunications (see Roberts et al, 1999) are supplemented by a need for policy-makers to understand something of the technical elements and implications of the latter:

The old-style planner talked about physical zoning, the balance of employment, housing and open space and traffic flows. The new style planner has to consider the configuration of electronic systems and Local Area Networks (LANs) and the provision of bandwidth to each urban area. The town planner dealt with the stocks and flows of vehicles. Today's public authorities have to face the stocks and flows of information (Howkins, 1987, p.427).

Building on our discussion of the conceptualisation of networks in the previous chapter might provide a way of bringing together the social, technical and territorial elements and processes inherent to the parallel shaping of urban places and telecommunications. Buijs summarises how '[a] network consists of junctions and the relations between these... Connections are the physical carriers of the relations. The aggregate of these connections is called the physical infrastructure' (Buijs, 1994, p.53). As it is connections which carry the relations between junctions in the network, then it is disconnections which demonstrate a lack of relations between certain junctions. Both are equally important, but only the connections in a network are 'visible' – disconnections cannot be seen as they are frequently very subtle, and are therefore more difficult to find and to analyse. More consideration and understanding of the disconnections in 'networks' is evidently needed, otherwise the 'delinking' of 'core' glocally connected spaces in global cities from contiguous, but less favoured, neighbouring spaces is likely to increasingly become the key territorial outcome of urban telecommunications developments (Castells, 1996; Graham and Marvin, 2001).

There are perhaps two types of disconnection which characterise the relationship between certain less favoured groups and spaces of cities and telecommunications infrastructures and services – a subtle, incidental form of disconnection in which the lack of access of a group or space to infrastructures is not strictly purposeful; and disconnection as 'excludability' in which there is a deliberate control of access to infrastructures (see also Graham and Marvin, 2001, pp.170-171, for their categorisation of disconnected users). Whilst national universal service policy has traditionally attempted to reduce levels of disconnection in telecommunications, the liberalisation or unbundling of telecommunications markets has begun to seriously problematise the future of such policy, which has been the domain of monopoly operators.

Madon (1997) observes how, in the information-based global economy, there is an uneven spatial development between the cities and regions of the developing world as well as between those of the developed nations. In addition, the development of a telecommunications infrastructure in India has assumed a twin track form 'whereby specialised networks and advanced data services are increasingly forged ahead and developed for business needs, perhaps for technology enclaves, while the goal of basic universal services and other social objectives assume a secondary place' (Madon, 1997, p.231). Indeed, the example of Bangalore's suburban technology park at Electronics City highlights the increasing expansion of global connections and local disconnections: 'Companies that locate within the park are insulated from the world outside by power generators, by the leasing of special telephone lines, and by an international-style work environment' (Madon, 1997, p.234).

This all suggests how, as Mattelart puts it, 'networks, embedded as they are in the international division of labour, organise space hierarchically and lead to an ever-widening gap between power centres and peripheral loci' (Mattelart, 2000, p.98). Similarly, Veltz (1996) talks about an 'archipelago economy', and Petrella of 'global techno-apartheid'. These notions all portray the same story:

> The convergence around distinct poles and the organisation of the world economy into networks linking these poles – to the detriment of the areas in between that are less well endowed and therefore more exposed to marginalisation and abandonment – carry a risk of splitting the world economy in two and creating a two-speed social geography (Mattelart, 2000, p.99).

It would therefore appear that our focus in this chapter on the nature and implications of global economic restructuring, the rescaling of political and economic governance, and territorial fragmentation in global cities is a highly important backdrop to urban telecommunications developments, and to broader questions of socio-territorial cohesion within a 'global information society'. This perspective becomes, thus, the third crucial foundational element to our theoretical – empirical explorations of the differing, parallel territorial contexts and specificities of telecommunications developments in Paris and London.

PART II
COMPARING
TELECOMMUNICATIONS
DEVELOPMENTS IN PARIS AND
LONDON

Introduction

> The contrast [of London] with Paris could not be more complete. There, the French state worked from the heart of the medieval city, enlarging its *enceinte* and modernising the fabric to make the city a symbol of national unity and indivisibility, and cultural *point d'équilibre* for the world at large. The history of Paris is a history of absorption and incorporation into a progressively higher organic unity. London's is a history of multiplications, not just of local governments but every aspect of metropolitan life... London sails as a flotilla with two, three or more flagships (Hebbert, 1998, p.7).

In the next part of the study, we build on the theoretical foundations outlined in Part One and focus on the essence of the study – case studies of urban telecommunications developments in Paris and London. Chapter 5 gives, firstly, a historical perspective on Paris and London from the three viewpoints most relevant to this study – telecommunications / technological, planning and governance, and cultures of territoriality. The three chapters that follow then offer an explanatory, analytical and comparative discussion of telecommunications developments in the two cities, which are investigated from the angle of the three respective strands to the research as already seen in the theoretical section earlier.

The individual telecommunications developments discussed and analysed in chapters 6-8 should not be read as being necessarily totally separate strategies or projects, as, in the case of both Paris and London, taken together they represent part of the inherently strategic regional landscape of telecommunications strategies in the two cities as a whole. They are, in this way, at least partly mutually constitutive.

It should be highlighted that the choice of strategies was also crucially influenced by the more simple factor of availability of material and information, and that the aim here is *not* to outline or analyse *all* the main telecommunications developments in Paris and London, but to investigate a few of the more illustrative ones and their most salient and revealing points according to the key strands and substrands of the research as a whole.

Chapter 5

The Historical Context of Telecommunications Territorialities in Paris and London

Introduction

In this chapter, it is time to begin to narrow the focus to the two cities within the two nations that are the main object of the study. This will allow us to deepen the level of investigation and analysis of developments in these cities.

The aim here is to 'set the scene' for the empirical discussion and analysis of contemporary telecommunications developments in Paris and London. It is important to situate the material in subsequent chapters in a more historical context.[28] This can be seen as the first layer of analysis necessary to an understanding of these (or any such) developments. It is easy to forget and neglect the fact that, however 'modern' and high-tech telecommunications infrastructures and services in global cities might be today, they have not just appeared suddenly and out of thin air, but have steadily evolved through a number of interacting historical conditions and developments. As Thrift argues:

> what must be re-embedded if we are to understand modern informational spaces and telematic cities, is any concerted sense of new electronic communications technologies as part of a long history of rich and often wayward social *practices* (including the interpretations of these practices) through which we have become '*socially acquainted*' with these technologies (Thrift, 1996b, p.1472).

These technologies are thus 'always only a mutable and mobile part' of wider social practices (Thrift, 1996b, p.1474). In this chapter, therefore, we discuss some of these historical conditions and developments that have helped to shape (and account for differences between) the current contexts of telecommunications developments in Paris and London. These take three forms – the evolution of communications technologies and telecommunications regulation; planning and governance practices; and national traditions of territoriality.

---

[28] As Castells (1996, p.21; p.412 on the space of flows in history) suggests.

Technological Evolution and Telecommunications Regulation

The first element in our historical background to the subsequent comparative empirical explorations of telecommunications developments in Paris and London is a brief look at how communications technologies have progressively evolved and been shaped over the last two centuries or so. As we shall see in the next few chapters, the technologies and infrastructures being intensively deployed and developed in global cities today are, on one level, merely the latest manifestations or descendants of those which formed the basis for the first round of global networking in the nineteenth century. The basic socio-economic and geopolitical specificities of this precursory networking have, likewise, sometimes endured to the present day, thus justifying further the following historical perspective.

A Brief Retrospective of the Emergence of Communications Technologies

Parallels have frequently been drawn between the telecommunications systems and networks of today and the old, pioneering forms of communication of the nineteenth century (Standage, 1998; Hugill, 1999; Mattelart, 2000). One such approximation can be made with the development of telegraph networks and then the smaller scale metropolitan pneumatic tube systems in the mid-nineteenth century.

France and Britain were at the centre of the development of early forms of communications technologies. With the construction of two of the biggest empires in the world, long distance communication became increasingly imperative in order to ensure political and economic hegemony and stability. As centres of empires then, as well as national capitals, Paris and London have long been key nodes in international communications networks.

France constructed the first optical and signal telegraph line in 1794[29] – a revolutionary communications system for a revolutionary nation (Mattelart, 1999).[30] The proclaimed advantages this would bring were inevitably much vaunted:

> Lakanal[31] wisely observed that the establishment of the telegraph was the best answer to publicists who thought France too big for a republic. Perhaps he foresaw that the telegraph would forge close links between the central authority and its representatives in the departments that had replaced the provinces (Wilson, 1976, p122; quoted in Hugill, 1999, p.25).

In a way, this turned out to be the case as France had a fairly widespread visual telegraph network by the mid-nineteenth century (Hugill, 1999, pp.25-26), with

---

[29] Its inventor was Claude Chappe (see Musso, 1997, chapter 7).

[30] Noam (1992) briefly discusses even earlier forms of state-dominated communications in France, going back to the thirteenth century messenger service of the University of Paris.

[31] A member of the committee which investigated the development of the telegraph in France at around the beginning of the nineteenth century.

550 stations covering some 5000 kilometres (Mattelart, 1999, p.176). However, this was all for administrative and military purposes, rather than public communication, up until the mid-nineteenth century (discounting the provincial dissemination of lottery results). As Mattelart argues, then, 'The gap between prophecies based on the democratic potential of networks and the trajectory of realpolitik in their establishment is thus a permanent aspect of communication history' (Mattelart, 1999, p.178). French law allotted fines for unauthorised communications and prevented the development of private networks. As Noam wrote prior to the recent liberalisation measures:

> Succeeding governments of varying political persuasions have never significantly deviated from this principle. The French government monopoly over telecommunications was thus established by law even before the introduction of electric telegraphy. When the latter arrived, the French government was far from enthusiastic (Noam, 1992, pp.134-135).

A similar lack of enthusiasm from the French government seemed to characterise, more recently, the slow process of partly privatising the national monopoly, the European agreements on market liberalisation by 1998, and the development of the global Internet (which threatened their national Minitel system).

Cable was laid underwater for the first time between Dover and Calais in 1851, thus linking the London Stock Exchange and the Paris Bourse (Mattelart, 1999). Undersea cable was held to be an illustration of Victorian hegemony (Mattelart, 2000). The advent of the electric telegraph brought efficient global telecommunications connections for the first time. Transatlantic cable arrived in the mid 1860s, and Britain had its own global network by the 1870s. Consequently:

> [b]y the end of the nineteenth century the globe had been criss-crossed using a technology only a little more than a half-century old. Moreover, most of this new information infrastructure was installed with private capital, necessitating little of the massive government spending and subsidies that had accompanied the equally spectacular growth of the railroads... The most complicated problem with the electric telegraph was not technical but geopolitical. Britain built the first global information infrastructure... The other polities never caught up... (Hugill, 1999, p.28).

Global networks enabled the creation of the International Telegraph Union, an organisation promoting common understanding and action among the states with the most advanced networks. It is interesting though that Britain was unable to join because its telegraph system was completely run by private companies (Mattelart, 1999, p.183). This regulatory difference between Britain and its European neighbouring states surfaced again in the last twenty years or so with the precursory liberalisation policy pursued by the former in its telecommunications markets, and the contrasting monopolies kept in place by other countries. Whereas, we shall see how this recent regulatory opposition between Britain and France has had an important influence on market strategies and network politics in and between the two countries, in the late nineteenth century, 'so tightly entwined were the logics of trade and diplomacy that the fact that the British undersea cable depended on private companies – unlike France, where it was placed under state control – changed nothing from a geostrategic point of view' (Mattelart, 2000, p.11). As we shall also see, in some ways, little has changed in global telecommunications – today, the globe is also criss-crossed by relatively new, state of the art fibre optic cable technologies, and this is also financed by the private capital of dozens of global operators. Equally, the contemporary telecommunications sector raises key geopolitical issues, although one of its key elements is certainly the question of whether or not the geopolitics of global telecommunications have become more of the domain of the operators (and on another level, individual global cities) than nation states. This question is obviously a crucial aspect of the case study research which follows in this chapter and chapters 6-8.

In London, the congestion of the telegraph network in the 1850s was associated with the increasing reliance on it of business users and the Stock Exchange to transmit important information (Standage, 1998, p.94). Just as today's business users look for ever quicker and more efficient communications networks, the London Stock Exchange at that time was not going to function unless a way was found to ease congestion on the networks. The solution was the creation of a pneumatic tube system powered by steam engines to transmit messages from the Stock Exchange to the Central Telegraph Office, where they were then telegraphed to their final destinations (Standage, 1998; Aldhous et al, 1995). The way in which this system prioritised, at first, the business market makes it, in many ways, a precursor to the metropolitan fibre optic telecommunications networks of today which are constructed by operators in the key business areas of global cities, and are aimed predominantly at the business market. The success of the pneumatic tube system in London led to similar systems in other British cities, and eventually in European cities such as Paris and Berlin. The message-carrying capacity of the systems was continually improving (in the same way as the capacity of fibre optic networks is today), and in some cases, networks had been extended to the extent that messages could be sent entirely by pneumatic tube without requiring telegraphing. This was the case in Paris, and led to an innovative pricing system whereby any message sent within the city cost the same, irrespective of destination and message length (Standage, 1998, p.100). In London, this communication system amazingly remained in existence until 1962, with more than 10,000 messages a day being sent around the network even in the late 1950s (Aldhous et al, 1995). Besides the metropolitan pneumatic tube systems, the global web of telegraph networks, submarine cables, and messengers in the 1870s made up what Standage (1998) calls 'the Victorian Internet', and it is not difficult to make a comparison between that web and the contemporary construction of global telecommunications networks by carrier operators.

In the 1860s, the London telegraph system was dominated by two companies, the London District Telegraph Company and the Universal Private Telegraph Company. Much as the telecommunications operators of today, they focused on different markets. The former worked within a four mile radius of King's Cross, while the latter aimed its service specifically at business users (Trench and

Hillman, 1993, p.177). The first appearance of the telephone in London in the 1880s is also an interesting story of power struggles and obfuscation. Nationalisation of the telegraph system by the Post Office (at an estimated cost of £8 million) did not precede the telephone by much, and in similar vein, in order to safeguard its investment in the telegraph system, the latter also tried to take over the new private telephone companies (Trench and Hillman, 1993, p.178). This was prevented, but the Post Office was still intent on obstructing the development of the telephone, and a set of restrictions was created which effectively brought an end to most of the telephone companies, with the exception of the National Telephone Company. A competitive environment emerged in which the Post Office ran the main lines, while the National Telephone Company held the local lines, but the need to open up more underground space turned the struggle to the advantage of the former:

> the Post Office would rent out underground space to the National Telephone Company; in return, the Post Office would acquire the plant of the Company. When the Company's charter ran out in 1911, the Post Office stepped in and took over the entire system (Trench and Hillman, 1993, p.179).

In any case, the development of the telephone in London revolved around business and commercial users for a significant period of time: 'That the wider public might have need of a telephone and that this was a potentially profitable market had still to be learnt' (Stein, 1999, p.62). As we shall see in chapter 6, there appear to be vague similarities between parts of this story and the current process of local loop unbundling in telecommunications networks, particularly with regard to the territorial division of markets (main and local lines), the obstinate and protective stance of the main operator, and the negotiations over control of underground space.

In France meanwhile, in contrast to both the situation in London, and its own previous stance on the development of the telegraph, the state initially positively encouraged the private sector to become involved in the development of the telephone, albeit with supervision of network construction, as a means of devolving some of the investment and risk (Noam, 1992, p.135). The prohibitive cost and slowness of the service though gradually led to the government regulating more and more the role of the private operators, until in 1889 it decided to take over the entire system itself in a process of nationalisation (Noam, 1992, p.136).[32] This nationalisation would subsequently dominate the French communications sector for the next one hundred years.

In providing this historical perspective on telecommunications network developments, it becomes possible to compare the late nineteenth century developments with more contemporary telecommunications. One of the most notable differences has been a shift from celebrating urban technological networks in the beginning of modernity to their subsequent underground burial during high-

---

[32] For a detailed comparison of the spatial development of telephone networks in France and Britain in the late nineteenth century, see de Gournay (1988).

modernity (Kaika and Swyngedouw, 2000). Recent progressions in the telecommunications market linked to demonopolisation and liberalisation on the one hand, and increasing technological sophistication and product proliferation on the other hand, may suggest another shift back, if not quite to the level of technological celebration and fetishism of before, then certainly to 'a renewed physical, social, political and discursive salience to urban networked infrastructures' (Graham, 2000b, p.185). Within the ways in which they changed relationships to, and experiences of, time and space (see Kern, 1983), or 'structures of feeling' (Thrift, 1996a), Kaika and Swyngedouw note how:

> Because of their significant role in the functioning of the modern capitalist city, networks of technology became *the* embodiment of progress during early modernity... Being excluded from the technological networks symbolised exclusion from the spheres of the powerful. Hence, the connection to the electricity or water networks of the city, or, similarly, the connection of one's home to a network of highways became a symbol of prestige and authority on the one hand and a terrain of controversies and power struggles on the other (Kaika and Swyngedouw, 2000, p.125).

The period of national Postal Telegraph and Telephone (PTT) monopolies in which the ideal of universal service became predominant diluted to a certain extent the symbolism of prestige and progress of the telephone because a majority of the population gained access to it. In recent years however, prestige and authority has become reattainable in telecommunications through mobile and broadband connections, for example, and at the same time, the levels of technological exclusion or disconnection are increasingly diverse. We shall see in subsequent chapters how the contemporary territorialities of telecommunications developments in Paris and London is a coalescence or a mosaic of prestige and authority on the one hand, as large corporations seek maximum global connection, and a terrain of power struggles on the other hand, as local, regional and national government bodies attempt to regulate and configure their territories and markets to meet both the economic development and social cohesion objectives that are the parallel goals of an 'information society'.[33]

The Era of Nationally Standardised PTTs in the UK and France

If we move the historical perspective on telecommunications in the UK and France forward from the times of great technological changes and progress, we enter a more recent period when telecommunications in both countries (and throughout the

---

[33] It is interesting, in turn, to compare this contemporary notion of 'information society' to how it was envisaged just after the Second World War by Norbert Wiener: 'An information society must be a society where information circulates freely. It is by definition incompatible with restriction and secrecy, with inequality of access and the transformation of circulation into commodity' (Mattelart, 1999, p.184). As we shall see, these last two processes in particular are highly characteristic 'presuppositions, mediums and outcomes' of the global information society of today, whether in an implicit or explicit manner. This would seem to turn the notion of 'information society', then, into something of a paradox.

western world) became more standardised as a recognisable service sector and an increasingly essential domestic and business utility.[34]

In both the UK and France, as we have seen, telecommunications infrastructures and services fell under the jurisdiction of the government departments of the Post Office, and were operated as public monopolies virtually from the late nineteenth century – early twentieth century onwards. In the French case, in particular, the task of improving and extending the national network was greatly complicated and delayed by the upheaval and destruction of two world wars (Noam, 1992).

By the 1950s and 1960s, however, the quality and reach of the telephone network and telephone services in both countries was subject to some criticism, so each government had to instigate plans for improvement. However, while investment plans were drawn up in Britain during the 1950s, and waiting lists gradually shortened, in France, the development of the telecommunications system was not viewed as a priority in post-war economic and industrial development until the late 1960s (Thatcher, 1999, p.41). As Noam notes, at this time, 'the bitter joke was that half of the country was waiting to get a telephone installed, and the other half was waiting for a dial tone' (Noam, 1992, p.141). The eventual result was a substantial modernisation and expansion plan implemented under the presidency of Giscard d'Estaing, the *Plan de Rattrapage Téléphonique*, which quadrupled the number of subscribers to the telephone network in just eight years (Noam, 1992, p142). This renewed state intervention in the telecommunications industry in the 1970s and the 1980s has been termed 'high-tech Colbertism' (Cohen, 1992).[35]

It was just before this time (in the early 1970s) that great technological changes in the telecommunications sector were beginning to take place, which would notably lead to the digitalisation of transmission and switching, and improved transmission methods via use of satellites, copper coaxial cable and then eventually optical fibre cable (Thatcher, 1999, pp.50-55). One of the most important results for the UK and French public monopoly operators of such methods of communications transmission was the increasing dissociation of cost from distance. Big differences in tariffs for long distance and local communications became harder to justify, and indirectly this placed additional pressure on the monopoly system because competition became more plausible. This was not necessarily seen as all bad news by the monopoly operators: 'For PTOs [Public Telecommunications Operators], the competitive dynamic pointed to pressure on domestic markets from new entrants, but opportunities for expansion abroad, both directly or through alliances with other operators' (Thatcher, 1999, p.86).

---

[34] Although demand for the telephone across Europe, and especially in France, up to the 1950s only augmented very slowly. Its greater expansion in the 1950s has been linked to the beginnings of suburban development and decreasing urban concentration and density (Ascher, 1995, pp.51-52).
[35] Colbert was a 17th century French statesman and advocate of mercantilism, or the intervention of the state in economic affairs to maximise exports.

In parallel with vast improvements to the national telephone network, a 1978 French report on '*L'informatisation de la société*' by Nora and Minc represented 'one of the first documents through which a major industrial country reflected on 'how best to handle the computerisation of society'... [According to the report], a network of computerised communication was the guarantee of a 'new global mode of social regulation'' (Mattelart, 1999, p.186; see also Musso, 1997, chapter 8). This document contributed to the adoption by the French government of an overall national industrial strategy focused on a kind of technological self-sufficiency with market protection of its own national champions as a priority.[36] This policy would, over at least the following decade, contrast greatly to that of the British government, which was beginning to look towards privatisation and market competition to improve and expand the national telecommunications sector.

Cable Policies in France and the UK in the 1980s

Within the different types and intensities of negotiations and actions relating to the restructuring of telecommunications in the UK and France in the 1980s, there was a common recognition of the potential importance of, and therefore a quite substantial policy focus upon, the development of cable networks.

In the UK, the government was aware of the importance of broadband cable networks for social and economic development, and so decided in 1982 to promote a cable-led 'communications revolution' (Greater London Council, 1985, p.395). The GLC, for example, was, however, against the market-led vision for introducing cable across the country, which was going to completely exclude local authorities from consultation. The 1985 London Industrial Strategy highlights the failure of the sector to establish itself at that time. In fact, key institutional and investment factors blocked the evolution of a national broadband network, which might have been rolled out by BT. Local franchises for private cable operators became the way forward instead (Thatcher, 1999). Nevertheless, as Hulsink observes, 'The government made clear that cable television systems and a future nationwide broadband infrastructure, as an alternative local network to BT's network, would increase the possibilities of competition in the provision of communications services and equipment' (Hulsink, 1998, p.147). As cable was envisaged as an alternative to the traditional telephone networks, the involvement of BT and Mercury in the Cable Programme was restricted. In 1990 it was decided that cable companies would be allowed to supply voice telephony in limited geographical areas, and that these companies could interconnect their individual networks to create a parallel national network to that of BT (Hulsink, 1998, p.148).

In the UK, then, cable has been seen as the preliminary step for introducing a form of competition within the local loop. In recent years, the cable services of

---

[36] Beyond noting the dominant role of the state in this regard, the wider concerns and implications of French technology policy in the late 1970s and 1980s lie well beyond the realm of this study (see Noam, 1992, for an account of the restructuring of the French telecommunications equipment industry in the 1980s to privilege the likes of Alcatel and Thomson).

companies such as NTL and Telewest have reached more and more homes and businesses. The 'local' nature of the services of cable operators has been emphasised in legislation, although this created a problem as 'they were originally obliged to represent themselves as local rather than national operators which inevitably made many potential customers suspicious of their ability to deliver a quality service' (Curwen, 1997, p.133).

In France meanwhile, the search by central government for important *grands projets* which could meet social, economic and political objectives within a framework involving both the public and private sectors brought about a focus on the telecommunications sector. In 1982 the *Plan Câble* was initiated, involving the deployment of an interactive broadband fibre optic network, which, at a cost of between FF 45-60 billion (£4-5 billion), would cable 6 million homes by 1992, and another one million homes a year thereafter (Morgan, 1989, p.41; Hulsink, 1998). It was also partly a means to encourage the decentralisation of administrative powers to the local and regional level, following the introduction of a territorial decentralisation policy in 1982 by the Socialist government. In contrast to the UK, though, the private sector was relatively absent. Investment for the cable networks came from a partnership between the DGT and local authorities:

> The establishment of a policy for cable, simultaneous with the laws on decentralisation, was to emphasise the role of local groups in putting the plan into action. The local mixed economy, which let territorial executives and private capital be associated, was to be the only formula authorised by the legislators, 'haunted' as it were by the risks of the privatisation of audiovisual and telecommunications as contained in the Plan. The development of the networks was to depend on elected officials (under whose initiative fell local programming) and the administration, which registered and dealt with local solicitations on a centralised level (Négrier, 1990, p.15).

It has been suggested that the *Plan Câble* was an attempt at increasing the monopoly position of the DGT in the French telecommunications sector, therefore again contrasting with the UK policy (Morgan, 1989). However, the project was a failure: '[it] suffered from an institutional separation between the construction and exploitation of the network and a reluctance on the part of local communities to engage in sponsoring the various cabling projects' (Hulsink, 1998, p.259). In the late 1980s, the deployment of cable networks in French *métropoles* was turned over to the private sector, in the shape of companies like Lyonnaise des Eaux, thus becoming one of the earliest elements in the opening up of the French telecommunications sector.

Nevertheless, while the *Plan Câble* may have been a relative failure in itself, the more recent development of telecommunications infrastructures in France is probably greatly related to its initial development of cable networks in the 1980s, on which the state and *collectivités locales* first worked together (IAURIF, 1998b; Négrier, 1990). Whereas the first cable networks in Ile-de-France were deployed in the mid 1980s, the number of points installed in the region was multiplied by 13 between 1987 and 1997: 'This growth in cable networks is a positive point for Ile-de-France. Cable is a real infrastructure, supporting numerous services in the domain of information, culture, leisure, and more recently, telecommunications' (IAURIF, 1998b, p.5, author's translation).

What the French *Plan Câble* did, possibly more than anything else, is illustrate (in contrast to the UK) 'the multiple representations of the territory of intervention of partners and on the methods, which are specifically French, of the territorialisation of networks' (Négrier, 1990, p.16). Négrier here cites Chantal de Gournay's distinction between France and the UK – USA in terms of network – territory relations: 'where the *Plan Câble* leans directly on the local government in order to establish the dynamics of cable, Great Britain takes 'extreme' care to avoid any intervention by territorialised authorities' (Négrier, 1990, p.20). These cross-national differences in the territorialisation of networks are clearly a key theme for this study, and we will return to consider its subsequent (re)shaping in France and the UK in the following chapters in relation to recent telecommunications developments in Paris and London.

Early Liberalisation in the UK and Subsequent Telecommunications Initiatives

As was discussed in chapter 2, the liberalisation of the telecommunications market in the UK (see table 5.1) was closely linked to the government's neo-liberal, market-led stance at that time. No other European country dared to forge such a position in relation to telecommunications. In telecommunications restructuring, therefore, the UK in a way played 'the role of policy laboratory for the world' (Garnham, 1990; quoted in Hulsink, 1998, p.111).

In the early 1980s, the UK government faced the contradiction of trying to promote a more competitive telecommunications environment, while also privatising BT which necessitated keeping it in a strong market position (Beunardeau and Phan, 1992). The government and large business users:

> argued that the (gradual) deregulation of domestic telecommunications would promote a favourable business environment, enhancing efficiency, choice, market responsiveness and technological innovation, to the benefit of both large and residential customers. Such a favourable business environment would have a positive effect on international trade in services and strengthen the position of the UK in the world by attracting international traffic and international businesses critically dependent upon adequate, flexible and efficient telecommunications facilities (Hulsink, 1998, pp.111-112).

It was also a response to the perceived out-of-date national infrastructure of BT and their poor performance, especially in terms of service delivery to large business users who wanted more advantageous telecommunications packages. The City of London, for example, was trying to modernise its computer and electronic communications systems and wanted a high quality telecommunications infrastructure put in place (Hulsink, 1998, p.132). The deregulation and liberalisation of financial markets in the City of London in 1986 – the Big Bang – was an important factor in ensuring the continued primacy of the City as a global financial centre (see King, 1990, pp.93-100), and tied up with this was the increased demand it led to for high quality telecommunications infrastructures and services.

Table 5.1 Principal dates in telecommunications liberalisation in the UK

1981 – The introduction of the first elements of competition and liberalisation by the Telecommunications Act.
The separation of the General Post Office (GPO) into British Telecom and the Post Office.
The privatisation of Cable & Wireless.
1983 – Mercury Communications is given long distance and international licences. They immediately decide to construct a new fibre optic network for the City of London. This leads eventually to BT having to upgrade its network to fibre in 1987. The BT / Mercury duopoly in fixed networks is established until 1990.
1984 – The Telecommunications Act privatises BT, ends its statutory monopoly, and introduces both a new licensing regime and a new telecommunications regulatory body, the Office of Telecommunications (Oftel), to regulate the liberalised market environment.
1990 – The duopoly of BT and Mercury is abolished in all markets except international services.
1991 – The White Paper finalises the move from managed to open competition.
Mobile operators and cable companies are permitted to provide fixed services.
1994 – Liberalisation of international simple resale.
December 1996 – The UK liberalises international facilities a year in advance of the EU deadline, leading to the end of the duopoly on international voice traffic.

The creation of the market duopoly and the privatisation of BT inevitably led to certain tensions. For example, Oftel had to intervene in the issue of the interconnection of BT and Mercury networks in 1985, without which Mercury would not have been in a position to significantly compete with BT. In addition, the GLC draft strategy for telecommunications in London in 1984 focused on the provision of a telephone service for all, and maximising social benefits from the technology, which were felt to be under threat from the implications of the privatisation of British Telecom (Greater London Council, 1984). Nevertheless, it has been argued that the seven year duopoly in the 1980s had seemingly little impact on the dominant position of BT (Beunardeau and Phan, 1992), even if it did force BT to modernise its networks as Mercury was offering an entirely digital network focused on large business users.

Following the abolition of the duopoly in 1990-91, and the subsequent licencing of a number of alternative competitive new entrants (such as Colt, WorldCom and Energis), in December 1996, the government decided to allow competition in the international voice traffic market, the final element to be liberalised, as an attempt at making the UK the hub through which the whole of Europe would route their communications to the USA (Hulsink, 1998, p.145). Consequently, by 1999 there were 125 PTOs offering domestic and international telecommunications services in Britain (Department of Trade and Industry, 1999, p.23). The government is still ensuring it has a significant role to play, however, partly through the shaping of advanced technological infrastructure and service opportunities for the benefit of the whole country. One of the key buzzwords in telecommunications in recent years has been 'broadband':

> The Government is committed to encouraging the rapid roll-out of broader-band networks throughout the country with the aim of ensuring that the whole population can benefit from new technologies and widening the choice of advanced commercial, entertainment and cultural services at affordable prices. The Government's liberalisation policy has already made the United Kingdom a world leader in creating the new information infrastructures required for the multimedia world of the 21st century. By the end of the decade, cable communications companies alone will have invested £10-12 billion in new networks. BT and the other telecommunications operators are likely to have at least matched that investment (Department of Trade and Industry, 1999, p.23).

We shall see in the following chapters, how the socio-territorial implications of broadband network deployment have become a key element in a reconfigured form of government action in the telecommunications domain, and how this 'reregulatory' role is an increasingly significant aspect, in parallel with competitive markets, of the overall landscape of telecommunications developments in London.

Recent Liberalisation in France and Moves Away from Dirigisme?

As was discussed in chapter 2, telecommunications policy in France in the 1980s and 1990s was closely linked to the government's *dirigiste* state intervention approach (see table 5.2). This contrasted strongly with developments in the UK. While the UK government was pursuing its strategy of liberalisation and privatisation, the French were trying to actually reinforce the position of their public monopoly, the DGT, which was operator, service provider and regulator all in one. The DGT even extended its influence into the domains of value-added services, cable networks and mobiles (Hulsink, 1998, p.226). Developments in French telecommunications exhibited therefore a potent state interventionism, in keeping with overall *dirigiste* and *étatiste* traditions: 'Pressures for reform came later than in Britain, due to the successes of the 1970s and the flexibility of public sector finance rules following new mechanisms for raising capital' (Thatcher, 1999, p.170).

At this time, expansion of the national telephone network was still continuing, following on from the *Plan de Rattrapage Téléphonique* of the 1970s, but more advanced measures, which would put France a step ahead of many other western countries, were also being put into place, such as digitisation, the Minitel videotex system and telematics services (Hulsink, 1998, p.249). The Plan Télématique had been outlined in the late 1970s – early 1980s, from which offshoots such as the Minitel, Integrated Services Digital Networks (ISDN) and Transpac, an innovative packet-switching data-network, grew (Humphreys, 1990, pp.216-222). The first of these offshoots turned out to be quite representative of the evolution of French telecommunications policy in the 1980s and 1990s.

Table 5.2 Principal dates in telecommunications liberalisation in France

1986-1988 – The hegemony and authority of the state administration, the DGT, as public operator, commercial service provider, executor of industrial policy and regulator, begins to be questioned.

1987 – Proposals to open up the value-added network service (VANS), cable television and mobile markets are outlined by the government.

1989 – SFR is granted a mobile licence to compete with France Télécom in this sector.

1990 – New legislation outlines 'a balance between public monopoly, controlled competition and full competition' (Hulsink, 1998, p.254). This conforms with EC decisions on liberalisation of certain markets, and the separation of operational and regulatory functions.

1991 – The postal and telecommunications sectors are finally separated, as France Télécom is designated as an *exploitant public*.

1993 – Full liberalisation of the value-added network service market is achieved.

1994 – The publication of the Théry report on the creation of advanced national broadband networks.

October 1994 – Bouygues obtains the third mobile network licence.

26 July 1996 – The 'Loi de réglementation des télécommunications' and 'Loi relative à l'entreprise France Télécom' (with France Télécom to become a national company from 1997) are passed.
The alternative network market is to be opened up immediately, and the voice telephony market by 1998.

January 1997 – An independent regulatory body, the Autorité de Régulation des Télécommunications (ART) is established.

October 1997 – Part privatisation of France Télécom, with 25% of shares sold on the stock market.

1998 – Liberalisation of the voice telephony sector completed.

The precursory nature of the Minitel system turned out in many ways to be to France's disadvantage with the growth of the Internet in the mid to late 1990s. Nonetheless, Minitel had been the first and, by far, the most successful videotex experiment in the world.[37] This success is mostly attributed to the commitment of

---

[37] The UK's Prestel service launched by the Post Office in 1979 proved quite quickly to be a failure, at least partly due to a lack of coordinated strategy between the parties involved (see Thatcher, 1999, pp.234-235).

the French state which envisaged the further development of the information society in France, and which, through France Télécom, gave out free terminals and subsidised the system for a number of years (Castells, 1996, pp.342-345). However, once the Internet started its global hegemony, the capacities of Minitel seemed limited in comparison, and although France Télécom came up with plans for a more advanced Minitel II, it was fairly clear that France was going to have to adapt to an information society based around the Internet: 'Outrun and encircled, even France – one of the world's largest and most powerful economies – stepped back from independent network development on a national and, indeed, a more limited transnational scale' (Schiller, 1999, p.80). Indeed, by 1998, the government had allocated a subsidy of FF1.5 billion to help increase the integration of the Internet into all levels of French society (Hulsink, 1998, p.257). The Minitel example of relative national independence in technological development has a historical precedent, as France also did not begin to use the Morse electric telegraph system at the same time as other countries in the nineteenth century, as it had its own version called the Foy-Bréguet system (Mattelart, 1994, p.8). Independence and particularity of national policy appear, then, to be a constant and continuing feature in the evolution of the French telecommunications sector.

The beginnings of liberalisation in French telecommunications appeared around 1987, with government proposals to open up the markets of value-added network services (VANS), cable networks and mobiles (Hulsink, 1998, p.252). In parallel with the political ruminations about the future pathway of the sector, and, as in the UK, technological developments in the late 1980s and early 1990s, such as electronic switching and optical fibre, were perceived as having the potential to bring many benefits.

France Télécom was established as a relatively autonomous public corporation from 1991 onwards, although its relations with the government were to remain closely tied, and partial privatisation was continually put off (Hulsink, 1998, pp.254-255). 1996 saw the passing of the *Loi de réglementation des télécommunications* dealing with competition and regulation. This opened up the market for alternative networks with immediate effect, and proposed the liberalisation of the voice telephony market for 1998, in accordance with EU legislation. Another law was passed in the same year which turned France Télécom into an *entreprise nationale*, although this was done with the provision that the state would still hold 51% of the company (Thatcher, 1999, pp.162-164). In addition, the ART was created in early 1997, and France Télécom was corporatised with the government selling 25% of its shares on the stock market (Thatcher, 1999, p.163).

One of the key elements in the new French telecommunications legislation was the drawing of a distinction between network operators and service providers, depending on ownership of copper or fibre transmission infrastructure (Maxwell, 1999). Although this distinction was meant to promote infrastructure deployment by new operators, it has created problems as these operators must decide which type of licence they want straightaway, and, in the case of network operators, must demonstrate their infrastructure investment strategy to the satisfaction of the ART (Maxwell, 1999).

Competition came, nevertheless, with a number of measures aiming to protect client interests and to make competition as healthy as possible: public service and universal service to be financed by all operators; the creation of a regulatory body; operators to obtain permission for construction work and pay a rental charge to the public sector; operators to define the conditions for the interconnection of their networks (Sipperec website). Since the opening of the telecommunications market, new operators have been most interested in the long distance and business parts of the market. Competition at a local level is rare or still non-existent, which leaves communes and the small businesses located there at a disadvantage (Sipperec website). Nevertheless, the French telecommunications market is now estimated to be worth around FF240 billion (36.58 billion euros) a year (e-territoires, 2001c).

Unlike the UK, then, where pressure for change in the telecommunications sector came internally from the government and large business users amongst others, it seems that external pressures, from the EC and large multinational companies together with technological developments like the Internet, have been some of the main influences in the gradual transformation of the French telecommunications sector (Hulsink, 1998, p.278). In addition, '[l]arge users were also increasingly demanding flexibility and international services, especially for data transmission, causing worries about the need to adapt France Télécom's institutional position' (Thatcher, 1999, p.155). Overall then, as Thatcher argues, 'Institutional alteration in France was slower, took place later and was more gradual than in Britain' (Thatcher, 1999, p.152).

Nevertheless, as in the UK, the French government have become keen to stress the advantages of ICT use among both businesses and the wider public. The 1998 government action plan for France in the information society (PAGSI) gave DATAR the responsibility of producing an annual report on regional disparities in access to the information society (DATAR, 1999). The first report focused more on teleservices for citizens than economic development initiatives, but this possibly reflects the orientation of DATAR:

> the priority today for *aménagement du territoire*[38] is to accelerate the development of new uses for telecommunications by the government and local authorities which will have both an exemplary nature and a positive effect on the modernisation of public services as well as reinforcing the attractiveness of territories and their processes (DATAR, 1999, p.7, author's translation).

The aim now would appear to be to find a way of recreating the innovation of the 1980s, within a new competitive context. As the Forum d'Études et de Développement des Infrastructures Alternatives (FEDIA – Forum for Studies and the Development of Alternative Infrastructures) argues within today's context:

---

[38] The key French notion and practice of *aménagement du territoire* will remain untranslated throughout this study because of the difficulty in finding an equivalent phrase in English. We will discuss the notion later in this chapter, but it has broad connotations of an explicit form of equitable territorial planning and management, which has traditionally most concerned attempts at balancing Paris – provincial and urban – rural development.

'The historical dynamic of French teleservices which developed with the telephone network and Transpac in the 1980s must find a veritable successor through the services connected to the broadband Internet' (FEDIA, 1999, author's translation).

Planning and Governance Practices

The second element in our historical background to the subsequent comparative empirical explorations of telecommunications developments in Paris and London is a consideration of the changing role of French and UK planning and governance processes and practices, in relation to the overall development of Paris and London. Whilst it is evident that the evolutions in national telecommunications policy and regulation of the previous section should have a significant bearing on urban telecommunications developments, it remains important, in research on the relationships between telecommunications and cities, not to underplay the importance of planning and governance practices and dynamics in the shaping of institutional responses to these potential relationships.

London and Paris as Capitals and Historical Centres

In providing a brief historical perspective on national planning and governance practices in relation to Paris and London, we must start with the most obvious element. Both Paris and London are the capital cities of their respective countries, and key historical centres, both within Europe, and in the more global empires that both countries built up, the resonance and remnants of which are still important for both cities (see King, 1990). Given this background, it is unsurprising that the planning and governance of Paris and London has nearly always been accorded priority status by their respective states, for the good of the cities themselves, the nations as a whole, and their positions on the world stage. The difference between this historical role, and the contemporary roles of, and mechanisms bound up with, the cities has been captured by Amin and Thrift with regard to the City of London: 'In its old incarnation, the City was the result of the local going global. In its new incarnation, the City is a result of the global going local' (Amin and Thrift, 1992, p.583).

Recent similarities and differences between Paris and London in planning and governance dynamics within the context of urban telecommunications developments can be illuminated by consideration and comparison of historical processes and projects in the two cities. There are, for example, significant similarities in each country in the relationships between state and capital in the mid 19th century compared to recent times, and subsequently significant historically rooted differences between the two countries. Peter Hall describes London and Paris in the mid 19th century, respectively as:

> the apotheosis of laissez-faire and the minimalist state, directed by the purest utilitarian principles... [and, in the case of the latter, as having] developed a quite different model of capitalism... a model in which the state itself acted as the motor of economic

development, using its prestige and its power to persuade private capital to accord with its plans. This was the first developmental state... (Hall, 1998, p.744).

Goodman (1999) shows how even in the early nineteenth century, London was a polarised capital with a rich West End and a poor East End. He demonstrates also that planning and infrastructure developments in London and Paris were influenced by the form of government in the two countries. Much more was spent on transforming Paris, as Napoleon III had drawn up a master plan to which all potential obstacles were quickly removed, and for which Georges Haussmann as prefect became responsible for putting into action, improving the water supply and sewer systems, building parks and developing straight, wide boulevards. The fact that the state, through its tax revenues, largely foot the bill for these developments in its capital suggests that Paris at this time was perhaps strategically more important and more representative of France as a whole than London was within Britain, where a similar level of expenditure for transformation of the capital would have been unthinkable (Goodman, 1999). Indeed, it has been argued that never in the history of urban planning has one city changed so greatly so rapidly as during Haussmann's transformations of Paris in the mid to late nineteenth century (Hall, 1998).

Much later, Patrick Abercrombie's Greater London Plan of 1944 and other subsequent regional plans outlined a hierarchical system of roads linking existing towns, a Green Belt surrounding Greater London, and a series of new towns beyond the Green Belt to take overspill population from the capital (Chant, 1999). Other ideas, including those for an integrated transport system, suffered from the absence of a regional planning authority for London, until the creation of the GLC in 1965, along with a new administrative structure of 32 boroughs (Chant, 1999). Chant shows how post-war regional planning for London has been most subject to political negotiation and change, as, for example, despite the establishment of the GLC, 'frequent changes in the political hue of the council prevented the continuity that large-scale regional planning requires... The abolition of the GLC in 1985 was symptomatic of a new political tenor, much opposed to regulation and planning' (Chant, 1999, p.184).

Meanwhile, Delouvrier outlined a new regional plan for Ile-de-France in 1965, which led to a more polycentric structure of five new towns and the business zone of La Défense, and new elements in transport infrastructure including the *péripherique* ring roads and the Réseau Express Régional (RER – regional express network) system (Sallez, 1998, p.101; Hall, 1998, pp.744-745), thereby 'doing for the suburbs of Paris what Haussmann had done one hundred years earlier for the city' (Hall, 1998, p.745).

In spite of obvious contextual differences between the two cities, urban developments in London and Paris in the last 150 years or so have been influencing each other to some degree:

First, the Paris of Haussmann was a model of decisive action to be aspired to; then, British attempts to decentralise London through garden-city satellites found favour in Paris. But there remained key differences, partly because of the intractability of two urban fabrics that were quite distinct historical palimpsests, and also because of differing traditions of government (Chant, 1999, p.221).

In this way, while the authorities of Paris and London could each look to the other for broad ideas in the planning and development of a major historical capital city, it was inevitably impossible for them to get away from the territorial specificities of their own city. As we shall continually see throughout the following chapters, this notion of crucial territorial specificities in Paris and London becomes absolutely intrinsic to an understanding of how cultures of regulation, planning and governance influence and are influenced by telecommunications developments in global cities in different ways.

A Retrospective of the Key Planning Projects of the London Docklands and La Défense

The key elements in French and UK planning and governance systems, and a pertinent further illustration of London and Paris as important capital cities and historical centres can be achieved through a brief exploration and broad comparison of respective major urban redevelopment schemes in the two cities.

The ongoing projects in the London Docklands and at La Défense to the west of central Paris are representative of state-led urban (re)development projects in many global cities, which have been necessary for each city to remain competitive as financial hubs with plentiful office space. In this respect, they can be viewed perhaps as emblematic of a state interventionist approach based on 'spatial selectivity', in which certain strategic places are prioritised for economic development over others, for wider territorial gain (Jones, 1997). This has been associated with the ways in which new informational technology demands and working practices in the late 1980s and early 1990s made much office space in global cities inappropriate and out of date for the needs of multinational banks and financial service companies.

These projects, which also include Battery Park City in New York and First Canadian Place in Toronto, are often compared to each other (see Zukin, 1992; Crilley, 1993). However, the comparison of the Docklands and La Défense projects has itself been suggested to be problematic from a temporal perspective (Daniels and Bobe, 1993). While the Docklands redevelopment was only firmly and materially initiated in the early 1980s,[39] with the creation of the London Docklands Development Corporation (LDDC) in 1981, which was given a remit due to last ten years, the La Défense scheme was started in 1958, with the creation of the Établissement Public d'Aménagement de La Défense (EPAD), which was given a thirty year remit, but which was only recently due to be officially wound up. Indeed, the relative age of the La Défense project was highlighted recently by Edmund White, when he wrote that: 'La Défense went directly from being

---

[39] Although numerous plans and policies for regenerating the area had been put forward throughout the 1970s (see Brownill, 1990; Hall, 1998, chapter 28; Foster, 1999).

futuristic to being passé without ever seeming like a normal feature of the present' (White, 2001, p.2).

Nevertheless, from the point of view of this study, and, in particular, its subsequent focus on the telecommunications networks of the two zones, it is highly relevant to compare the two developments in a general sense, as their major similarity is perhaps the way in which they both occupy a fairly peripheral location in relation to the traditional core of the global city, while, in turn, creating a major urban axis of economic development between this core and their periphery. Indeed, Fainstein (1994, p.207) suggests that the UK government often cited the French government's success with the La Défense development as an argument in favour of the Docklands project. The establishment of this axis has, in turn, become at once a 'presupposition, medium and outcome' of telecommunications infrastructures in both cities, as we shall see in chapter 8, when the juxtaposition and parallel importance of new and old networked urban spaces in Paris and London is explored further.

Before the redevelopment projects of the Docklands and La Défense were undertaken, the two areas probably could not have been much different. In the latter case, the villages of Puteaux and Courbevoie were traditional agricultural communities, before slowly succumbing to suburban development from the late 19th century onwards (Lacaze, 1994). The London Docklands, in contrast, were a key site of the industrial revolution:

> In the 19th century when the Docks were originally built they represented a massive revolution in communication that enabled the industrial revolution to take place. With the waterways serving as the chief means of transportation and communication, not merely with distant territories, but within the City itself (Ward, 1987, p.291).

It was the dereliction and run-down state of the Docklands area following the decline of its traditional industries, together with the overwhelming need for office space in London, that prompted fresh government regeneration thinking in the early 1980s.[40] The area was seen by the Conservative government as an ideal location for testing out their new 'enterprise zone' initiative (Hebbert, 1998). The Conservatives adopted a market-led, or leverage planning (Brindley, Rydin and Stoker, 1996), approach, in which the requirements of the private sector were privileged at virtually whatever cost. Local public needs would supposedly devolve through a 'trickle down' effect (Coupland, 1992; Crilley, 1993).

The origins of the La Défense project, meanwhile, were in government proposals for an architectural competition in the 1950s which would 'regenerate the axis, between the porte Maillot and the La Défense roundabout, 'for the honour of the heros of the First World War'' (Lacaze, 1994, p.113, author's translation). This was followed by the more practical necessity, as in London, to create

---

[40] The state of the Docklands was argued by the government to be a national problem as much as a local one, because of both its location within London, and the importance of London to Britain as a whole (Brindley, Rydin and Stoker, 1996, p.102).

additional office space in the Paris region, but 'without jeopardising the scale and architecture of inner Paris' (Savitch, 1988, p.147).

Both developments were placed under the auspices of quasi-government authorities – the LDDC and the EPAD. The latter though had allegedly less autonomy and received more government interference than the former (Daniels and Bobe, 1993). The role of the LDDC was to 'prime the pump' to encourage private capital investment: 'The powers available to the corporation allow it to 'carry out any business or undertaking for the purpose of regeneration', although no clear definition of regeneration has been published. This gives the LDDC almost limitless powers' (Coupland, 1992, p.153).

This differing level of government interference was not necessarily completely to the respective benefit and detriment of the two projects in the way we might think: 'In Paris, the French government restricted development to help the huge La Défense project; but the British government did nothing to stop the City of London wrecking Canary Wharf' (Hall, 1998, p.926). Michael Hebbert suggests how 'the Canary Wharf development team, mostly North American, was astonished at the government's policy indifference, the diffusion of responsibility and the communication gaps that existed within and between the Departments of the Environment and Transport' (Hebbert, 1998, p.121).

This difference can be suggested to be all about conscious intra-urban competition in London, and the greater priority of inter-urban competition in Paris. The fact was that the rivalry between the Docklands development and the City of London as a key office centre was (and is) much more intense than that between La Défense and central Paris. Equally, in contrast to the French government, the UK government evidently seemed to think that this rivalry was beneficial to London's overall competitiveness as a financial centre because the market was big enough for both of them:

> This overproduction [of office space] has to be seen as the product of the fragmentation of the metropolis into a patchwork quilt of competing intra-urban growth coalitions, each eager to establish or consolidate its status as a financial district. When Docklands gained Canary Wharf, for example, there was a consciously chosen strategy in the City of London to weaken conservation regulations, raise plot ratios, increase the level of permissions and generally pursue anything that would ward off competition from Docklands (Crilley, 1993, pp.135-137; see also Hebbert, 1998).

In the early 1990s, however, both the City and the Docklands suffered from high office vacancy rates, reductions in employment in the financial and business services sector, and the bankruptcy of many developments and companies.

La Défense, meanwhile, is a veritable hub of international investment and multinational companies, which not only 'slackened' the border of the central business area in Paris and prevented the actual city from being overrun with tall office towers, but offered Paris a means to compete with the highly concentrated City of London: 'a more abundant offer, more moderate prices, more diverse possibilities in setting up, such assets are not to be ignored in the international competition between large cities' (Lacaze, 1994, p.139, author's translation). It is now 'the main weapon waged by the French capital in international competition' (Burgel, 1997, p.115). In fact, the 1994 regional Schéma Directeur, written by the state's regional Préfecture, set out plans to double the size of La Défense, as a response to the developmental constraints in the centre of Paris for office construction (Llewelyn-Davies et al, 1996, p.35).

Indeed, the way in which all levels of government were determined to assist the development of the *quartier* is illustrated by the response to the perceived disadvantage of La Défense not being part of the territory of Paris, which is a major consideration for some potential investors looking for a prestigious Parisian address: 'How could this be remedied without creating impossible political problems? Simply by asking the Minister in charge of the postal service to create a specific postal address, a Cedex, to be called Paris – La Défense' (Lacaze, 1994, p.119, author's translation). There were similar efforts made for the Docklands, but whilst the area was included in the 0171 telephone code of central London, other proposed changes to its postcode (from E14 to D1, thereby eliminating the 'East') and its public transport travel zone (from Travelcard Zone 2 to Travelcard Zone 1) failed (Hebbert, 1998, pp.194-195).

Whilst an Enterprise Zone was created on the Isle of Dogs in Tower Hamlets and Newham, in which there were supposed to be no restrictive planning controls and where businesses became exempt from paying rates for ten years, the construction of local housing declined abruptly and the LDDC even took over some local authority land for redevelopment (Coupland, 1992). As Fainstein observed: 'Instead of viewing the territory under its planning control as embedded within the Docklands boroughs, the LDDC pictured the riverbank as a new vibrant core for the whole metropolis' (Fainstein, 1994, p.194).

One of the major problems, or 'the Achilles heel', of the Docklands development (Daniels and Bobe, 1993) concerned infrastructure provision. Despite the eventual creation of the Docklands Light Railway, it became clear that the public transport and road networks should have been vastly improved before the main development was started, and were far from adequate for the number of people travelling to and from work in the area. Canary Wharf provided 40,000-60,000 office jobs alone. The finance problems and subsequent construction delays associated with the extension of the Jubilee underground line have been well documented in the national press. Transport strategies in the Docklands were 'targeted exclusively at making Canary Wharf a viable business location at the expense of strategic transport planning for the metropolis as a whole' (Crilley, 1993, pp.134-135). In contrast, it was ensured that La Défense was well connected to public transport networks, through the extension of the Métro, the provision of a stop on line A of the RER, and a tram system linking the zone with Boulogne-Billancourt and Issy-les-Moulineaux to the south.

Of course, the most severe problem with the Docklands development was the collapse of the Olympia & York (O & Y) development company in the early 1990s, which threatened the future of the whole Canary Wharf scheme. It is interesting that part of the reason for the problems of the company with Canary Wharf has been held out to be its lack of understanding of the traditional (and continuing) importance of the City of London as a *place* for finance and business, from which companies are extremely reluctant to move:

> Whether or not it was possible in the British context to overcome the appeal of central London, its failure to do so placed O & Y at a serious disadvantage in its competition with property-owners within the City of London, especially once the office glut there eroded the previous cost differential between the City and the Docklands (Fainstein, 1994, p.202).

In any case, as Newman and Thornley appraise: 'The Canary Wharf story shows how in the 1980s central government intervened in local planning in order to create the environment for the market to determine priorities' (Newman and Thornley, 1996, p.139).

Ten years ago or so, in the fresh light of the property market slump and Olympia & York going into receivership, opinions about the Docklands redevelopments were (necessarily) somewhat harsher than they might be today:

> Docklands started as a story of hope; a dream of opening up the area to meet the needs and aspirations of the East-Enders who had lived there for generations. Once hijacked by the private sector developers in league with a new market-led government-sponsored approach, it rapidly turned into a nightmare of deregulated planning and massive overdevelopment. The huge glass- and marble-clad offices have little of relevance for the local community, and represent a long-term monument to how 'regeneration' can become a disaster in less than a decade (Coupland, 1992, p.161).

Canary Wharf was seen as a 'grandiose set piece', an 'autonomous entity with concern only for what is immediately adjacent or directly contributive to their rental value' (Crilley, 1993, p.127). This is held out to be symptomatic of '...capital's continuing subordination of urban space to the logic of profit' (Crilley, 1993, p.159). In contrast, the local, least empowered groups were left to restore social issues to the Canary Wharf development agenda, although there was evidence in the late 1980s of a more socially conscious agenda being put in place (albeit temporarily) through negotiation between the LDDC, Olympia & York, and local representation groups (Foster, 1999, Chapter 6).

Canary Wharf may have wanted 'to avoid the 'project look' of modernist urban renewal such as La Défense in Paris' (Crilley, 1993, p.148), but it has not avoided substantial criticism of its placing global finance and big business well ahead of local redevelopment and social cohesion. The UK government has arguably come under most fire for stubbornly persisting with a market-led approach, even when the problems of the market became most overwhelming. For Fainstein:

> its latter-day puritanism about letting the market do its work contradicted the project's state-based origin. If the government initially had been content simply to await private initiatives, at most there would have been a gradual development of the Docklands, moving slowly outward from the boundary with the City (Fainstein, 1994, p.211).

The Docklands is now seeing the undertaking of a new phase of redevelopment, with notably the renewed construction of two office buildings next to the Canary Wharf Tower: 'Despite the pessimistic predictions that Canary Wharf would be a white elephant, it is thriving with activity and those working or simply visiting get out in large numbers here' (Foster, 1999, p.354). The relocation of Citibank, the largest foreign bank in the City, to the Docklands in the mid 1990s has been viewed as a prime example of the turnaround in its fortunes (Hebbert, 1998). The LDDC has been wound up, and responsibility for local development has returned to the hands of the local authorities, but Foster's conclusion remains one dominated by the continuing neglect of the needs of the local community (Foster, 1999).

In these respects again, the comparison with La Défense may not be especially pertinent, because, on the one hand, the French project was not as closely associated with social cohesion objectives to parallel economic development ones as the Docklands scheme was supposed to be,[41] so that the inevitable big business – poor neighbourhood socio-geographical juxtapositions were never as clearly visible as in the Docklands. On the other hand, state intervention in the project to ensure its success was continually forthcoming – the infrastructure was provided from the start; as we saw, the government heavily restricted similar development in central Paris; public funding was made available when the project hit problems during the economic crisis of the 1970s; empty office space was filled by government officials (Fainstein, 1994). In short, 'while the French used private-sector resources to create a whole new section of Paris, they did not rely on the market to regulate the flow of those resources' (Fainstein, 1994, p.211).

The Spatial Restructuring of State Socio-economic Regulation in France and the UK

We can suggest, then, that the development projects undertaken at La Défense and in the Docklands are, to a certain extent, emblematic of the relative dominance of the state or the market with regard to planning and economic development policies and strategies in France and the UK.

The French state has traditionally kept close control over planning and economic development policy, through departments such as the DATAR and the Ministère de l'Équipement. Market developments have been steered, to a certain extent, according to state policy and *volonté* (Keating, 1991a; Newman and Thornley, 1996; Harding and Le Galès, 1997). *Départements* in France, for example, are headed by a government-appointed prefect, who provides government oversight of policies and public funding in the area. In the Ile-de-France region, the Institut d'Aménagement et d'Urbanisme de la Région Ile-de-France (IAURIF – the institute of planning and urbanism of the Ile-de-France region), as the planning arm

---

[41] In spite of the fact that high density housing was an important element in the overall project, as at the Parc estate (Noin and White, 1997).

of the Conseil Régional, assembles representatives from national government and all levels of local government.[42]

The UK state, meanwhile, has done likewise on one level, but has steadily become more open to the requirements of the market and the private sector, as opposed to those of the traditional welfare state, and, to a certain extent, within a neo-liberal *laissez-faire* approach, has oriented elements of planning and economic development policy in accordance with these requirements (Keating, 1991a; Newman and Thornley, 1996; Harding and Le Galès, 1997). As Newman and Thornley suggest: 'The recent emergence of public-private cooperation at local level gives planning in the 1990s a different character to the centralised, deregulated model of the mid-1980s' (Newman and Thornley, 1996, p.145). However, this neo-liberal, centralised approach was very much a strategy of Conservative governments, particularly under Thatcher. Before this, a common approach to steering the space economy involved 'carrots' (in the north) and 'sticks' (in the south east). There was huge public intervention, but more in bridging the failures of the market than in shaping a particular form of national territorial development.

The differing balance between market orientation and state regulation in the UK and France is also bound up in diverging intergovernmental systems in the two countries since at least the early 1980s. While the Conservative government in the UK kept administrative control highly centralised, the French Socialist government introduced decentralisation measures to devolve certain powers to regional and local level. However, this situation is tempered by the fact that the UK government was intent on forging important links and relations with the private sector, while the French state was bidding specifically to improve levels of intergovernmental relations. As Newman and Thornley observe, in a London context: 'Centralisation of decision-making [in London] has been a dominant factor since the abolition of the GLC, and the private sector has taken an increasingly important role in strategic planning and promoting the capital as a 'world' city' (Newman and Thornley, 1996, p.128). Indeed, in the late 1980s and 1990s, on the one hand, it was the Corporation of London, the local authority for the City and the voice of its business community, which often represented London at international gatherings or general meetings, whilst on the other hand, the government attempted to increasingly coordinate its policy role in London, leading to the creation of the cross-departmental Government Office for London (GoL) in 1993 (Hebbert, 1998). In France, one of the key mechanisms in decentralised governance has been the introduction of strategic *contrats de plan*, the content of which is subject to negotiation between the state, the regions, and the local level. These plans are informed by regional *schémas directeurs*, which are negotiated by regional bodies of the state and the Conseil Régional (Llewelyn-Davies et al, 1996, p.34).

---

[42] The precedent to IAURIF, the Institut d'Aménagement et d'Urbanisme de la Région Parisienne (IAURP) was created in 1960 during the presidency of Charles de Gaulle, becoming the IAURIF following the change in the name of the region of the capital from Région Parisienne to Région Ile-de-France in 1976 (see IAURIF, 2001a).

The particularity of Paris and London (as global cities) in national state regulation
We should not, however, underestimate the particularity of Paris and London within these national systems of state regulation. Both cities are national capitals, the powerhouses of their national economies,[43] important historical centres, and key centres of command and control functions or global cities, as we shall see here. Subsequently, central government in France and the UK remains, like it nearly always has done, more involved in the development of their capitals than any other city or area of their territories. For example, Brechet has discussed how:

> In Ile-de-France the state has retained significant power in regard to regional planning. Decentralisation did not in any way modify the manner of drawing up the regional master plan; it remained exclusively under state authority (Brechet, 1994; quoted in Noin and White, 1997, p.63).

The relationship between the state and the local authorities of the Paris region has usually been characterised by opposing points of view, which have been difficult to mediate, and 'reciprocal distrust' (Lacaze, 1994, p.33). This is perhaps unsurprising, given, as Guy Burgel suggests, that: 'Paris has remained essentially a personal battlefield for private investment, prestigious improvements and presidential enrichment' (Burgel, 1997, p.115). Because of this, it has only had an elected mayor since 1977, and given the length of time Jacques Chirac occupied this office (1977-1995), it is safe to suggest that the French President and the wider French state still takes a significant role in the affairs and the development of the city. As Noin and White sum up:

> Within France, therefore, Paris plays a complex set of roles. It is an extremely dominant capital, and yet the agglomeration itself is also dominated to a considerable degree by the same state interests that have underpinned the evolution of that dominance. And the agglomeration does not represent a single monolithic power-block within the French polity, but is instead divided within itself both in administrative and in wider psychological terms (Noin and White, 1997, p.8).

With regard to London, the recent reinstatement of a regional body of government, the Greater London Authority (GLA), even with its limited powers in certain policy areas, is part of the recent focus of the current Labour government on reconfiguring the level of (de)centralisation of the UK state and on subsequently improving intergovernmental relations.[44] But, as with the role of the state in Paris above, it is also a recognition of the status and (national) importance of London as a global city, and of how, until recently, it was virtually the only global city without a mayor and strategic level of government, which had increased the level

---

[43] The economy of London, for example, has been estimated to be worth £75 billion, or 20% of the total GDP of the UK (Hebbert, 1998, p.132).

[44] However, a regional level of government for London was not meant to reduce the role of other levels of government: 'The GLA model reflected Labour's desire to restore London-wide government, but in a way which did not imply any threats to power for the boroughs' (Tomaney, 2001, p.244).

of fragmentation in the governance of the city (Newman and Thornley, 1997). The decision to introduce a directly elected mayor for London has, then, been viewed as an example of wider current shifts from government to governance (Tomaney, 2001).

The establishment of a Mayor and Assembly mean, however, that London must now function as a region as well as a global city (Local Futures Group, 1999b, p.24). According to the Local Futures Group report (1999b, p.14), the London Development Agency (LDA), which was the precursor to the GLA, had to deal with two aspects to the competitiveness of the capital:

- Improve London's *external* competitiveness in a national, European and even global context of inter-regional competition.
- Improve the competitiveness of weaker local and sub-regional economies *within* London, where both cohesion and growth matter to sustainable development.

The role of central government in the former objective has increased in recent years, both as a general trend to intensified inter-urban competition in Europe has emerged, and as it has become clear that the competitive advantages that London has traditionally held, in particular the dominance of the City as Europe's pre-eminent financial centre, are perhaps being eroded by the likes of Frankfurt and Paris. These are inevitably also key concerns at the local level for the Corporation of London (Pete Large and Pauline Irwin, Corporation of London Economic Development Unit, interview with author, August 2000).

In comparison, it is evident from the focus of debate and policy orientation that it is felt that Paris – Ile-de-France must start functioning as a global city as effectively as it does as a region: 'Paris's problem is to confirm its position as one of the members of the private club of very great world capitals, with New York, London and Tokyo' (Carrez, 1991; quoted in Llewelyn-Davies et al, 1996, p.37). At the end of the 1980s, for example, some French planners and agencies saw the development of an Ile-de-France megapolis as the only way to ensure France was not left marginalised within Europe (Lipietz, 1995). For Noin and White, though, there are strategic difficulties with such a vision:

> The principal problem posed by the administrative structure of the agglomeration is not...the exceptional position of the City of Paris (nearly 20 times the population of the next largest municipality), but the absence of an overseeing authority for the whole agglomeration (Noin and White, 1997, p.36).

Paris does not have a *communauté urbaine* like Lyon or Bordeaux. It has actually been seen as 'un état dans l'état' – a state within the state (Renaud, 1993). As Noin and White continue:

> the realisation of important projects which necessitate the agreement of the state, the region and the communes concerned is thus slow and littered with potential pitfalls. Rejecting both the excessive centralisation of the past and the political fragmentation of

today, there is now a necessity to find a middle route which respects useful decentralisation whilst also ensuring the indispensable coordination of planned actions at the spatial scale of the agglomeration as a whole (Noin and White, 1997, p.36).

In a study of one element of these planned actions – promotional strategies – Chevrant-Breton (1997) suggests that Paris is most interested in attracting firms which are in keeping with the status of the city, such as high value added firms. The French particularity of having the same organisation responsible for both planning and attracting inward investment is also questioned (Chevrant-Breton, 1997, p.152). As Llewelyn-Davies et al observe:

> The Départements and the Communes are more interested in attracting enterprises, because enterprises create jobs and generate a much needed tax base. In that sense, they compete with each other; in any event, they have no strategic view. Any enterprise (provided it does not pollute) is seen as worth retaining or attracting (Llewelyn-Davies et al, 1996, p.34).

The Llewelyn-Davies et al report goes on to note three main reasons why central government is not involved in developing specific policies for the Paris region: *aménagement du territoire* attempts instead to reduce disparities between Paris and the rest of France; decentralisation means that it is the Conseil Régional which is seen as being responsible for the development of Paris; the government does not want to interfere in 'industrial choices' either, ie through determining regional growth sectors (Llewelyn-Davies et al, 1996, p.34).

Meanwhile, the governance situation in London remains just as complex. Newman and Thornley summarise how:

> Following the fragmentation of London planning in the late 1980s new public and private, city-wide and subregional interests have developed. A complex pattern of agencies has been formed and at the London-wide level the relationships between public and private interests are intricate and in a state of flux (Newman and Thornley, 1996, p.151; see also Newman and Thornley, 1997).

The development of the new London government illustrates the conflicts between central and local levels of government in the UK (see Newman, 2000).[45] There is a distinct lack of independence of the London mayor from central government, as in the lack of tax raising powers, or the obligatory continuing influence of national planning policy in setting the framework within which the mayor must formulate a Spatial Development Strategy (SDS) for London (Tomaney, 2001).

Llewelyn-Davies et al summarise, then, the possible comparison between the two cities: 'Paris is, with London, perhaps the only 'all-features' world city in its

---

[45] The continuing tensions and controversies surrounding the formulation of a transport strategy for London, and particularly the vast difference of opinion between the government and Ken Livingstone over the proposed public-private partnership approach to modernising the Underground system, illustrate this very well.

range of economic, social, political and financial attributes; but it is perhaps less powerful all round' (Llewelyn-Davies et al, 1996, p.37).

National Traditions of Territoriality

The third element in our historical background to the subsequent comparative empirical explorations of telecommunications developments in Paris and London is a consideration of differing national cultures of territoriality. This is important on at least two counts. Firstly, the ways in which different national, regional and local actors in each country understand and relate their practices and actions with the notion of 'territory' (and in turn the interactions between these various understandings) shape, and are in turn shaped by, urban telecommunications developments. Secondly, the links between the construction and deployment of infrastructure networks and the notion, configuration and signification of territoriality have, as we shall see, always been inherently close, at least in French contexts. There seems, therefore, to be significant grounds for exploring the ways in which telecommunications infrastructure networks are closely tied to national territorial ideals and traditions.

'Aménagement du territoire' in France

'When Paris coughs, France sneezes' (traditional French saying).

The French notion and practice of *aménagement du territoire*, broadly meaning territorial planning, is almost unique. There are a variety of factors for this, including the inherent Republican principle of egality, combined with a large national territory characterised in many regions by low population density, but one of the chief ones seems to be related to cultures of mobility. The mobility of work and therefore of people in North America has been opposed to the relative 'sedentary' culture of Europe, and of France in particular (DATAR, 2000, p.8). The French are held to be:

> sensitive to the build-up of patrimony; it is a cultural and societal choice: being allowed to live and work in one's home region. Consequently, in order to compensate for this low mobility of work, authorities are constrained to initiating transfers of income and public funds, with the aim of reducing disparities. This is one of the fundamental aspects of 'French' *aménagement du territoire* policy (DATAR, 2000, p.8, author's translation).

The term might date back semantically only a few decades, but its origins and basic ideas can be traced back as far as Napoleonic times. The French Revolution gave us the powerful nation state: 'International communication emerged with modern nationalism, which established the territory as the basis of sovereignty and of an imaginary community' (Mattelart, 2000, p.1). Mattelart indeed suggests that the emergence of new communications technologies helped formulate the earliest notion of *aménagement du territoire*:

The preelectric telegraph aided the French state, as it emerged from the Revolution of 1789, in its project of mastering space. It was an element in a unified territorial scheme. A coherent vision of the national territory gave form to regulations assuring the flow of merchandise and people (Mattelart, 1994, p.4).

The more recent political idea(1) of *aménagement du territoire* in France developed in the immediate post-war period with state focus on national reconstruction. By the early 1960s, it had the rather ambitious parallel objectives of modernising the territory, renovating or 'shaking up' the procedures of the Jacobin State, and changing society (Wachter, 2000, pp.81-82). As Serge Wachter makes out: 'The aim was to introduce new administrative working methods, breaking with the inertia and the rigidity of ministerial divisions and their crystallisation in the devolved terminals of the State' (Wachter, 2000, p.82, author's translation). A new government agency, the DATAR, was created in 1963, expressly to explore and act within the domain of *aménagement du territoire*. The whole concept, however, was frequently reduced to the Paris – province debate and their uneven development (IAURIF, 2001a).[46] The imbalance between Paris and other French cities and regions should, nevertheless, not be underestimated, and was the key concern for the French government from the end of World War II, leading to policies such as the '*métropoles d'équilibre*' in the late 1960s (Sallez, 1998, p.102). The aim here was to promote the development of key provincial French cities, especially through infrastructure deployment, thus 'rebalancing' overall national territoriality.

The decentralisation project of the 1980s could seemingly not help but dilute the nature of *aménagement du territoire*, with agendas and responsibilities passing from the state to regional and local levels of government. It was no longer clear which geographical scale was prioritised in *aménagement du territoire*: 'Did its task consist of helping local development, regional action, national solidarity, or the three at the same time?' (Wachter, 2000, p.84, author's translation).

One of the main attributes of *aménagement du territoire* in the 1990s was its reconfiguration towards a recognition that territorial differences might actually bring benefits:

> We are now increasingly seeing in it a creative scope for organisation, or an active structure, and no longer a simple space that we use to welcome potential investors or to construct an infrastructure...Policies of *aménagement* are discovering that real 'territory' is not functional 'space' (DATAR, 2000, p.11, author's translation).

As a result of this, competition between regions was now recognised as a valid pathway to follow (Wachter, 2000, p.87). Another aspect of *aménagement du territoire* that came back into focus was urban policy, following many years of relatively more interest in rural areas (Wachter, 2000, p.89). The main reason for this was the changing nature of scale – the French urban system could no longer be viewed as a system in itself, but had to be viewed in relation to the European scale,

---

[46] This had been identified as early as 1947 in Jean-François Gravier's book 'Paris et le désert français', which was very influential on French territorial thinking for many subsequent years.

which led to the realisation that, of all French cities, only Paris was capable of occupying a major role in the European urban system (Wachter, 2000, p.89). The Paris – province debate has subsequently been downplayed, although this has led to a dual and paradoxical focus for French territorial planning, of both increasing the competitiveness of Paris within Europe and the world, and promoting decentralised territorial development away from Paris whenever possible (see Lipietz, 1995). The role of the DATAR agency remains crucial for the latter policy (see DATAR, 2000), but the benefits of a strongly competitive capital for the whole of France are being increasingly highlighted (Burgel, 1997).[47]

Aménagement du territoire in the 21st century must therefore relate to every scale from the local to the global: 'more and more, it has to mobilise and coordinate an increasing number of levels of decisions and wishes – public and private – to take on and carry out its programmes of action' (Wachter, 2000, p.95, author's translation). Wachter sees it as becoming a case of *ménagement* instead of *aménagement du territoire*: 'to open latitudes of choice, to create effects of integration, and to improve decentralised coordination' (Wachter, 2000, p.101, author's translation). As the national DATAR agency and regional prefecture of Ile-de-France discuss:

> The territorial approach is being modified: the regional or the inter-regional are settling themselves into determined spatial delimitations; the metropolitan areas of today have no well-established limits, analytical scales vary according to the questions being asked. To study Ile-de-France is also to study the Parisian urban region up to the 'Francilien' border, and its connections with cities in the Parisian Basin. In short, the urban poses clearly the question of governance to *aménagement du territoire*, in other words, of the mobilisation of numerous actors and at varying scales in a dynamic of common projects (DATAR and Préfecture d'Ile-de-France, 1999, p.16, author's translation).

The French territory is therefore bound up with three structuring logics – institutional (its historical administrative division), functional (its relations with multiple economic processes), and patrimonial (its everyday agrarian roots, opposing 'the France of Roquefort to that of Coca-Cola') (DATAR, 2000, p.12). Thus, through parallel consideration of these logics and recent experience of French territorial development, the DATAR recently proposed a future perspective for the *aménagement* of France up to 2020, based on a polycentric and networked model of national territoriality.

---

[47] This seems to be a shift back from the 'anti-megapolis' discourse which became quite dominant in the early 1990s, when notably Charles Pasqua, the Minister of the Interior and Territorial Management, argued against substantial further growth of the Paris region. Much of the debate focused around trying to agree on the size of a suitable prospective population of the Paris region in 2015, and thus the level of spatial and physical development that would be needed (Lipietz, 1995, pp.148-149).

The Lack of an Equivalent in the UK?

The French concept and policy of *aménagement du territoire* is undoubtedly unique in the world. Few other countries have devoted such effort into thinking 'territorially' about how to manage and organise national development.[48] Few other countries have governments as regularly involved in publishing documents or entire books about 'territory' or 'territoriality' as the French government through the DATAR agency (see DATAR, 1992; DATAR, 1993; Wachter et al, 2000; DATAR, 2000).

In the UK, territorial thinking, planning and action has been nowhere near as complex or prolific. Nevertheless, the planning guidance documents (development plans and regional planning guidance) of the UK in the 1990s have been seen as perhaps working towards a similar set of relations between the government and the regions as *aménagement du territoire* partly implies (Wachter, 2000, p.99). However, there would seem to be a strong link between the political tradition of the UK 'based on pragmatism, incremental change, and disdain for grand principles' (Keating and Midwinter, 1994, p.180), and the lack of innovative strategic territorial policy. This may also be a strong question of the relationship between London, as both national capital and global city, and the rest of the country, which again is a familiar question to the French. However, equitable territorial development in the UK has never really been on the political agenda in the last two or three decades. Instead, Thatcherism has been viewed as characterised by the construction of 'a clear two-nation strategy present from 1979' (Jones, 1997, p.850). As Jones summarises: 'The Thatcherite neoliberal accumulation strategy was and still is based on promoting free-market ethics, deregulation, privatisation, recommodification of the public sector, tax cuts, and favouring the South East and London as its financial centre' (Jones, 1997, p.850). Consequently, possibly the principal territorial notion and mechanism in play in the UK in the last two decades has been the North – South divide: 'a product of the spatial selectivity ascribed to the South East within Thatcherite accumulation strategies and the spatial selectivity denied to the North, which was gradually severed from industrial assistance and forced down a route of intense industrial restructuring' (Jones, 1997, p.851).

Indeed, from a London point of view, Peter Taylor has argued that 'London would operate just as well (in fact, probably more effectively) as a city-state without the encumbrance of the rest of the United Kingdom' (Taylor, 1997, p.766). This view also says very little for the state of territorial cohesion and policy in the UK. He goes further as well: 'The interests of the London city region do not coincide with the interests of the rest of England. That is why the United Kingdom is not big enough for London and England' (Taylor, 1997, p.769). London's global city role is crucial, then, not just for the city but for the whole nation. This is a point of view shared by some of the regional authorities of London (Colin Jenkins, GLA, personal communication, June 2001).

---

[48] The Netherlands can be seen as a rare exception.

As we shall see in the next chapters, this intrinsic territorial focus in France and the lack of something similar in the UK have inevitably greatly influenced telecommunications developments in Paris and London.

Infrastructure and Territory

We saw earlier in the chapter something of how infrastructure networks necessarily have or imply important territorial elements and configurations. This goes back as far as the development of the visual telegraph network in nineteenth century France, which created a far greater level of connection and communication between the central state and the previously far-flung provincial departments. It was important for the Republic to be seen as one closely linked territory. The traditional territorial focus of French national development has, then, seemingly nearly always been considered in parallel with infrastructure networks (Offner and Pumain, 1996; Mattelart, 1999).

The French word for network, *réseau*, has a long and complex etymology in the French language. It has only had a meaning related to communications systems for around 200 years. Prior to this, it had developed both biological and military analogies (Mattelart, 1999, pp.170-171). Saint Simon (1760-1825) was one of the key orchestrators of extending the dimension of the meaning of *réseau* (Mattelart, 1999, p.174; Musso, 1997). It fitted in with both 'Enlightenment ideals of exchange as a creator of values' (Mattelart, 1999, p.174) in the search for the goals of Reason, and the recognition that French national unification at the time still needed to be fully achieved.

In the 18th and 19th centuries indeed, comparisons between the advanced communications networks of Britain and the embryonic *réseaux* in France were frequent: 'the image of a map criss-crossed by free-flowing communication networks inspired many French authors of social utopias during the nineteenth century, who made explicit reference to the perfection of the English model' (Mattelart, 1999, p.175). The regional integration of the British telephone system in the late nineteenth century contrasted strongly with the centralisation of the French network and its inevitable vast areas of provincial disconnection. The remarks of a French telecommunications adviser, quoted by de Gournay, illustrate the differences, and the early recognition of the need to shape telecommunications and territoriality in parallel:

> We have just seen that the organisation of the French telephone system has not yet enabled us to link many suburban areas with urban networks, nor have we been able to group different telephone networks within the same region. Abroad, some countries are ahead of us in this area. The principal reason for this is that their telephone concessions are not strictly limited to the city proper. Generally, the concession covers a rather large zone which includes neighbouring small towns and the suburbs, in other words, the whole agglomeration surrounding the centre of the concession; sometimes, it is even spread out over a whole region, in which case all the regional networks are part of the same concession (quoted in de Gournay, 1988, p.331).

These remarks highlight the ways in which communications networks and territorial development are very much bound together. This remains the case today, as is demonstrated quite simply by the distinctions made in the telecommunications industry between local and wide area networks, or metropolitan area networks and pan-European networks. The ways in which territoriality and scale are thus bound up in telecommunications developments in Paris and London will be explored in more detail in the following chapters.

Concluding Remarks

In this chapter, we have provided a broad, historical perspective on French and UK telecommunications, planning and territoriality in relation to Paris and London, as a means of setting the scene and providing some important context for the discussion and analysis of telecommunications developments in the following chapters. This perspective serves as a reminder that many of the primary issues and elements of contemporary telecommunications developments in global cities have crucial historical roots and backgrounds, thus refuting any suggestion that they are completely new developments with little or no relation to any contextual precedent. What we have found here is that the contemporary landscape of telecommunications developments in global cities is, on one level, made up of a series of interacting historical processes and practices in the domains of technological evolution and telecommunications regulation, planning and governance practices, and national traditions of territoriality. With this study having a substantial comparative element, we have also importantly shown that many of these differing historical processes and practices have varied, sometimes quite substantially, between France and the UK. As we shall see in the subsequent case study chapters, these differing historical processes and practices have influenced the differing contemporary landscape of telecommunications developments in Paris and London.

It is now time to move into an actual empirical exploration of our case studies, and investigate both the continuing influence of the processes discussed in this chapter on contemporary developments, and whether these telecommunications developments in Paris and London are emblematic of, or partly configured by, new and / or different dynamics of technological evolution and telecommunications regulation, planning and governance practices, and national views of territoriality. As already explained, the case study chapters are structured according to the three principal cross-cutting strands of my research in order to privilege the comparative aspect of the study in greater depth. The similarities and the differences between the many elements and implications of telecommunications developments in Paris and London should, then, become hopefully much clearer.

Chapter 6

National Influences on Telecommunications Developments in Global Cities

Introduction

The different elements in the analysis of the territorialities of telecommunications developments in Paris and London have been divided and spread across several chapters, in order to approach them from a slightly different, more thematic perspective which aims to bring out as effectively as possible the salient strands of both cases. In the previous chapter, therefore, we opened discussion of the case studies by providing a historical overview of important developments and processes in telecommunications, planning and governance, and views of territoriality, in which the empirical material on Paris and London needs to be grounded. It is now time, therefore, in the next three chapters, to discuss, analyse and compare these telecommunications developments to explore what they demonstrate about the relationships between global cities and telecommunications in a more detailed perspective. Each of these chapters follows on largely, but not exclusively, from the three respective theoretical chapters in Part One, according to our research strands. The choice of these strands illustrates and brings out the focal interplay between differing geographical scales that characterises contemporary urban telecommunications developments (and many urban developments in general) in a supposedly globalising era. Hence, in this chapter, we investigate first the extent of national level influences on telecommunications developments in Paris and London.

We need to explore and compare the nature of these influences in two ways. Firstly, the influence of national telecommunications policies and regulatory practices. Secondly, the influence of national intergovernmental systems, and urban and economic development policies. Identifying, analysing and comparing the level of influence of both these broad national policy fields on telecommunications developments in global cities serves the purpose of specifically investigating the basis behind the current popular rhetoric and focus on how the convergence of globalising processes and supranational actions and local processes and subnational agendas has perhaps somehow squeezed out the national level, and state policy and decision-making in particular. In other words, we are exploring here the ways in which processes and practices from the national scalar level influence and are influenced by telecommunications developments in Paris and London, as the first layer of the territorialities of these developments, which are bound up with parallel and mutually constitutive scalar processes and practices.

The Influence of National Telecommunications Policies and Regulatory Practices in Telecommunications Developments in Paris and London

It is to be expected that national telecommunications policies and regulatory practices should have a quite substantial general influence in the make-up of the landscape of telecommunications infrastructures and services in global cities such as Paris and London. What is perhaps not so clear is the level of their continuing influence in shaping predominantly liberalised markets, and the deployment of infrastructures and services in global cities, where these infrastructures are firmly and deliberately territorialised to promote and increase global – local connections in ways which apparently bypass the national policy level. It is these elements and practices which we focus on in this section.

The Differing Continuing Influences of Lengthy Monopoly in France and Precursory Liberalisation in the UK

As we have already seen in both chapters 2 and 5, the recent history of telecommunications policy and regulation is substantially different in France and in the UK. In the UK, there has already been a relatively long period of liberalisation and competition, the beginnings of which already date back about twenty years, whereas in France, although some limited forms of competition were introduced from the late 1980s onwards, the overwhelming dynamic has been one of a protection of the monopoly until the obligation of EU-wide liberalisation in the mid 1990s. The liberalisation of the UK telecommunications market was part of a conscious government strategy to boost the UK, and especially London, as a business centre. Changes in the French telecommunications market had more to do with European Union policy-making to bring competitive markets into being throughout all member states by 1998, and were carried out with reluctance and reservation from some quarters. Within these differing national contexts, then, we might expect telecommunications developments in Paris and London, up to the last couple of years at least, to be firmly intertwined with the dominant monopolistic environment of France and the dominant competitive environment of the UK respectively. We can explore and compare the extent of the influence of these differing national contexts by investigating the development of the Téléport Paris – Ile-de-France and the shaping of the City of London telecommunications market between around the mid 1980s and the mid 1990s.

Téléport Paris – Ile-de-France I: configuring a restrictive national regulatory context for regional territorial development Like many cities and regions around the world in the late 1980s and early 1990s (see Richardson, Gillespie and Cornford, 1994; IAURIF, 1998a), the public sector response to the emergence and recognition of telecommunications as a tool for territorial development in Ile-de-France took the form of the implementation of a teleport.

The origins of the Téléport Paris – Ile-de-France date back to the mid 1980s, with the creation by IAURIF in 1987 of the *Association pour l'étude d'un téléport en Ile-de-France*:

> IAURIF wasn't asked to reflect on a teleport project, it was IAURIF which had a small division of people working on the theme of information technologies, who were taking notice a little of what was happening, and they saw in the 1980s the development of an international movement for opening the telecommunications market to competition, well in the United States, and then in Britain. Then there was the evolution of technologies and also of satellite communications technologies at the same time. So they saw the first concept of the teleport appear quite rapidly... (Daniel Thépin, IAURIF, interview with author, September 2000).

Fairly rapidly, proposals became ambitious in scope, taking in five extensive sites around the Ile-de-France region – Marne-la-Vallée to the east of Paris, Roissy to the north, La Défense to the west, and the Plateau de Saclay and Saint-Quentin-en-Yvelines to the south. By including the *villes nouvelles* in the project, the Téléport was thus building on continuing regional territorial policy, first outlined in the Schéma Directeur of 1965. In addition, each of these sites was to be the focus for a specific type of activity:

> Activities were split up which meant that specific telecommunications services could be developed for a category of users. So, there was an EDI [Electronic Data Interchange] service for the transport sector at Roissy. Roissy is an important pole for the transport sector. On the Plateau de Saclay, it's research. At La Défense, it was business zone functions. At Marne-la-Vallée, it was the image sector (Daniel Thépin, IAURIF, interview with author, September 2000).

The list of members belonging to the *Association Téléport Paris – Ile-de-France*, which had succeeded the earlier association, consisted of representatives from different local and regional authorities, as well as the directors of large private companies, and representatives from local universities. Four 'founding principles' of the Téléport development were originally outlined in 1987 (IAURIF, 1998a, p.59, author's translation):

- To build an innovative project, making use of communications technologies, which is part of regional development plans and contributes to making Ile-de-France a key site in new global networks of communication.
- To develop a teleport on a regional scale which builds on the main development areas in Ile-de-France and takes into account their specificities in constituting teleport development sites.
- To offer services to enterprises which are attractive to multinationals and a development support for small and medium enterprises.
- To promote study themes which reflect the competences of association members, and lead to projects which make use of applications from both new technologies and more traditional sectors.

The basis for the development of the Téléport in the first place was, therefore, a perceived need for another layer of infrastructures and services in the region to supplement those of France Télécom (or the DGT as it was in the mid 1980s). As Emilio Tempia of the Direction Régionale d'Equipement en Ile-de-France (DREIF – regional infrastructure directorate of Ile-de-France) stated, 'it was a way for the Ile-de-France region to work against the monopoly at the international level' (Emilio Tempia, DREIF, interview with author, September 2000). The implication here, then, is that creating a regional telecommunications project which connected specific local sites to global networks via satellites could be a way to bypass the national monopoly of France Télécom, which was stifling the competitiveness of the region in terms of telecommunications.

Whilst the DGT was a member of the original working group 'perspectives for electronic communication in Ile-de-France in 2000' set up in the early 1980s, whose work directly inspired the proposal of IAURIF for a teleport in 1986 (IAURIF, 1998a, p.57), by the time of the first partnership of regional actors led by the Conseil Régional, the *Association pour l'étude d'un téléport en Ile-de-France*, in 1987, the national operator had disappeared from direct involvement in the project. This partnership eventually included other telecommunications companies instead – Alcatel, Cegetel and Telcité, the telecommunications arm of the RATP (IAURIF, 1998a, p.57). This lack of involvement from DGT / France Télécom fits with the implicit objective of the Téléport to create alternative networks (and much less expensive communications costs) to serve and link the five regional sites, something which the national operator evidently did not approve of (Daniel Thépin, IAURIF, interview with author, September 2000). This tense link between France Télécom and the Téléport development is something we will return to in chapter 7 on the regional strategy of the incumbent operator. However, the question at the time of the development was 'has the spectre of a Trojan horse teleport of deregulation put off the public operator?' Catherine Viasnoff of the teleport association defended the project: 'We have never evoked the question of preferential tariffs with France Télécom. We position ourselves as a federative or unifying platform which complements the offer of France Télécom. We're not doing business, but driving things forward: in other words, attracting new businesses and preventing others leaving. As planners, we're reasoning for the long term' (quoted in Bessières, 1989, author's translation).

Because of its monopoly, however, infrastructure deployment at first had to involve France Télécom. Trying to shape a telecommunications project in France in the late 1980s and early 1990s in a monopoly context without the full cooperation of the monopoly operator would have been problematic to say the least. This was to prove part of the reason for the eventual downfall of the Téléport.

In any case, linked to the Téléport strategy was the overall strategic concern of the Conseil Régional to improve access for businesses throughout the region, but particularly those in the *grande couronne*, to medium-band and broadband networks. In 1991, therefore, they partly financed the Francilienne des Télécoms initiative in which France Télécom was to accelerate fibre optic deployment throughout the *grande couronne*. At the time, this was a substantial investment for the Conseil Régional (120 million francs) in an initiative with important territorial cohesion objectives, despite the subsequent downplaying of the resulting network, which we will see in the next chapter. Indeed, for the Conseil Régional and France Télécom, it was 'the first French electronic superhighway', totalling 900 kilometres of cable and 26,000 kilometres of fibre optics, thereby supporting information flows of more than 2 Gbits/s (Butor and Parfait, 1994, p.58).

In addition to the Francilienne, France Télécom had to be involved in providing local networks, *Réseaux Optiques Flexibles*, such as those connecting up the various poles at Marne-la-Vallée, 'given the regulatory situation in force at that point' (IAURIF, 1998a, p.61, author's translation). So, the local Réseaux Optiques Flexibles and the Francilienne des Télécoms regional loop of France Télécom were eventually, by the mid 1990s, juxtaposed in the overall Téléport project by a regional 'electronic highway', the Réseau Électronique Express Régional (R2ER) of Telcité, linking Marne-la-Vallée to La Défense, as well as the networks of Cegetel, SANEF (Société des Autoroutes du Nord Est de la France) and Media Marne (IAURIF, 1998a, p.61). By this time, it was clear that, in trying to reconfigure a restrictive national regulatory context for regional territorial development in the Téléport project, the Conseil Régional and IAURIF had to have recourse to France Télécom and their networks, and that the aim of meeting specific telecommunications needs in specific sectors of the region through promoting alternative, less expensive networks only became possible with the prospect of competitive markets in the mid 1990s. This was the continuing influence of monopoly in the Téléport Paris – Ile-de-France.

Networking the City of London I: configuring a competitive telecommunications market for global finance As we saw in chapter 5, the square mile of the City of London, as both a national and international centre of business and finance, has a relatively long history as a focus for the deployment of communications technology networks:

> Over the last 140 years this small space has been one of the key centres of a globally extensive web of telecommunications; what was once the telegraphic heart of the Empire of British commercial capital is now one of the electronic hearts of an international imperium of commercial capital (Thrift, 1996b, p.1473).

In the 1980s, however, the Square Mile had to face up to a dual competitive dynamic – the threat posed by other financial centres around the world, particularly in Europe, which were suggested to be closing the gap on the City in certain markets, and a shift towards a demonopolised national telecommunications sector, following the US lead. We are obviously most interested here in the latter dynamic and its influence on the territorial deployment of alternative networks, but we cannot ignore the way in which the former is also bound up in the reinforcement of the City as a networked urban space, particularly in the deregulatory response of City authorities, who allowed foreign banks and businesses to compete more in City markets, thereby instigating a 'Big Bang', which just increased the demand for telecommunications.[49] The development of London's telecommunications infrastructure was therefore helped intrinsically by financial market liberalisation in the City:

> For instance, BT invested more than £40 million in a 32km advanced fibre optic network for the City in preparation for the Big Bang in 1987 and continues to invest around £400 million per annum in telecommunications infrastructure in London. This continued dynamism is illustrated by the fact that in the first half of 1998, the City's share of BT's new ISDN line installations in the UK was about 10%, whilst its share of the country's total employment was just over 1% (Local Futures Group, 1999a, p.9).

Indeed, we noted in previous chapters that the large business users of the City were instrumental in influencing the decisions of the UK government to open up the telecommunications market to a duopoly and to privatise British Telecom.[50] These actions were supposed to lead to an increase in the standard of BT's networks and services, which had been subject to criticism in the early 1980s. Certainly the emergence of Mercury as a rival carrier prompted activity on the part of BT, particularly given that the former had decided to specifically target and serve the large telecommunications consumers of the City:

> In the month that Mercury formally applied for a licence (in June 1981) BT announced plans to build a second network in the City of London. Over the next 18 months BT accelerated its plans to offer a nationwide network for its digital leased line services, started to restructure tariffs on analogue leased lines (an area in which Mercury did not compete) and cut tariffs on long distance calls between major cities (Ireland, 1994, pp.37-38).

For the City of London, then, the presence of a second network in the form of Mercury was beneficial in itself, but perhaps even more for its influence on BT strategy: 'Mercury's decision to target the City represented a significant competitive threat to BT – although business users represented only one fifth of

---

[49] As Ireland summarises:

Over 1986-1987 private circuits grew fivefold as trading on the stock exchange floor was replaced by telephone trading, fuelling demand for point to point analogue circuits. The creation of large-scale dealing rooms for foreign exchange and securities trading stimulated demand for PABXs [Private Automated Branch Exchanges] and dealerboards. This was accompanied by a property boom to accommodate both the expansion of existing City firms and the entry of new international banks. Property developments brought additional benefits in communications terms since banks were able to build in flexibility to meet future telecoms requirements. In old buildings the ducting in the risers offers little additional space for telecoms cabling and there is not always room to build the false floors necessary to accommodate the mesh of cables required to service a dealing room. Extra space and the use of new technologies such as structured cabling allow banks to reconfigure buildings and businesses much more quickly and cheaply than would otherwise be the case (Ireland, 1994, p.38).

[50] Although Morgan (1989, p.24) notes that these business users never directly demanded either of these actions.

BT's customers, they accounted for more than 60 per cent of its income' (Ireland, 1994, p.38). BT was forced to upgrade their network serving the City to fibre to compete with Mercury, so the BT City Fibre Network was completed in 1987.

In this way, with the new networks and services of Mercury, and the deliberate response of BT in their restructuring, precursory competition in UK telecommunications brought about by the Thatcher government substantially influenced telecommunications developments in the City of London. This was the beginning of a market-led approach to infrastructure deployment in the City, that would be extended even further after the lifting of the duopoly restrictions, and the move to an open form of competition (allowing the entry of other operators) in the UK telecommunications market in 1991.

With this increasingly competitive context, the potential benefits to the City of London and its vast numbers of bulk telecommunications users were clearly conceivable but not necessarily completely evident. At around this time, the Corporation of London was concerned with a number of issues broadly related to the 'infrastructure' supporting the financial and commercial activities of the City, of which 'the use of technology' was one. They therefore funded the City Research Project to investigate more fully the competitive position of the financial services sector in London (London Business School, 1995), through the publication of a series of papers focusing on these specific issues. Jenny Ireland's report on 'The importance of telecommunications to London as an international financial centre' came out in September 1994 (Ireland, 1994).

The report was keen to point out even basic observations about the role of telecommunications in the City, such as their critical importance to foreign exchange, securities dealing, broking and international banking (Ireland, 1994, p.5): 'Financial firms' use of high bandwidth services and wide area networks is matched only by academia' (Ireland, 1994, pp.9-10). Ireland compares the telecommunications environment of London and the UK to the other major financial centres of France, Germany, Japan, and the USA (although without much specific discussion of individual cities in these countries) in terms of technology, deregulation, tariffs, and network reliability. The key issue highlighted was the fact that, although London had had a competitive advantage over other financial centres in the telecommunications domain, this advantage was starting to erode away as deregulation and competition set in elsewhere.

Nevertheless, Ireland (1994, pp.48-49) noted the entry into the London telecommunications market in the early 1990s of new operators such as COLT, WorldCom and Energis,[51] all of whom were prioritising the building of fibre optic rings in and around the City. In summary, she was able to conclude:

> As a major international financial centre London is an attractive target for telecoms providers. Financial firms can now choose between providers in all three segments – the local loop, long distance and international traffic. Increased competition in the local

---

[51] The telecommunications subsidiary of the National Grid launched in 1993, which therefore uses electricity transmission infrastructure for the deployment of its fibre network.

loop in particular will lower telecoms costs and force BT and Mercury to be more innovative (Ireland, 1994, p.51).

The development of fibre networks by operators seemed to reduce the need for public sector involvement in the telecommunications infrastructure sector: 'Although the development is piecemeal it seems likely that the UK will have a national fibre network in place within the next 5-10 years and it should not be necessary for the government to fund additional network investment' (Ireland, 1994, p.17). Certainly in the City of London, then, the development of telecommunications infrastructures and services in the last twenty years has been left largely to the free market.

It is clear, then, that a competitive telecommunications market supports the City of London as a global financial centre, but, beyond this, it also equally clear that the City is probably *the* key element in the development of a 'knowledge-driven information economy' not just in London, but for the UK as a whole (Local Futures Group, 1999a). The relations between ICTs and territorial development are, then, relevant at and between multiple scalar levels.

The socio-territorial legacy of PTT monopolies: the continuing need for universal service   A grasp of the sheer breadth and diversity of activities and processes bound up in the telecommunications sector today can be gained just by comparing transnational joint ventures and partnerships with the continuing local obligations to provide a universal service to all communities and territories of a nation. The creation of universal service was commonly associated with the era of the PTT monopolies, but has continued into the liberalisation era relatively unchanged, apart from now being primarily the responsibility of an incumbent which must compete in an open market as opposed to its monopoly of the past.

In France, the 'uniformity' of universal service has been tied up with the idea of *service public* and its typically French 'egalitarian' connotations (see Gille, 1995). Universal service is the provision of a fixed-line telephone service available to all, plus a number of other elements such as the compilation of a directory and the provision of public call boxes. France Télécom, being the incumbent public operator, is entrusted with fulfilling these obligations, although the costs are partly paid for through the interconnection charges other operators pay for use of the network of France Télécom.

The scope of universal service in the UK covered similar elements, and the funding of universal service via interconnection charges also took place in the UK up to 1997. At this point, however, Oftel decided that the costs to BT of providing universal service were not sufficiently high to warrant indirect contribution from other operators, and indeed, that BT benefitted in some ways from it. BT must now fund the various elements of universal service itself (Thatcher, 1999, p.210).

In France, recent debate about universal telecommunications service had begun to focus on whether the new information society law would change the notion at all, and particularly whether it would include broadband Internet access within it. However, universal broadband is problematised by high cost, European legislation, and the liberalisation process which requires keeping market entry possible.

The key aspect of national universal service policy in France and the UK is perhaps the way in which it has reinforced the roles of France Télécom and BT, the former public monopoly operators, in ensuring a balanced socio-territorial development of telecommunications provision. This would seem to be increasingly important in a competitive context, even within Paris and London, as alternative operators are focusing their activities more and more on the most profitable users and geographical areas. While these users and areas are to be found more often within Paris and London than in other parts of France and the UK, both cities have also got some of the poorest areas or communities in their respective nations. How the notion of universal service evolves in relation to new telecommunications markets such as broadband and 3G mobile will therefore be absolutely crucial for ensuring that balanced socio-territorial development continues to be promoted.

The Differing Influences of the Shift to Liberalised Markets in France and the Maturity of Liberalisation in the UK

Discussion and analysis of the telecommunications sector in France and the UK in recent years needs to take into account national policy and regulatory contexts which are, in many ways, as divergent as the earlier ones of the previous section. Much has been made of the supposed convergence of this policy and regulation across Europe since 1998, but there remains a key difference in this study insofar as France only opened the main part of their telecommunications market in 1998, while the UK has been experiencing varying forms of competitive markets since the early 1980s. In this way, whilst competitive telecommunications markets in the UK are quite mature, competition in France has only relatively just begun and the many different actors in French telecommunications are still 'finding their feet'. In the previous section, we considered the relations between telecommunications developments in Paris and London and differing monopoly and competition contexts in France and the UK respectively. In this section, we need to explore the relations between telecommunications developments and differing national contexts of new and mature forms of competition. We begin once again with a continuing analysis of the Téléport Paris – Ile-de-France and the City of London telecommunications market.

Téléport Paris – Ile-de-France II: regulatory difficulties and redundancy with the prospect of liberalisation While the UK was ploughing ahead with the gradual liberalisation of its telecommunications markets, the closed regulatory environment in France tended to slow telecommunications developments down. If anything, the relations between the proposed Téléport developments and the French regulatory context became increasingly important through the 1990s. Difficulties intensified, largely because of the restrictions on deployment of local telecommunications infrastructures and services in the context of the protected monopoly of France Télécom. The Téléport Paris – Ile-de-France contained a number of different initiatives based on telecommunications on a number of different sites around the region, 'but in each case, the regulatory situation didn't allow us to put networks in place, at least before 1995...' (Daniel Thépin, IAURIF, interview with author, September 2000).

Most of the proposed developments on the sites around the region never got built, with one or two exceptions, as Daniel Thépin of IAURIF explains:

> On the other hand, one thing which was really done was the antenna site in Marne-la-Vallée, which was located in proximity to the RER. The site actually touched the RER because that allowed the direct connection of the different sites. Therefore, the Téléport *has* created an association, a group, which was bought out afterwards, taken up by other groups, which means that there *is* an antenna site at Lognes in Marne-la-Vallée. There are problems with the site though, because of the regulation and France Télécom... Anyway, there is an antenna site, but... ARTESI[52] still has a very very symbolic percentage of capital in the group, but because it was them which created and launched it, they have inherited that from the Téléport if you like, but it's not at all their activity... There was also the project for the network at La Défense, and the Conseil Régional financed that in partnership with EPAD. The fibre network connects up all the office blocks, almost all the office blocks at La Défense, and is rented to operators who want to serve the site. It's not necessarily a good thing though because suddenly the operators didn't agree with the situation, saying it was too expensive to rent. They'd prefer to deply their own fibre to the office blocks at La Défense (Daniel Thépin, IAURIF, interview with author, September 2000).

Overall, however, when EU regulations dictated the opening of the French telecommunications market to competition by 1998, the Téléport Paris – Ile-de-France had 'lost its *raison d'être*' (Emilio Tempia, DREIF, interview with author, September 2000), and would be overtaken by the logic of the market. As Yannick Landais of the former Téléport association ARTESI stated: 'With competition, there emerged a sufficient number of operators and telecommunications offers in the region. It was time to let the market do its work' (Yannick Landais, ARTESI, interview with author, October 2000). Daniel Thépin of IAURIF summarises the reasons why the strategy as a whole never really got off the ground:

> In spite of all the proposals and activities, the Téléport finally wasn't done – we can say that much. Politically, there wasn't a strong will or desire to do it because it wasn't clear enough how. Plus, the operator was pretty much opposed... Then, in particular, there was a moment where we began to be caught up by the evolution of regulation in France, and it's true that now there's no real reason for the authority to put in place a telecommunications network, at least not aimed at specific sectors, as the operators are there, and it's a deregulated and competitive industry, and situating ourselves in it... private initiatives, directly it's not possible, and indeed in a way it's the opposite. The Téléport was that we needed a certain number of specifics to put in place particular conditions. It's true that operators have located in the most interesting zones of Ile-de-France, and the role of the planner is instead to look at what's happening elsewhere other than these zones which are a little privileged. Then, the Internet massively called

---

[52] The current *Agence Régionale des Technologies de l'Information* in the Ile-de-France region, which was originally the association of the Conseil Régional set up for the development of the Téléport.

into question the structuring of value added services, and now the notion of services aimed at research companies, not that they no longer exist but that they are no longer defined like that... Before, it was all very innovative, and that in itself could justify an action giving a push in the right direction from a planner. Now, it's in the market, and we have to try to ensure that we have compared the development as best as possible, but we're no longer in an anticipation phase, and in addition telecommunications sales have accelerated enormously... (Daniel Thépin, IAURIF, interview with author, September 2000).

Despite its status as a kind of preliminary 'bridge between the global, regional and local levels' (IAURIF, 1998a, p.129, author's translation), the Téléport Paris – Ile-de-France petered out, then, partly because of the problems in both justifying it to a suspicious France Télécom, who withdrew their support, and, perhaps relatedly, obtaining political support at the regional level. Here, an understanding of the importance of a regional telecommunications project was not really forthcoming, especially for a project which might have been viewed as needlessly standing on the toes of the national operator. This can be linked back to the French political culture and, in particular, its close intergovernmental relations and the importance of the 'cumulation of mandates' of politicians, which means that local and regional developments can influence their possible national mandates. The lack of institutional and political will was followed by the evolving regulatory situation. Nevertheless, for ARTESI: 'The Téléport opened the doors for other developments in the telecommunications domain in the region. It very much showed the way. Perhaps particularly in leading to the offers of other competitive operators' (Yannick Landais, ARTESI, interview with author, October 2000).[53] Indeed, while very few of the specific material elements proposed in the Téléport actually saw the light of day, the five regional sites which became its focus all continued to develop infrastructures and services as a shift to liberalised markets became evident, and the founding principles served to highlight the importance of partnerships and a strategic regional approach to the telecommunications domain. The shift towards liberalised markets in France thus changed the nature of telecommunications developments in Paris, and re-problematised the role of territorial authorities in this domain.

Networking the City of London II: the overwhelming dominance of the market In the 1980s, with just the two network carriers, BT and Mercury, telecommunications developments in the City of London had already been

---

[53] ARTESI has steadily broadened its horizons to dealing with the development of all new technologies in the region, with particular regard to users. Thus, ARTESI had its roots as the *Association Téléport Paris – Ile-de-France*, before it became the *Agence Régionale pour le développement du Téléport et de la Société de l'Information en Ile-de-France*, and now it is just the *Agence Régionale des Technologies de l'Information*. Consequently, with a change of president around the beginning of 1999, ARTESI developed a reconfigured focus around ICT uses: 'we abandoned the axis revolving around supporting businesses in ICT development. With businesses, the machine is well on the road' (Yannick Landais, ARTESI, interview with author, October 2000).

A Tale of Two Global Cities

dominated by the market, albeit as a result of the national policy which created the competitive environment in the first place. When a further round of liberalisation was instigated in 1991 with the abolition of this duopoly (apart from in international services) and a shift to an open form of competition, the emergence of other carriers increased this dominance. We can question, though, whether the increasing maturity of liberalised telecommunications markets in the City has been bound up with an even less necessary territorial regulatory role for the public sector, notably the Corporation of London, the local authority of the Square Mile. The City Research Project of the mid 1990s, including the Ireland report on telecommunications, seems to symbolise a continuing concern of the Corporation with reaffirming the competitiveness of the City in the face of growing pressure from elsewhere, particularly continental Europe. This competitiveness objective, thus, defines a broad territorial regulatory role for the Corporation, cross-cutting the whole of its services and functions. The territorial implications of telecommunications developments in the City appeared limited because the free market was seen to be working in this domain, as a result of liberalisation. The affirmation of this in the Ireland report of 1994, and its conclusion that other European cities would take several years to reach the quality and quantity levels of the City in telecommunications provision, negated any need for the Corporation to intervene in this domain, at least for a few years, and confirmed the hypothesis that the increasing maturity of liberalised markets greatly influences territorial provision of telecommunications infrastructures and services.

Since the City Research Project, the Corporation of London has largely continued its *laissez-faire* stance to the telecommunications market. As the Corporation of London's Economic Development Unit made clear:

> We don't deal with telecommunications very much... We rather take it as given that if a company in the City had a problem with the telecoms offer or could not get what they wanted, they would tell us, and this doesn't seem to happen... (Pete Large and Pauline Irwin, Corporation of London Economic Development Unit, interview with author, August 2000).

Through annual surveys such as those of Healey & Baker, it has had confirmation of the leading position London holds in Europe in the perception of the quality of its telecommunications infrastructure. Part of this lack of material action from public sector actors in the telecommunications domain in London can be related to their lack of basic information about the activities of operators. The Corporation of London has little idea about the actual quality or extent of the infrastructure serving the Square Mile:

> None of these companies will tell us very much because it's so commercially sensitive to give that information. What we'd really like to know is how many fibre optic cables and rings there are beneath us, and it's nearly impossible to find that out... We just don't know. They just won't say... I think if you wanted to dig a cable, you have to get permission off a huge number of different people who have the right. I think there's a list of about 14 people you have to get in contact with. You have to get permission to dig the hole from the Corporation's Technical Services Department, and then you're

required to get in contact, I think, with about 14 different people – gas people, telecoms people, about 6 different telecoms companies, to make sure that when you dig your hole, you're not going to cut into any of their cables. But the Corporation doesn't even have a confidential map or model of what's underground... Which does seem a bit mad. I mean we couldn't believe it when we found that out... It's a bit of a sort of a Holy Grail, this idea of getting a map of what's underneath the City. Everyone you speak to knows someone who's got one, but when you speak to that person they say 'I haven't got one myself, but I can get one, I think, from the bloke down the corridor' (Pete Large and Pauline Irwin, Corporation of London Economic Development Unit, interview with author, August 2000).

This means that the local authority for the Square Mile is rather left *presuming* that there are no problems with the quality of telecommunications networks in the City:

We don't really know about the quality of the telecommunications infrastructure. We wouldn't have any hesitation in saying 'yes, it is', but... as we said before, if a company came in, between all the different agencies we'd make sure they get exactly what they need. And in that respect, it's a world class city with world class telecommunications infrastructure. But we can't tell you how many fibre optic cables there are underneath the ground. Just have to take our word for it... We know it's true, but we've just got no evidence, and that's fine... And, you know, by speaking with the people that are here we know that that's right. That that's true. That it is world class... That's the sort of thing if it was a problem would crop up – exactly right... And nobody's left because they couldn't get the telecoms or IT locations that they needed. As far as we know... (Pete Large and Pauline Irwin, Corporation of London Economic Development Unit, interview with author, August 2000).

This all illustrates perfectly the absolute dominance of market operators in the telecommunications domain in the City. The local authority does not appear to have the right to know who has what kind of infrastructure and where, within, or rather under, its own territorial jurisdiction. As the Kerbes report verifies: 'it is difficult to be sure who offers telecoms services to whom in the City at any point in time... There is also only limited information on where ducting has been laid in the City...' (Kerbes, 2001, pp.12-13).

Nevertheless, whilst the Corporation may not have been involved in direct material telecommunications market intervention in the City, we can suggest that the City Research Report in the mid 1990s, and two key subsequent pieces of research on IT and telecommunications (Local Futures Group, 1999a; Kerbes, 2001) are examples of an explicit policy of discursively constructing and packaging telecommunications developments in the City of London as part of their overall competitiveness and attractiveness agenda. This is perhaps an overlooked type of territorial regulation to which we will return in the next chapter, fitting in as it does with the second strand of our research. We can make the point here, however, that discursively promoting territorially fixed, market-led telecommunications developments in this way (while simultaneously not having detailed knowledge of them) demonstrates, or is even the end product of, to a certain extent, the profound influence of a national telecommunications policy oriented towards competition and liberalisation over the last twenty years or so, and the increasing maturity of this competition.

The territoriality of cross-national convergence in telecommunications regulation: the unbundling of the local loop and the deployment of DSL   Across much of Europe, the majority of telecommunications markets were opened up to competition by 1998. The continuing exception has been in access to the local part of telecommunications networks, between exchanges and the customer, variously called the 'last mile', the 'last kilometre', or the 'copper pair', which remained in the control of incumbent operators, meaning that their competitors have always had to interconnect their own networks to those of the incumbent (and thereby pay the determined interconnection charge to the incumbent) in order to reach their clients. In the last couple of years, however, a Europe-wide process of 'unbundling' the local loop has been initiated, albeit extremely slowly, with the objective of opening up the final monopolistic segment of the telecommunications market, and therefore, in particular, promote the development of widespread and competitive broadband services through varying types of Digital Subscriber Lines (DSL).

With its focus on the 'last mile' of networks, creating potentially fully contiguous relations between all operators and their clients, and the negotiations over juxtaposed or 'co-located' operator equipment and infrastructure in the local exchanges of incumbents, the unbundling of the local loop is perhaps the clearest example of the inherently territorial interactions and *enjeux* between telecommunications developments and national telecommunications policy and regulation. Given the intensity of telecommunications networks and services already present in the key sectors of global cities, it is not surprising that unbundling and the subsequent deployment of DSL have been most heavily pushed in France and the UK in relation to Paris and London, as competitive operators have shown most interest in offering their own services via the 'last mile' of networks in and around the City of London, the Docklands, central Paris and La Défense. The process has not run smoothly, however, in both countries, with this potential direct territorial access to users being constructed through a series of negotiations between the incumbent, the regulator, the state and competitive operators.

The French regulator, the ART, has stated the key *enjeu* of the unbundling process: 'Access to the copper pair is… considered to be the only way to ensure that the new entrant has complete freedom to define services and form a relationship with the customer' (Jean-Michel Hubert, ART, quoted in Taaffe, 1999, author's translation). Alternative operators want direct access to their clients, and this equates to the territorial contiguity or tangency of their own networks with the terminal (architecture) of their clients. This is the equivalent in the telecommunications industry of face-to-face communication in global financial centres, and for the benefit of both parties is seen as just as important. It is, then, all about the opening up of telecommunications developments to total territorial competition, and subsequently, possible increased territorial cohesion in terms of deployment of infrastructures and services. However, it has already become apparent that this latter point might be difficult to achieve, given the lack of interest from competing operators in the UK and France in the peripherally located, and often dispersed, exchanges that BT and France Télécom opened up first. In the UK though, Oftel allegedly offered some of the more popular exchanges (Rob Tapping, BT, interview with author, January 2001). Nevertheless, it is primarily access to the key market spaces that other operators would like to see (the City of London, the Docklands, central Paris, La Défense), which is perhaps not surprising given the focus of many of these operators solely on business clients. In sum, it is clear that there are difficult problems to be resolved within this territorially-based process:

> The process of unbundling has been painfully protracted. David Edmonds, the Oftel director-general, has described dealing with BT as 'trench warfare'. Decidedly low-tech arguments about how the limited space in exchanges is allocated have rumbled on. Oftel has been accused of going soft on BT and BT has been accused of dragging its feet. Among the issues still to be resolved is BT's insistence that other companies' equipment must be housed in separate rooms in its exchanges (Teather, 2001).[54]

Given the length and uncertainty of these negotiations and procedures, it is unsurprising that many companies have recently withdrawn their interest in co-location in these exchanges, and therefore in accessing the 'last mile' of the network at all. One plus point from this has been noted by Teather:

> Ironically, the number of companies pulling out of the process has actually made local loop unbundling more attractive for the remaining few. Less competition for space means it will be easier to build a regional or national network of DSL lines (Teather, 2001).

Nevertheless, the fact remains that unbundling was supposed to increase territorial competition in the telecommunications sector, rather than reduce or refocus it.

While both France and the UK have endured delays to the unbundling of their local loops, the national regulatory authorities took initially different courses of action in relation to the ADSL services of the respective incumbent operators. The ART in France controversially allowed France Télécom to launch ADSL services in parts of Paris before alternative operators were given a chance to develop wholesale offers. The ADSL service of France Télécom, Netissimo, was therefore available in central Paris from mid 2000, and its deployment was due to cover the departments of Hauts de Seine, Seine Saint Denis and Val de Marne by the end of 2000. In the UK, however, Oftel prevented BT from deploying their ADSL services before alternative operators were able to rent ADSL lines (from March 2000). In both cases though, slow unbundling has generally been viewed as granting the incumbents a head start on their competitors (Emmanuel Tricaud,

---

[54] According to the e-business advisor of the GLA, 'at BT's current rate of DSL deployment, it would take over 900 years to reach every home in the UK' (Jenkins, 2001a).

Colt, interview with author, November 2000).[55] The incumbents inevitably see it differently. As the external relations director of France Télécom explained:

> What is certain is that the more we localise the telecommunications market, the more we create the risk of a differentiation, even an inegality, with the development of new technologies... When we reduce massively our long distance tariffs, we are participating in *aménagement du territoire*. Meanwhile the strategy of our competitors consists of concentrating on the densest zones (Gérard Moine, quoted in La Gazette des Communes, 2000, p.10, author's translation).

The French Government was, initially, fairly indifferent to the prospect of the unbundling of the local loop, in contrast to the position of the regulator, the ART. This illustrates both how 'you must really distinguish the regulatory body from the decision maker in France' (Pierre Blanc, consultant, in Local Loop Report, 1999), and the continuing concern of the state to protect France Télécom (in which it still holds 51% of the capital). For Daniel Thépin of IAURIF, the prospect of open competition through the unbundling process signifies how rapidly the telecommunications sector has evolved in France: 'it's an element that we would never have imagined possible even just a year ago [in 1999]' (Daniel Thépin, IAURIF, interview with author, September 2000). This contrasts with the situation in the UK, where the regulator is part of the government (albeit acting as a relatively independent agency), and there is very limited concern with protecting a fully privatised BT. This is a key difference, then, in terms of the territorial implications of national telecommunications policy, and demonstrates a cautious French approach to newly competitive markets, compared with the maturity of competition in the UK and the experience of BT in shaping its strategy within this.[56]

---

[55] In fact, many of the recent events involving DSL development and France Télécom have brought widespread criticism. The French operator decided to charge competitors a high price for its DSL network planning information and maps, which was seen as 'a classic piece of incumbent obstructionism' by one consultant (quoted in Taaffe, 2000). According to other operators, they were also trying to overcharge on unbundled line rental tariffs, which remained high even after the ART reduced them a little. The feeling was that the slowness and obstruction tactics employed by France Télécom on local loop unbundling – for example, by June 2001, France Télécom had received four official orders from the ART to speed up various aspects of the unbundling process (e-territoires, 2001b) – were an attempt at guaranteeing the incumbent an ADSL monopoly, as at that time other operators could still only offer their customers France Télécom's wholesale ADSL service (Taaffe, 2001).

[56] Unbundling in France was eventually set to happen from 1 January 2001, but France Télécom was seeking something in return for giving up its fight to keep its monopoly on the local loop, notably a much reduced control of its tariffs by the government and the ART (Les Echos, 2000b). The weakening of the French government's reservations about unbundling (being the major shareholder in France Télécom) seemed to be strongly linked to a realisation of its economic and political *enjeu* or importance. In other words, that the broadband Internet connections that would result were absolutely essential for building an 'information society' in France (Le Figaro, 2000).

Unbundling also illustrates a typically paradoxical situation in keeping with the transnational nature of the telecommunications sector. The restraining measures of France Télécom with regard to unbundling the local loop in France contrast with the way in which the same operator is leading the fight for unbundling in other European countries. For example, France Télécom has a 25% stake in NTL, which provides an alternative Internet access offer to that of BT in the UK (AFOPT, 2000, p.34). The operator has also been involved in the MetroHoldings joint venture with Deutsche Telekom and Energis, which has constructed three Synchronous Digital Hierarchy (SDH) broadband fibre optic rings in London, at Broadgate, Aldgate and Canary Wharf. In addition, it has been noted how this international strategy of France Télécom is paradoxically paid for from its own profits on the local loop monopoly in France (AFOPT, 2000, p.34). This is certainly an extreme example of the differing territorial strategies of a single operator following the liberalisation measures of the European telecommunications sector as a whole.

The widespread, competitive deployment of ADSL relies on a completion of this local loop unbundling process, but this, and the deployment of wireless local loop networks (which has made further progress in France than the UK) are the two main new types of infrastructure, via which both governments hope socio-territorial access to broadband can be expanded, for example to specific local zones in the Paris and London regions. In spite of France Télécom's obstinate stance on unbundling, the process may, for example, 'lead eventually to medium-band connections being available in, say, 5 million buildings rather than the current position of broadband being available in only a few thousand buildings' (Emmanuel Tricaud, Colt, interview with author, November 2000).

We have seen here, then, that this dual process of unbundling the local loop in France and the UK and deploying DSL services is a crucial element within the whole landscape of telecommunications developments in Paris and London. Whilst its current early stages prevent much exploration of its territorial interactions and implications in specific areas or developments in either city, we should be aware of its place within territorial strategies of operators and its possible importance for the future shaping of the telecommunications developments we are looking at in this study in both cities. For example, it is already a major component of national and regional strategies for increasing the deployment of, and access to, broadband infrastructures. Nevertheless, the e-business advisor to the GLA struck a cautionary note with regard to DSL development in London in relation to the broader strategies of regional and central government:

All BT local London Exchanges are DSL enabled, but at £40 per month (residential) and more than £100 per month (business or multi-user), it is not affordable! DSL over copper is only an interim solution. If we are to reach, say, 10 Mbits per household by 2010, then we have to get fibre into the local loop. Can BT afford to invest in this alone? (Colin Jenkins, GLA, personal communication, June 2001).

In this section, however, unbundling and DSL deployment have already illustrated, on the one hand, a certain shift towards a convergence between French and UK telecommunications policy (for example, the parallel process in both countries of unbundling itself), yet on the other hand, continuing differences or divergence in these national policies due to differing levels of maturity of competition in national telecommunications markets (such as the level of leeway attributed to each incumbent in both the speed of opening up local exchanges, and in permission to deploy ADSL services before their competitors).

Harmonising technological and regulatory developments: the example of the wireless local loop in France   The recent emergence of wireless local loop networks is viewed by regulatory authorities, operators and local authorities as a technological solution which could increase local market competition and territorial broadband access. This debate, and development of the actual infrastructure required, has though been more prevalent in France than in the UK.[57]

The attribution of operator licences for wireless local loop activity was carried out by the ART using specific criteria. Beyond the usual technical and financial capabilities of the operators, the ART was also interested in the following (OTV, 2001c, author's translation):

- The capacity for increasing competition in the local loop.
- The scale and speed of deployment.
- The coherence of the project.
- The optimisation of use of the spectrum.
- The contribution to employment in France and in Europe.
- The contribution to environmental protection.

The externalities of the wireless networks were therefore as important in the decision of the regulatory authority as the characteristics of the operators and the networks themselves. Two operators were chosen for national licences on the frequence bands 26 Ghz and 3.5 Ghz (FirstMark and Fortel), and two others for each of the 22 regions of France on 3.5 Ghz only (La Tribune, 2000b). In Ile-de-France, this was Broadnet and Landtel.[58] The promotion of competition is the main objective: 'With the wireless local loop, alternative operators have their first real chance to compete with France Télécom on the local market for telecommunications and fast Internet access' (La Tribune, 2000b, author's translation). The main advantages of the wireless local loop are that it is simpler than fibre optics and more supple than ADSL, as well as being cheaper than either

---

[57] Although a spectrum auction did take place in the UK in November 2000, after which six operators were given licences 'to provide Broadband Fixed Wireless Access at 28 GHz'. This infrastructure deployment will allegedly cover 57% of the UK population (Office of the e-Envoy, 2001, p.27).

[58] The operators given a licence in France for deployment of wireless local loop networks are all new entrants in the French telecommunications market. In addition, the four operators mentioned here with licences (national or regional) concerning Ile-de-France are all North American, and are constructing their territorial strategies on an international basis (OTV, 2001c).

to put in place (Jean Nunez, Cegetel, interview with author, September 2000; La Tribune, 2000b). Connection between each user and the base station of the operator takes place via a small antenna. Engineering costs (which represent a very substantial part of the cost of deploying fixed telecommunications infrastructures) are therefore much lower.

The wireless local loop however serves a strictly limited geographical area. The antenna of the client must be in the line of sight of the transmitting antenna which is in turn linked to the backbone network. Given this territorial restriction on deployment and access, the French regulatory authority, the ART, made the point of still choosing administrative regions as the means of delimiting the boundaries of operator licences. *Aménagement du territoire* remains an important consideration, especially as the operators view the deployment of this technology in cities of less than 30,000 inhabitants as unprofitable. Whilst more than 60% of the population of Ile-de-France is likely to have access to the wireless local loop, in more rural provincial regions of France the figure will be less than 10% (OTV, 2001c). The main clientèle is suggested as being small and medium enterprises (SMEs) in the main metropoles.

For alternative operators the main advantage of the wireless local loop is its complete independence from the network of the incumbent, unlike the fixed local loop. One of the principal problems though is asserting the competitiveness of this technology in urban markets such as Paris which are already 'saturated' by operators offering fibre optic, cable and DSL networks and services. Thus, Emmanuel Tricaud of Colt in Paris argued the wireless local loop to be 'at the limit of being a gadget' (Emmanuel Tricaud, Colt, interview with author, November 2000), although Valerie Aillaud of the Chambre de Commerce et d'Industrie de Paris (CCIP – Paris chamber of commerce and industry), in contrast, suggested that big things were expected of the wireless local loop, in terms of creating more choice in local markets where France Télécom still holds a monopoly (Valerie Aillaud, CCIP, interview with author, March 2001). Equally, Jean-Philippe Walryck of the Association Française des Opérateurs Privés en Télécommunications (AFOPT – French association of private telecommunications operators) suggests that the development of the wireless local loop together with ADSL deployment 'is starting at last to open up the competition game' (Jean-Philippe Walryck, AFOPT, interview with author, November 2000). We can perhaps see here, then, an apparent reason for the greater development of wireless networks in France than in the UK. It appears as though they are being viewed as an additional form of technological infrastructure which can be used to increase the level of territorial competition in response to relatively recent market liberalisation, where the monopoly of the incumbent is proving difficult to bypass. Wireless is perhaps not receiving so much attention in the UK because of the perceived maturity of territorially competitive telecommunications markets.

There are a number of other elements in the landscape of telecommunications developments in Paris and London, which also relate to the theme of this section, but because they resonate even more with themes discussed in the rest of this chapter, they are explored in these later sections. This includes the particular difficulties which emerged in Paris, and in France in general, in the shift from monopoly to competition, such as the ambiguity of notions of publicity procedure, infrastructure deficiency and the length of time for recouping investments in relation to the interventions of territorial authorities in the telecommunications domain (see chapter 7). There is also a key example of the difficulties involved in moving from a monopoly to a liberalised and competitive national telecommunications market – the conflict relating to the local loop serving the La Défense business sector – which will be investigated further in chapter 8.

Changing Policy and Regulatory Influences on Operator Activities

There are inevitable difficulties in moving from monopoly to competition in national telecommunications markets. These include the changing nature, and territorial basis, of operator strategy and activities. From having just one operator in the market to influence the development of telecommunications infrastructures and services, there becomes many, which means that telecommunications developments are now shaped increasingly by a form of 'network politics', in which different operators are constantly negotiating, shaping, constructing and packaging their territorial strategies and activities in relation to those of other operators, as well as in relation to a host of other contexts and specificities. This is especially the case in France where, as we have seen, competition has emerged generally much faster and more intensely since 1998 than in the UK which has steadily opened up its markets over nearly 20 years. Scores of licensed operators in both countries are rapidly bringing down tariffs in local, national and international voice markets. This would appear to create difficulties for incumbents in particular, as they must adapt their previously 'standardised' strategies to maximise their own competitiveness in territorial markets. On the other hand, however, the Yankee Group Europe suggests that 'although incumbents like Deutsche Telekom, France Télécom and BT certainly feel the effects of competition and falling prices, the impact has probably been even greater on the competitive carriers themselves.' Table 6.1 offers a brief comparison of the differing market and regulatory influences on the two types of operator in the telecommunications sector today. Incumbents have, after all, still managed to retain a general market hegemony – 96% in the local market, 88% in long distance, 81% in the international market (Yankee Group Europe, 2001, pp.8-9).

Telecommunications developments in Paris and London have, then, been influenced by the changing roles and actions of the historical operators in the two countries. We will explore more closely the nature and extent of their specific territorial strategies for the Paris and London regions in the next chapter, but here, it is interesting to compare more general changes in the actions of both BT and France Télécom between the periods of monopoly and of competition in Britain and France, and then to compare cross-national similarities and differences in the two periods.

The implications of 3G, or third generation mobile network roll-out, for example, are common to both BT and France Télécom, as they have spent a lot of money on developing a dominant presence in the European mobile market. In both cases, this appears to be related to their incumbent status, which is perhaps felt to dictate that they must be present in all the major telecommunications markets. Thus, BT acquired a third generation licence at great cost in the UK government's trumped-up 'auction', while France Télécom bought up Orange. Rolling out the new mobile networks will also be a massive financial burden. Both incumbents have subsequent debt problems which are substantially influencing overall corporate strategy.

Table 6.1 Comparing incumbents and competitive operators

Incumbent	Alternative
Slow decision making	Generally more rapid decision making
Inability to divest consumer business easily	Flexibility to reshape business according to market requirements
Heavy debt load increased by 3G license costs	Generally speaking, no 3G commitments; debt load generated by network build-out
Anticipated new revenue streams from new mobile services	May use MVNO to gain access to 3G revenues
Can often offer a more complete, less expensive service bundle than competitors	Bundling for consumers is limited; for businesses, many lack mobile
Stretching the law with subsidies and delays	Frustration over incumbent delays and special deals

Source: The Yankee Group Europe, 2001, p. 13.

The current reorganisation of BT, brought about by changing market and regulatory influences, has a definite influence on the broad landscape of telecommunications in London, mainly because of the budgetary restrictions it implies. The different parts of BT's operations have been split into separate sections (mobile, Internet etc). Nevertheless, despite the increasingly competitive telecommunications market, particularly in London, BT is still undoubtedly the main player. However, the reorganisation of the former monopoly is leading to capital spending restrictions, which is a relatively major change in policy from the past, when large investments in infrastructures and services would barely be thought twice about: 'The planning programme was more reflective before, and there was less care taken with spending... it's more reactive now, we have a leaner organisation, which is very rigorous in what it does... it's a change from 'mañana to now'' (Rob Tapping, BT, interview with author, January 2001). In other words, before, BT would carefully consider its possible actions for a while, in a way that has become nearly impossible in the heated, fast-changing competitive environment of today. These restrictions are bringing about a focus on 'value for money', as a kind of justification for investments. This is why BT is placing important emphasis on its fault volume reduction programme. Nevertheless, the need to take a look at commercial needs is also stressed, which means 'trying to maximise where demand is highest' (Rob Tapping, BT, interview with author, January 2001).

There is a key division between reactive and pro-active types of work for the main operators (Rob Tapping, BT, interview with author, January 2001; David Ellis, BT, interview with author, January 2001). With the increasing speed of change in the telecommunications industry, much of their work is becoming more reactive: 'making quick movements to respond to technical and market developments' (David Ellis, BT, interview with author, January 2001). To be pro-active now, it is clearly necessary to be incredibly quick off the mark in identifying new areas or sectors of demand.

Market developments in one sector of the telecommunications industry, such as mobile, are also likely to influence developments in other sectors, primarily through the strategies of incumbents. Broadband network deployment in the UK is seen to be suffering in particular in this regard. As the e-business advisor to the GLA has argued:

> The development of innovative broadband services requires that broadband access is available to a significant number of potential users; the roll out of broadband access requires that broadband services are available for adoption by potential users. The industry is unlikely to invest in broadband infrastructure given the current market conditions which have been brought about, in part, by the need to raise £22 billion to purchase third generation mobile licences. This level of debt funding has had major implications to industry players beyond the successful bidding companies. Clearly, if the broadband access / services dilemma is to be overcome, *some form of intervention is required and this requires further consideration at national level* (Jenkins, 2001b, original emphasis).

Here then, Colin Jenkins is talking about 'current market conditions' at the national level in the UK. The central government policy of auctioning off 3G licences to the highest bidders has therefore had an impact on operator strategies in the UK and the national telecommunications market as a whole, which is now influencing the development of territorial broadband access. Broadband looks like being one element of the telecommunications sector which might not be able to be left to the market, even in London, as we shall see.

Nevertheless, the Yankee Group Europe see many advantages for BT and France Télécom: 'Europe's incumbents are helped by the fact that they own the local loop, and are better placed than carriers like AT&T and WorldCom to provide bundled services to consumers and corporates' (Yankee Group Europe, 2001, p.9). Certainly, operators in the two countries follow closely the market positions of, and imposed regulations on, each other, which illustrates that they see the importance of taking into account, at least to a certain extent, Europe-wide regulatory influences on national regulatory practices. For example, when France Télécom presented their wholesale or interconnection tariffs for 2001, they argued that these should take into account the tariffs in other European countries:

> We shouldn't start from the idea that these tariffs should be reduced in France quicker than elsewhere. Admittedly, BT proposes tariffs which are apparently inferior by 30 to

50%, but, as the ART recalled, the UK opened its market more than ten years ago. Whatsmore, because of UK geography and demography, BT benefits from economies of scale, size and density. If we were to extrapolate the tariffs of France Télécom on to the UK, they would be comparable to those of BT (Gérard Moine, France Télécom, quoted in Les Echos, 2000b, author's translation).

The cross-national influence on France Télécom strategy extends also to its *collectivités locales* directorate, as Irène Le Roch illustrates:

There's a common scope now in European telecommunications. We have contacts through the Telecities network, we have contacts through our subsidiaries located in European countries, which give us reports on the state of play there, and to whom I ask questions about how authorities in Sweden, Germany, Denmark or Italy can function with or deal with certain problems. As France Télécom's operations abroad are not too concerned with *collectivités locales* aspects, but concentrate more on the business aspect, at the moment we haven't really taken into account this aspect in the strategy of the company. On the other hand, for me, this serves to explain internally to France Télécom how in other countries, *collectivités locales* work on these problems and how eventually French *collectivités locales* could very well be shown how to take up the same prerogatives as Germany or Sweden for example, because those are the countries where *collectivités locales* are extremely active in the telecoms area... As you've been to the OTV [Observatoire des Télécommunications dans la Ville – the observatory of telecommunications in cities], you'll have seen all the publications they do for the *collectivités*, whereas a few years ago, they weren't at all interested in these aspects. Now it's become unavoidable. It's so unavoidable that in the next municipal elections, which will be held in March 2001, it's unthinkable that *collectivités locales* won't include at least a little project focused around the information society (Irène Le Roch, France Télécom, interview with author, September 2000).

In France, the continuing influence and 'protection' of France Télécom remains great, with it still being majority owned by the state to the tune of 51%. Le Monde recently recalled how any possible privatisation of France Télécom in the future would require a change in the current law, which prevents the state from surrendering overall control (Rocco, 2001). This protection needs to be placed within a wider national context of overall state protection in France for many key public services or utilities (in contrast to the British government's policy in the 1980s of selling them off[59]). We can also suggest a slightly paradoxical situation here, in which France Télécom seems to maintain a closer relationship to the French state than the regulatory body, the ART, does. This contrasts sharply with Britain where Oftel is a non-ministerial government body (but which retains a certain independence from the government) and BT now a fully privatised operator (although many staff at Oftel are former employees of BT).[60] Still, the continuing close relationship between the French government and France Télécom has no

---

[59] See Guy, Graham and Marvin, 1997.
[60] A request was made to Oftel for an interview, but 'Oftel does not generally grant interviews to discuss strategies of operators and how our regulation influences them' (Rachel Reeve, Oftel, personal communication).

equivalent in the UK. Even if the government does still possess a few 'golden shares' in Cable & Wireless, and they have a theoretical right to veto elements in the latter's corporate strategy, it is felt that they would never use this veto (Matt Cochrane, Cable & Wireless, interview with author, January 2001).

The closeness of the relationship between the French state and France Télécom has not, however, always stopped tensions between the two. Indeed, the latter has paradoxically considered legal appeals on a couple of occasions against the decrees of its main shareholder – firstly, in 1997 relating to rules on interconnection and its schedule of conditions; secondly, in 2000 relating to local loop unbundling (Les Echos, 2000b).

The involvement of France Télécom in the creation of an advisory agency for territorial authorities in the telecommunications domain constitutes a major difference with BT, largely relating to the formerly contrasting regulatory environments. France Télécom, as a public monopoly, clearly had more of a 'duty' to territorial authorities than BT, which as a private operator, would not have been necessarily willing to invest unprofitably in such an agency.

The Observatoire des Télécommunications dans la Ville (OTV) was set up in 1991 by France Télécom with a broad objective 'to inform local politicians about the role of ICTs in local development' (Marie-Hélène Bonjean, OTV, interview with author, June 2000). It can be seen as part of the *service public* obligation of the operator (Christophe Barthelet, OTV, interview with author, June 2000). Although funded entirely by the monopoly operator, it was not aimed at the commercial side of the telecommunications market at all. This reflects partly how France Télécom has separate sections which deal with businesses and *collectivités locales*, thereby apparently separating the economic and the social, and the private and the public (Irène Le Roch, France Télécom, interview with author, September 2000). A council for the association was set up on which France Télécom had only one vote. Previously, however, it was the chief executive of France Télécom who chose the president of the OTV, but since the prospect of competition in the market, it has been the whole council of the OTV which has decided its president (Marie-Hélène Bonjean, OTV, interview with author, June 2000). Marie-Hélène Bonjean noted three roles for the OTV in their relations with *collectivités locales* (Marie-Hélène Bonjean, OTV, interview with author, June 2000):

- *Collectivités locales* can ask the OTV for advice and information on possible strategies.
- If the OTV sees an interesting development, they can make an invitation to the relevant *collectivité locale* to join their working group.
- The OTV produces an information bulletin, a detailed website, and a number of dossiers and reports on telecommunications developments, which are specifically 'non-technical and simplified' for the benefit of local politicians.

The OTV has had to change its stance somewhat since liberalisation. It has always been funded entirely by France Télécom, but now tries to stay neutral in its role of informing and advising local politicians and *collectivités locales* about developments in telecommunications: 'we're realistic. We try to think of everything, and if that means citing the offers of Cégétel or Bouygues...' (Marie-Hélène Bonjean, OTV, interview with author, June 2000).

In addition to these close relations between some operators and territorial authorities, partnerships between operators, particularly the new entrants, are also increasingly common. In France, the Association Française des Opérateurs Privés en Télécommunications (AFOPT) was set up at the time of the liberalisation of the French telecommunications market to join together for their mutual benefit many of the relatively new operators who wished to compete with France Télécom.[61] AFOPT, however, holds frequent discussions with the historical operator, and they enjoy 'good commercial relations' (Jean-Philippe Walryck, AFOPT, interview with author, November 2000). It is also worth remembering that operators are frequently the biggest customers of each other. For example, Cable & Wireless and BT usually buy voice 'minutes' off each other rather than having the hassle of always deploying big cables everywhere (Matt Cochrane, Cable & Wireless, interview with author, January 2001).

All these general elements and transformations reflect, then, important changes in operator strategies and activities in the light of differing changes in national telecommunications policy and regulation in both France and the UK. These differing 'national influences' on the strategies of France Télécom and BT, in particular, form part of the background to their more territorially focused strategies in regions such as Paris and London. It is these territorial strategies that we will explore further in the next chapter.

Changing Policy and Regulatory Contexts and Reconfigured Roles for the State

The recent broad shifts from monopoly to competition in France, or in the increasing maturity of competition in telecommunications markets in the UK, are bound up with changing roles for the state, and for regional and local authorities in the telecommunications domain. In both France and the UK, recent telecommunications policy has seemingly become part of a more widespread national policy relating to the development of 'the information society' in both countries. Both states have, for example, highlighted how the Internet, the most visible form of ICT infrastructure, can be utilised as a means for social, economic, political and cultural development.

The reconfiguration of the role of the state: the French 'projet de loi sur la société de l'information' and the UK e-envoy's broadband vision The French government has outlined three main objectives or roles for itself in the telecommunications sector (Christian Pierret, secretary of state for industry; quoted in OTV, 2001d, author's translation):

---

[61] In 2001, AFOPT joined with another operator partnership, AOST (Association des Opérateurs de Services de Télécommunications), to form AFORSTélécom (Association Française des Operateurs de Réseaux et Services Télécom), which includes 28 of the major competitors to France Télécom (e-territoires, 2001c).

- Promoting the development of broadband networks to facilitate the development of innovative and attractive services.
- Guaranteeing access for everybody everywhere to these services.
- Adapting the regulatory context in order to ensure its efficiency and modernity.

Indeed, the French Prime Minister Lionel Jospin has himself reaffirmed the crucial role of the state in the telecommunications sector: 'In a wider perspective, the impetus given by the Government to this strategic sector seems to me to exemplify what the role of a modern State can be in a market economy' (Jospin, 1999, author's translation). This role is suggested to be constituted by the following five elements (Jospin, 1999, author's translation):

- A State which looks ahead.
- A State which knows how to give impetus without substituting itself for social actors.
- A State which leads by example.
- A State which assumes its responsibilities in the functioning of a market economy.
- A State which guarantees the principals and keeps the values that society has chosen.

Jospin elaborates on this final element:

> A State which attends to the respect, in digital space, of the essential values of liberty, egality and fraternity. The expansion of information technologies must not cut a 'digital divide'. The Internet must not foster new inequalities in the access to knowledge. It is the role of public service to ensure the balanced development of these technologies on the national territory and equal access for all to the essential content of these networks (Jospin, 1999, author's translation).

Whilst we should be careful in reading too much into such statements, it does illustrate very clearly how the state is still much intent on developing its role with regard to telecommunications developments, thereby refuting arguments which de-emphasise the possibility of state intervention in this sector in favour of more global logics. In addition, it is particularly interesting that, despite the suggested advanced or revolutionary nature of telecommunications technologies, the Prime Minister chooses to recall the traditional French notions of liberty, egality and fraternity to underline the continuing role of the French State in this sector.

As part of all this, the French government has taken a lead on the national development of broadband networks (see Bourdier, 2000). In July 2001, following a first meeting of the Interministerial Committee for Territorial Development (CIADT), it was announced that FF1.5 billion would be made available over five years for ensuring complete territorial broadband access at a reasonable price by 2005, by 'accompanying' the projects of local and regional authorities. This investment was to come from the state investment bank, the Caisse des Dépôts et

Consignations (CDC), which was also to offer its increasing expertise in the ICT domain to assist authorities with the development of their initiatives. In addition, a proposal was put forward to use the widespread state-owned electricity infrastructure as a means to deploy fibre optic networks in underserved zones[62] (Comité Interministériel d'Aménagement et de Développement du Territoire, 2001). Whilst this offer now exists, it has rarely been taken up so far because of the prohibitive rental price for the dark fibre.

This is, then, all part of the reconfigured role of the state in French telecommunications. As Irène Le Roch of France Télécom in Paris summarises:

> The state intervenes, nevertheless, in regulation. It's not the regulatory authority, but even so, it's the state which gives authorisations. It's the final authority, and then the French state remains for now the proprietor of France Télécom, the public operator. We can tell ourselves that in a few years, things are going to diminish, the share of the state has diminished, but for now, it's still considered that a public operator like France Télécom is not a bad thing for the management of the French state. This proves as well that the French model at the level of the public economy is not so bad as that, compared to countries where everything is liberalised... We have never been completely public. For a few years now, we've introduced a lot of private things into the public economy, but it's also all this economic, political and administrative architecture... The state is still a representative, and is supposed to be a mediator and to be the balancing authority. It redistributes the finance after all. It won't disappear (Irène Le Roch, France Télécom, interview with author, September 2000).

The UK government has also simultaneously shaped its own project for national broadband development (Office of the e-Envoy, 2001), but this is unlikely to include public sector-led network deployment, although some financing has been made available. In common with traditional differences in national telecommunications policy between France and the UK, while the former is considering a state-led approach, the UK favours shaping a market-led approach, although the government has realised that promoting broadband network provision in peripheral or rural areas will probably necessitate some form of intervention, or 'pump-priming' of the market, on its part:

> It seems likely that in due course the market should, possibly using a variety of technologies, be in a position to ensure that most people in this country are able to receive higher bandwidth services at a price that is affordable for the majority. The combination of competition and regulation (where necessary) should drive prices downwards, leading to the hoped for virtuous circle of falling prices stimulating increased demand in turn leading to further price reductions and so on. This is what has happened in the mobile market and a similar pattern is likely in the market for broadband services (Office of the e-Envoy, 2001, p.22).

Nevertheless, as part of its leadership role, the government has set up a Broadband Stakeholder Group, bringing together the major players from the public

---

[62] By enveloping the fibre around high tension electrical lines, deployment costs can be two or three times less than for underground cables.

and private sectors. This seems to be a recognition of the increasingly important socio-territorial implications of broadband infrastructures, and constitutes a clear transformation from the more traditional recent UK government approach to telecommunications of promoting market-led developments. As stated above, whilst competition in the broadband market may continue to lead the way, there is now equal emphasis on the probable need for some kind of territorial regulation. The e-business advisor to the GLA has underlined how this will probably be required even within the London region (Colin Jenkins, GLA, personal communication, June 2001; Jenkins, 2001b). Territoriality and broadband telecommunications are, thus, becoming increasingly intricately linked in London and the UK as a whole, which is a significant policy shift that explicitly recognises the differing socio-territorial implications of parallel technological, market and regulatory developments.

The place of regulation within French and UK central government French and UK telecommunications policy also differs slightly in terms of the structure of their regulatory bodies in relation to central government. Being a non-ministerial government department with a staff of civil servants, Oftel in the UK is seemingly closer to the government than the ART in France,[63] which operates officially as a separate independent body, although there are clear links between it and the Ministère de l'Economie, des Finances et de l'Industrie. The ART is subordinate to the Ministère in so far as it is the latter which has the final word on anti-competitive behaviour and key regulatory decisions, albeit following the recommendations of the ART. However, in the same way as Oftel, the ART stresses its role as a national regulator with identical spheres of activity in all French regions (Lorraine Margherita, ART, personal communication, February 2001). Perhaps equally crucial to the economic development possibilities of telecommunications infrastructures and services in Paris and London is the way in which central government organises responsibility among departments for telecommunications / IT – economic development issues. In both France and the UK, this would seem to lie within the remit of several ministries or departments, which is perhaps unsurprising as telecommunications and IT have very broad implications, but the slight difference between the two countries occurs in that the French government has a specific telecommunications department, DiGITIP (Direction Générale de l'Industrie, des Technologies de l'Information et des Postes – the general industry, information technologies and postal service directorate), within the Ministère de l'Economie, des Finances et de l'Industrie, whereas in the UK government, if we ignore the regulatory side which Oftel takes care of, telecommunications would appear to fall primarily under the responsibility of the

---

[63] Hulsink has noted, however, that Oftel was set up in order to 'insulate regulatory activities from short-term political pressures and government interference…[thereby becoming] an independent specialised agency acting relatively independently from parliamentary and ministerial controls' (Hulsink, 1998, p.163). The ART, meanwhile, is obliged to act as a neutral agency, because of the impossibility for the government to be both regulator and majority stakeholder in France Télécom (ART website).

Department of Trade and Industry (DTI), which controls, for example, the issuing of licences. The recent creation of the Office of the e-Envoy, linked to the DTI, suggests a move towards a closer, more detailed structure for the telecommunications domain within central government in the UK, as in France.[64]

The Role of Intergovernmental Systems, and Urban and Economic Development Policies in Telecommunications Developments in Paris and London

In the previous section, we saw how telecommunications developments in global cities such as Paris and London have to be seen within a context of national telecommunications policy and regulatory practices. In this section, we will illustrate how it is equally important to place them within our second set of national policy contexts, of national systems of governance and urban and economic development policies.

The Influence of Contrasting French and UK Systems of Governance

In previous chapters, we have already discussed the ways in which France and the UK have differing systems of governance, varying in political cultures and traditions, and in the emphasis placed on intergovernmental relations. In the last twenty years or so, France has been generally characterised by a certain degree of decentralisation of some powers to regional and local government, but with continuing strong intergovernmental relations and negotiations. As Daniel Thépin of IAURIF summarises in relation to the Ile-de-France region:

> As Ile-de-France is the capital region, the State is very very present. The State is much more present in the *aménagement du territoire* of Ile-de-France than perhaps in other regions. It's not the same kind of thing... In turn, the Ville de Paris has an enormous budget, which is more important than the Conseil Régional. The Conseil Général for the Hauts de Seine also has an enormous budget, which is more important than the Conseil Régional. Therefore, it's like that, there isn't any hierarchy between the different levels of local power. There's no authority... they can work together, but practically, it's not always obvious, so a lot of negotiation has to go on (Daniel Thépin, IAURIF, interview with author, September 2000).

The UK, meanwhile, generally tended to recentralise powers which had been devolved to lower levels of government, and is seen to have a weaker set of intergovernmental relations, with much top-down policy initiation from central government. In recent years, though, these arrangements in UK governance have begun to change a little, with, notably, the re-introduction of a regional level of government. In this section, we need to consider the influence of these differing

---

[64] Here, there is a perhaps not unrelated point relating to the proliferation of documentation about telecommunications and IT policy which is available from both the ART and the French government, whereas in the British case, documentation is harder to get hold of.

national governance systems on telecommunications developments in Paris and London, building on our initial discussions at the end of chapter 5.

National systems of governance and broadband roll-out in Paris and London We suggested in the previous chapter that the decentralisation measures of the French state in the 1980s, which devolved some economic development intervention to regional and local authorities might be seen in the ways in which the Conseil Régional d'Ile-de-France and IAURIF took control of shaping much of the Téléport initiative in the late 1980s and early 1990s. This regional government intervention in the telecommunications domain for territorial development and competitiveness purposes has continued in recent years. Both the Conseil Régional and IAURIF still have departments or officials concerned with IT and telecommunications, which formulate discursive and material strategies for regional government intervention. IAURIF has held two important seminars in the last eighteen months concerned with the regional territorial implications of IT and telecommunications. One of the most important current interventions at the regional level in Ile-de-France, however, is in shaping and funding the roll-out of broadband infrastructure, particularly that for the various research institutes of the region. While this fully concerns central government, especially in relation to increasing the competitiveness of the Paris capital region, we can suggest that the construction of broadband networks for research is being mostly shaped by a negotiation process between the institutes themselves and the Conseil Régional, and is therefore an illustration of the decentralised economic and territorial development powers present in the French system of governance. We will look further at the implications of these broadband research infrastructures in Ile-de-France in the next chapter, when we focus more on the changing territorial regulation of regional government intervention in telecommunications developments in Paris and London.

The roll-out of broadband infrastructure is equally a primary concern in the London region, as we suggested in the previous section. Here, again, we can argue that the construction of a territorial intervention strategy at the regional level fits in with the generally centralised nature of the UK system of governance, in so far as the GLA response to broadband requirements, in the form of the reports of the e-business adviser (Jenkins, 2001a; 2001b), have been strongly shaped and influenced by central government policy on 'Broadband Britain' (see, for example, Office of the e-Envoy, 2001).[65] Central government has the leading strategic role in broadband deployment at the moment, and has highlighted the importance of the London region leading the way or setting the example in its development. It is up to the GLA, in collaboration with other agencies and local authorities, to work

---

[65] The recent changes in the UK governance system have also influenced, to some extent, operator activity. For example, BT has realigned its UK strategy to fit in with new RDA boundaries: 'BT is determined to play its full part in regional development, working in partnership with the RDAs and other regional players, to meet the challenges that the new regional agenda has set for us all' (Sue Davidson, BT, Foreword to Local Futures Group, 1999b).

within this national context in shaping a specific territorial strategy for the London region. Economic governance, in this respect, still appears to be more centralised in the UK than in France.

Intergovernmental relations in the London Connects initiative, the work of France Télécom, and at Issy-les-Moulineaux Many London boroughs started interacting on a small scale in the telecommunications and IT domain a few years ago. A 1991 report for the London Research Centre and BT outlined possible roles for local authorities in the city in the domain of electronic communications: "The local/regional planning tier, by exerting the influence it has yet to learn it holds, should become a key arbiter of the social and economic impact of the information revolution" (Yeomans, 1991). This domain has received renewed impetus recently with the setting up of the London Connects initiative in 2000.

London Connects started out as a one-day conference held in October 2000 which was trying to promote the creation of a firm information society strategy in the capital. There were several broad objectives for the project (London Connects, 2000): 'It aims to ensure that London citizens get the best from the Information Society by:

- raising awareness of the issues
- demonstrating good practice
- demonstrating technologies and applications
- establishing London-wide networks
- starting the process of establishing an IS strategy for London – linking public, private and voluntary sectors
- starting the process of developing first a communication network and gateway for London using the world wide web, and then creating an ongoing London wide network tackling the organisational issues involved in joining up service delivery'.

'Citizens' are defined in the widest possible sense, so that this is not purely a social cohesion *or* an economic development strategy for London, but both in parallel, which is essential given the number of groups involved and their widely varying objectives.

The 'pioneers' of the strategy were three of the most forward-thinking boroughs in London with regard to ICTs, namely Camden, Lewisham and Newham, together with the Society of London IT Managers (SLIM), but a variety of public, private and voluntary sector groups have been encouraged to participate. It was, however, very much in response to a set of particular issues and contexts, some of which were set by central government, as a result of its concern with maximising the social and economic benefits of the 'information age'. The government had published its IT strategy in April 2000, and in particular, a series of targets for the delivery of electronic services by all levels of government. In addition, a number of funding programmes from the government such as Invest to Save and the Single Regeneration Budget were apparently means for ICT proposals to find money: 'These could provide funding for some of the infrastructure needed to deliver electronic services if used in a more coherent way, by being better coordinated at a political and management level' (London Connects, 2000). There were equally a set of wider contextual issues, such as the contradiction that London was seen as the centre of the UK's 'knowledge economy', but has several of the most deprived areas in the country, and the lack of a common technical infrastructure, which made communication and interaction very hard (London Connects, 2000).

From this centralised policy foundation, however, a partnership-based strategy has emerged, which relies on the input of both public and private sectors, and interaction between central, regional and local levels of government (London Connects, 2001). In this latter respect, it is relatively innovative for an IT / telecommunications strategy in London, or indeed in the UK, because, as we have seen, intergovernmental relations in the formulation and implementation of urban and economic development policies in the UK have generally been held to be quite weak in the past, with a top-down approach being usually dominant. Equally, it also has to be seen against the backdrop of the traditionally fragmented political identity of London as a whole, in which metropolitan representation was frequently assured by just the Corporation of London, with no borough involvement whatsoever (Antoinette Moussalli, Lewisham, interview with author, January 2001). Instead, in London Connects, on one level it was the local boroughs who were first to construct the actual strategy foundation. As Janice Morphet of the DETR (Department of the Environment, Transport and the Regions) argued: 'Local authorities in London don't have a long history of working together, so it's a big leap to London Connects with everybody working together, but it's a process of building up' (Janice Morphet, DETR, interview with author, January 2001).

Nevertheless, within this context of the need for all London boroughs to achieve the e-government targets set by central government by 2005, the boroughs of Newham, Camden and Lewisham saw an opportunity for collaboration and inter-borough assistance, as well as the possibility of a more strategic regional framework for developing IT and telecommunications in London: 'although Newham, Lewisham and Camden don't geographically touch, we can still build on the need for links between them. In any case, there is a great need for more of a regional infrastructure for London as a whole' (Alasdair Mangham and Andrew Stephens, Camden, interview with author, January 2001). A broader London-wide focus to e-government was welcomed by Alasdair Mangham and Andrew Stephens:

> This kind of idea makes sense, insofar as it promotes economies of scale, and allows the sharing of ideas and infrastructure. Services can be delivered in a seamless fashion. There's less focus on geographical boundaries, and less concern with own budgets and political structures. The problem comes when boroughs have to give things up to fit in with the wider strategy. Plus the GLA has no budget, except for transport (Alasdair Mangham and Andrew Stephens, Camden, interview with author, January 2001).

Still, for Antoinette Moussalli of Lewisham, 'the GLA enables London to be a joined-up capital' and the London Connects strategy in which it is heavily involved:

> has the possibility of creating thriving communities as well as thriving business, with shared responsibilities. It is very clear where responsibilities lie. The problem will come if power starts being overplayed somewhere, but for now, there seems to be mutual trust between boroughs and authorities (Antoinette Moussalli, Lewisham, interview with author, January 2001).

The formulation of a strategy in London and the UK for the electronic delivery of government services by 2005 can therefore be held up as a good example of the importance of intergovernmental relations in the IT domain for targetted territorial development: 'the questions that need answering for this initiative are being developed by local and central government together... The DETR is undertaking a pro-active policy and regulatory role. We're not trying to assess local authorities on their capacity to deliver IT services' (Janice Morphet, DETR, interview with author, January 2001). Only time will tell whether London Connects represents fully a more intergovernmental and partnership-based approach to social and economic governance in London and the UK as a whole.

France Télécom is a good example of an incumbent operator whose activities have been closely tied in with national governance arrangements. There is a longer history of this in France than for BT in the UK, because France Télécom was, for much longer, the public monopoly, with its expectation that they would be able to advise on the strategies of local and regional authorities in France in the telecommunications domain. At this time, this was almost a kind of central – local intergovernmental relations, given that the monopoly operator (formerly the DGT) was run by the state. This advisory role was partly the reason for the setting up of the OTV agency in 1991. Although the government has since loosened its control by selling some of its shares, France Télécom still has this advisory role, largely through its directorate of public affairs, which, in the new competitive context:

> maintains with *collectivités locales* a certain number of relations, partnerships and contacts, to ensure that France Télécom, at the institutional level, can still be the representative of reference, because we are, after all, the public operator, and we're present everywhere, so with regard to this, we have tasks which go beyond our business activity in the strictest sense... We work both at the national level, and we support the regional directorates of France Télécom who themselves work at the local level, but who don't necessarily have the analytical competencies or the vision of the problems as a whole, so we work with them permanently (Irène Le Roch, France Télécom, interview with author, September 2000).

French intergovernmental relations are important for the ways in which France Télécom can work with *collectivités locales* in the Paris region, and elsewhere in France. The national and local directorates of the operator fulfill differing duties in terms of relations with differing levels of government. As Irène Le Roch outlines:

It's true that we have some privileged relationships. For example, I work with certain regional councils, because they don't want to work with local representatives [of France Télécom]. It's not that they don't want to work with local people. We work with the local people with the regional councils, because the local people are considered to be 'reducers' or 'simplifiers', and it's always more chic for a big politician to have direct contact with the national directorate, so there you go. It's like that, as part of the political culture of this country. Perhaps in federal countries like Germany, it's not as pronounced, but in traditionally centralised countries, where there is this game at the same time of local and national politicians, because what we call a big politician in France is someone with a mandate of mayor, or deputy, or senator, or sometimes minister. Therefore, we understand that there is a political *enjeu* in which we, with our function as the external relations directorate, do this type of work (Irène Le Roch, France Télécom, interview with author, September 2000).

Here, we can see the influence of elements of the national policy context we discussed in chapter 2, particularly the emphasis in France on overlapping levels of government and the cumulation of mandates by politicians, which is a very different policy context than exists in the UK. The possibilities for local and regional government intervention in the telecommunications domain, that we saw above for broadband infrastructures, and which devolve to a certain extent from the decentralisation measures in French governance in the 1980s, have been held to have further territorial implications as well:

It's true that in France with decentralisation, we gave a lot of power to *collectivités locales*, whether they be regions, *départements* or towns. This decentralisation has provoked a little bit of a loss of the sense of national interest, and perhaps in France, the state has abandoned a little its role of giving impetus, and then local politicians have a slight tendency to see problems at their level, and finally not for the whole of the territory (Irène Le Roch, France Télécom, interview with author, September 2000).

Issy-les-Moulineaux, in the south west inner *couronne* of Paris with around 50,000 inhabitants, is certainly a commune in the Paris region, which has made the most of these decentralisation measures in the ICT domain, winning several awards as a result. Building on this increased local power, the traditionally large numbers of ICT companies located there, and the experience and drive of its mayor André Santini, a former national Minister of Communications, Issy has been able to develop an intensive and widespread territorial ICT strategy. This latter factor appears illustrative of the greater power and influence of local French politicians compared to their UK counterparts. M. Santini has indeed become a leading voice in the Global Cities Dialogue group of 44 cities from around the world, which exchange ideas and strategies for combatting the digital divide (Florence Abily, Issy Média, interview with author, July 2000).

M. Santini encouraged a *plan local d'information* to be adopted there as early as 1996, which included the development of the commune website and proposed ways of making local services more integrated and efficient through ICTs (Florence Abily, Issy Média, interview with author, July 2000): 'This global step has allowed us to move forward more rapidly [...] The website, for example, is at once a local information tool, a vector of communication, and, especially, a supply of services for residents and businesses' (Eric Legale, quoted in OTV, 2001d, author's translation).

The emergence of telecommunications and IT strategies at Issy-les-Moulineaux has thus been characterised by a relatively unilateral approach: 'The inertia, or indeed the opposition, shown by the partners of Issy-les-Moulineaux has forced us to act alone and to turn to international partners.' The commune has identified the reluctance of the *préfecture* of the department of Hauts-de-Seine with regard to this domain, which might show that 'There is a lack of political will on the part of the State' (Eric Legale, quoted in OTV, 2001d, author's translation). This approach has not prevented Issy from developing an ICT strategy, which is 'particularly striking because it illustrates well the combination of dimensions and tools which define a veritable ICT policy' (OTV, 2000, author's translation).

Telecommunications infrastructures and services play a major role in helping to attract companies to Issy-les-Moulineaux: 'If we don't offer businesses in the ICT sector the infrastructures which allow them to access networks, notably broadband Internet, quickly and easily, there's not much point because the businesses will leave' (Eric Legale, quoted in IAURIF, 2001b, author's translation). Boucles Locales Entreprises (company local loops) were therefore constructed virtually as soon as liberalisation permitted in 1998. The commune did not want a multitude of engineering projects going on continually, so it decided to mutualise the use of cable 'pathways'. Cegetel was put in charge of the engineering work, but the other operators contributed to the financing of the work. Each operator could then insert their own fibre and remain in charge of their own networks (Le Fil MC des Télécoms, 22 September 1997). France Télécom infrastructure (1700 kilometres of fibre optics at a cost of more than 13 million francs) remains, but has been joined by a host of private operators including Colt, WorldCom, Cégétel, Bouygues and Level 3. Business areas in the commune are therefore virtually completely connected: 'Having known how to follow the latest innovations, Issy-les-Moulineaux has become a real crossroads of competitive networks aimed at companies' (L'Eco d'Issy, Eté 2000, p.11, author's translation). Issy now has:

> the largest possible palette, before the wireless local loop, so that our businesses can benefit from an offer, in terms of tariffs of course, but also in terms of services which are needed for their activities. This is really essential because it is this which will be the attraction of Ile-de-France for many years to come. As we all know, many provincial cities complain about not having telecommunications networks for years (Eric Legale, quoted in IAURIF, 2001b, author's translation).

The commune has two ambitious new planning projects in this domain. Firstly, the increasing numbers of IT firms interested in locating in Issy led to plans for the old station to be transformed into a 'cyber-nursery' with quality IT, telecommunications and Internet infrastructure and facilities. Eight companies will be chosen to move in to the premises and to take advantage of the services offered therein (L'Eco d'Issy, Eté 2000, p.9). Secondly, the 'Digital City of the Third Millennium' project, planned to be completed after 2003, is to be located in the Fort d'Issy, and consists of 12 hectares with 1000 flats equipped with multimedia technology and broadband Internet access. A quarter of the flats are for social welfare, and others for very small businesses. Environmental considerations are to be a priority and vehicle access restricted (L'Eco d'Issy, Eté 2000, p.11). For the commune, 'this group of dwellings prefigures the city of tomorrow as much in terms of technical, technological and environmental innovations as in human terms' (Ville d'Issy-les-Moulineaux document, 1999, author's translation).

We can see here, then, the influence of intergovernmental relations, and national systems of governance as a whole, within particular telecommunications developments and contexts of the Paris and London regions. It can be suggested that these have usually been more visibly important in developments in Paris and France, but that now there is some evidence of how emerging reconfigured governance arrangements in the UK and in London are beginning, in strategies such as regional broadband deployment and London Connects, to shape quite intrinsically telecommunications developments in London.

The Influence of Urban and Economic Development Policies

Within the national systems of governance discussed in the previous section, differing national urban and economic development policies are formulated and implemented, and their influence on telecommunications developments in Paris and London needs to be likewise explored. This is especially important in the light of the fact that telecommunications infrastructures and services are viewed within both theoretical and policy perspectives to be an increasingly essential part of urban planning and economic development practice as a whole.

Telecommunications and planning practice and legislation Whilst national urban policies and planning practices in France and the UK have traditionally neglected telecommunications networks, and viewed infrastructure more frequently as concerning transport or other utilities, there is some evidence that this state of affairs is beginning to change. Recent French planning laws, for example, have integrated telecommunications issues into overall territorial development. As Vinchon suggests:

> Telecommunications now appear in urban planning and territorial development documents (Plan d'Occupation des Sols, Schéma Directeur, Loi d'Orientation sur l'Aménagement et le Développement du Territoire du 4 février 1995...), and have become the object of directly spatial interventions (Vinchon, 1998b, p.6).

Nevertheless, the role of the main French government planning department, the Ministère de l'Équipement, in the telecommunications domain is virtually negligible (Emilio Tempia, DREIF, interview with author, September 2000). Still, these evolving French planning regulations greatly contrast with British planning documents which talk about telecommunications almost solely in terms of the location of masts and the visual impact of dishes.

As a result of this, in London, the extent of the role of telecommunications in the new Spatial Development Strategy (SDS) of the Mayor is far from clear. The report of the e-business advisor to the Mayor and the GLA suggests more consideration of the relationship between telecommunications and planning: 'The ways in which planning agreements / S.106 agreements[66] can facilitate the development of new technology on a consistent basis also needs to be explored' (Jenkins, 2001b, p.11). However, these possible 'ways' are not outlined in the report. The SDS is supposed to include the 'positive step' of 'ensuring that the basic duct, building entry, and building cabling for the e-infrastructure is included in all new major planning applications' (Colin Jenkins, GLA, personal communication, June 2001). Still, following this, we can suggest that there may be possibilities for planning authorities in London to regulate land use in favour of the development of ICTs through the inclusion of certain 'provisions' in these agreements, which might, for example, take the form of ensuring construction or renovation of buildings includes their cabling or wiring to the relevant local broadband networks. This is beginning to take shape already, as the e-business advisor to the GLA also notes:

> A number of proposals in the GLA's consultative document 'Towards the London Plan' are designed to increase the availability of broadband access. There is a requirement for multiple duct nests and building cable entry to be included in all new major developments, for example, and for all new buildings to be e-enabled. This proposal represents best practice in the construction industry for office developments, but is new for residential developments... The approach presents some challenges, particularly in the e-enabling of buildings. In business accomodation this requirement has generally been met by providing an equipment area, cable risers between floors and floor ducting – the exact configuration being designed at occupancy (Jenkins, 2001b, p.11).

Up to now, then, there is clearly more experience of configuring office developments and business areas for telecommunications developments than residential developments and areas. The possible ways of achieving the configuring of the latter are unclear:

> For residential accomodation, a more prescriptive, and therefore less flexible, approach will be necessary. Here, the issue of developing technology becomes more important. The current approach would be to equip every room with small-bore ducting or Cat5 cabling. However, the emergence of power line telephony or domestic wireless LANs

---

[66] S.106 refers to Section 106 of the Town Planning Act 1990, in which 'a section 106 agreement is an agreement between a local planning authority and person interested in land, which restricts or regulates the use of land either permanently or for a specified period. It may contain such incidental and consequential provisions (including provisions of a financial nature) as appear to the local planning authority to be necessary or expedient' (quoted in Jenkins, 2001b, p.19).

(Local Area Networks) may make this obsolete, so further liaison with the construction and ICT industries will be necessary (Jenkins, 2001b, p.12).

Actual local authority intervention in telecommunications in France is a relatively new phenomenon. Certainly the traditional view of engineers and management at the former DGT was that they were the experts in conceiving, planning and deploying networks, and that local authorities were not: 'Even if some historical examples show the positive role of local officials in the development of the telephone, the 'technician's perception' of the network and its administrative organisation was centralist and resolutely 'anti-local'' (Négrier, 1994, p.99). French planning practice remains focused on physical or material developments. As the director of the DATAR agency argues:

> While officials demand a great deal of assistance in terms of roads, university facilities... their demands for new technologies are limited at 1% [of the total investment laid out in the contrats de plan État-région]. France still finds itself in a culture of amenities, of 'concrete'. The state and the territorial authorities are guilty of not having invested more on new technologies during the period 1990-1999 (Jean-Louis Guigou, DATAR; quoted in OTV, 2001d, author's translation).

Despite the presence of consideration for telecommunications in some official French planning documents, these mechanisms also illustrate in some cases the confrontation between new forms of networked urban space and old unadapted planning procedures. For example, the last modification of the Plan d'Occupation des Sols (POS) for Paris was in 1994,[67] and it is not at all adapted to the context or influence of ICTs. One of the best examples of this is that ICT businesses are not allowed to occupy the ground floor of buildings in commercial zones, despite the lack of available office space for these businesses in central Paris as a whole (Françoise Courtois-Martignoni, Mairie de Paris, interview with author, October 2000; Matheron, 2000, p.162; p.181). There is also a kind of confrontation between new forms of networked urban space and old historical urban space. For example, the development of the wireless local loop in the centre of Paris has strong opposition from many groups. In fact, placing antennas on buildings is forbidden in Paris for visual and aesthetic reasons. In spite of the potential advantages of this technology, Paris and its historical buildings must continue to be seen above all as a '*belle ville*' (Françoise Courtois-Martignoni, Mairie de Paris, interview with author, October 2000). Indeed, there is unquestionably a shortage of traditional buildings in Paris adapted to ICT activity (Françoise Courtois-Martignoni, Mairie de Paris, interview with author, October 2000). As Matheron explains:

> In Paris, there are numerous office buildings from the Haussmann era. The architectural characteristics of these buildings lend themselves quite poorly to the requirements of ICT companies. The problem of relatively reduced floor space is added to ceiling heights which are often too small to allow the deployment of IT infrastructure, and

---

[67] The POS is due to be replaced by a broader Plan Local d'Urbanisme (PLU).

sometimes even, insufficient maximum floor loads for heavy IT equipment (Matheron, 2000, p.91, author's translation).

We can also note here the influence of global processes on the relationship between telecommunications development and urban and economic development policy and practice. For example, in addition to the general problems facing the Téléport Paris – Ile-de-France as a whole during the 1990s, the development of one of the specific elements in the project, the proposed Médiacentre to be located in Marne-la-Vallée, was finally prevented by the collapse of international property markets (Daniel Thépin, IAURIF, interview with author, September 2000).

The principal way in which local authorities in France have begun to act in the telecommunications domain has been in the deployment of dark fibre infrastructure, and this is a particular feature of the situation in the Paris region. The Sipperec intercommunal initiative illustrates this quite well. The agency sees their work as explicitly linked to planning issues, for example, as it develops networks using dark fibre, which means, amongst other things, that the fibre only has to be laid once and does not need much maintenance (Gwenaelle Lalaux, Sipperec, interview with author, June 2000). In addition, Sipperec promotes its function as revolving inherently around planning:

> If Sipperec decided not to become a telecommunications operator itself, it was in order to assume more efficiently the role of planner which is its responsibility to the local authorities. And, within this planning mechanism which must include economic development and the reduction of social fragmentation, telecommunications are a major element (Sipperec website, author's translation).

Given this self-ascribed planning role, it is not surprising, then, that Sipperec (together with AVICAM[68] and two territorial authorities) have called on the French government to include telecommunications more in planning policy:

> Dark fibre infrastructures, as a territorially distinguishing element, must be considered as a tool for an urban and territorial policy, of which the modes of implementation are left to the discretion of the levels directly concerned: agglomerations, departments, regions. Urban policy can and must use the means of the 21st century (real time, interactivity, mobility) and not run after the obsolete ones of the 20th century (Sipperec et al, 2000, p.27, author's translation).

Building on our discussions here, we focus on the Sipperec initiative further in the next section, as a perfect illustration of the difficulties which French planning and territorial legislation created with regard to the interventions of authorities in telecommunications developments. Nevertheless, in conclusion to this part, it is necessary to underline how little national planning practice and legislation in the UK appears to be bound up in telecommunications developments, particularly compared to the situation in France. This would seem to be linked to the dominant

---

[68] The Association des Villes pour le Câble et le Multimedia (the association of cities for cable and multimedia).

market-led approach defined for the telecommunications sector by the government, that we looked at earlier, and the subsequent difficulties or inabilities of territorial authorities in London and the UK as a whole to define a role for themselves in this domain. Authorities in London are beginning to develop strategies around telecommunications and IT, but these do not appear to have the firm 'planning' link that strategies in the Paris region such as Sipperec have. In particular, the newly competitive context in France has encouraged local authorities to take up the deployment of dark fibre to boost competition in their local markets. There is not the same regulatory context in the UK, and therefore not the same type of intervention by local authorities.

French urban and administrative policy and territorial authority intervention in telecommunications: an initially restricting mechanism   If the French state or government is restricted now to a largely regulatory role with regard to urban telecommunications strategies (Emilio Tempia, DREIF, interview with author, September 2000), this remains nonetheless a highly important role, with a great influence on the actions of both telecommunications operators and local communes. For the ART, economic and social objectives, including *aménagement du territoire*, take priority over competition in the law on telecommunications regulation (Chinaud, 1999).

Local and regional authority intervention in telecommunications developments in France has nevertheless been severely restrained by national administrative policy and law. The postal and telecommunications sectors code makes it clear that the rules of competition must be abided by, with all operators having rights of passage on the public domain. In addition, a further complication for the development of cable networks in particular is the different regulations regarding networks carrying audiovisual services and networks carrying telecommunications services (Chinaud, 1999). Such complex uncertainties often delay or prevent communes from acting in the domain of telecommunications infrastructures. Nevertheless:

> that broadband is obvious and especially a necessity today seems to be largely admitted by local communes [...] Undoubtedly because they became conscious very early on of the essential role of new technologies in the planning of their territory, and because they also took note of the Internet phenomenon, communes are mobilising themselves (Payen, 2000, pp.10-11, author's translation).

Broadband is not only needed to manage the public services of a municipality, but especially to attract companies to locate there: 'Companies need more and more bandwidth and only locate where this bandwidth is available' (Richard Lalande, président de l'AFOPT, in Payen, 2000, p.10, author's translation).

The most restrictive mechanism in French urban and administrative policy for territorial authority intervention in telecommunications, though, devolved from an apparently paradoxical situation, whereby while local communes were given the right to promote competition in the sector on their territories from the postal and telecommunications sectors code, the code for communes, amongst others, was not so clearcut regarding such action, particularly following the Loi Voynet of June 1999, which attempted to link telecommunications, *aménagement du territoire* and sustainable development. One particular article in the Loi Voynet caused a great deal of discussion and frustration. Article 17 of Law 99-533 of 25 June 1999, regarding guidance for territorial planning and sustainable development, outlines that:

> territorial authorities can, provided the offer of broadband telecommunications services or networks that they ask for is not provided by market actors at an affordable price or does not respond to technical and quality levels that they expect, create infrastructures destined to support telecommunications networks in order to place them at the disposition of operators who request them... The decision to create a telecommunications infrastructure can only take place after the instigation of a publicity procedure permitting the recognition of a deficiency and the evaluation of the needs of operators susceptible to use the proposed infrastructures (quoted in La Gazette des Communes, 2000, p.7, author's translation).[69]

The wording of this important law is far from clear and open to different interpretation, and does not advance very far the cause of the *collectivité locale* in France in the domain of telecommunications. Already, France is one of only four countries in the EU (with Greece, Ireland and Portugal) which does not grant power to *collectivités locales* in the telecommunications domain (see table 6.2). In the UK by contrast, they have the power to deploy dark fibre where they wish and to have their own operator licence, as in the case of Kingston (La Gazette des Communes, 2000, p.8).

Specifically, the laying and running of dark fibre by communes in France is not forbidden. However, renting such infrastructure to an operator is equated with economic and commercial activity, so was only permitted when a deficiency (*'constat de carence'*) in private territorial initiatives had been shown in a 'publicity procedure'. The local authority was also obliged to 'amortise' its infrastructure investments within eight years, which led to an artificial increase in the tariff at which they rented the infrastructure to operators. Their investment was mostly for engineering work for the laying of dark fibre, which would usually take a much longer period to amortise. As one official observed: 'Such a short period for amortisation, for such a heavy investment complicates, or indeed forbids, the realisation of these projects' (Olivier Abuli, Assemblée des communautés de France, quoted in e-territoires, 2000, author's translation). Indeed, in response to the Loi Voynet, regarding the need to demonstrate infrastructure deficiency, the city of Nancy and the Sipperec initiative are two rare examples of French territorial authorities initiating plans for the deployment of dark fibre (Jean-Philippe Walryck, AFOPT, interview with author, November 2000). In this regard, we can

---

[69] The *Loi d'orientation pour l'aménagement et le développement durable du territoire* of 25 June 1999 also outlined the compilation of nine *schémas de services collectifs* which would shape national territorial policy in the future. A *schéma de services collectifs de l'information et de la communication* was to be one of these (see Ministère de l'Aménagement du Territoire et de l'Environnement, 2000).

look further at the Sipperec project which has undertaken to develop a broadband
metropolitan network for the benefit of its member communes in the inner
couronne of Paris.

**Table 6.2   Cross-national differences in the roles permitted for territorial
authorities in telecommunications in the EU**

Country	Rental of dark fibre	Operator licence
Austria	YES	YES
Belgium	YES	YES
Denmark	YES	YES
Finland	YES	YES
France	NO	NO
Germany	YES	YES
Greece	NO	NO
Ireland	NO	NO
Italy	YES	NO
Luxembourg	YES	YES
Netherlands	YES	YES
Portugal	NO	NO
Spain	YES	YES
Sweden	YES	YES
UK	YES	YES
Total (15 countries)	**11 countries**	**10 countries**

Source: Sipperec, 2000, p. 30.

*Sipperec and the Territorial 'Deficiency' of Private Telecommunications
Infrastructures in the Inner Couronne of Paris*

The Syndicat Intercommunal de la Périphérie de Paris pour l'Électricité et les
Réseaux de Communication (Sipperec – the intercommunal syndicate of the Paris
periphery for electricity and communication networks) is an agency in the Ile-de-
France region which has aimed to advise and assist communes on the outskirts of
central Paris to equip themselves with electricity and telecommunications
infrastructure networks. It is itself structured a little like a municipality, with a
president, elected officials or councillors, advisers and technicians. These latter
develop proposals for the activities of the agency which the elected officials accept
or decline (Gwenaelle Lalaux, Sipperec, interview with author, June 2000).
    Although it dates from 1924, it has only been involved in the
telecommunications field since 1997, just before liberalisation. Sipperec tries to
ensure communes take full advantage of the newly competitive market, through
obtaining a more competitive offer for territorial coverage than that of France

Télécom. This was particularly important once it became clear that the principal beneficiaries of the competitive market were the main business sectors in Ile-de-France, and that local competition elsewhere would be limited (Gwenaelle Lalaux, Sipperec, interview with author, June 2000). One objective was to cable the communes which had not been cabled during the duration of the *Plan Câble* in the 1970s-1980s.[70] This fits in with the overall aim of ensuring 'that nobody is forgotten, whether they are inhabitants or small and medium enterprises which might be inclined to relocate to communes which do have the networks they require' (Gwenaelle Lalaux, Sipperec, interview with author, June 2000). This would have both economic and social implications for communes, because they rely on companies for their revenue from the *taxe professionnelle*, which in turn means they can keep taxes for their inhabitants lower. Lower revenue from the *taxe professionnelle* would mean higher taxes for inhabitants. Sipperec is therefore about 'egality, equality and local development, serving a collective, municipal interest. It's not about money' (Gwenaelle Lalaux, Sipperec, interview with author, June 2000).

80 communes in the inner *couronne* of Paris are now part of the initiative, which has revolved around the construction of infrastructure *plaques*. Lyonnaise Communications was chosen in October 1999 to cable 28 communes of the inner suburbs in 2 blocks and in parallel – north (550 kilometres) and south (1100 kilometres). The investment required from the operator was around 1 billion francs. This was the biggest cable network project undertaken in France since the start of the *Plan Câble*. The first part of the work was finished during 2000, but it was expected to take a further year or eighteen months to connect every commune (Sipperec website). The concession concerns analogue and digital television, Internet access, a telephony service, and specific offers for individuals, professionals, schools and town halls. Coverage of the population will be 100%. Sipperec emphasised how the communes and users would benefit from the competition among companies the project generated: 'The importance of the proposed market (28 communes, 1,000,000 inhabitants, 490,000 potential points) and the willingness of towns to work together have been the determining factors in getting particularly competitive offers' (Sipperec website, author's translation).

Besides the cable network, Sipperec has also undertaken to promote the construction of a metropolitan broadband telecommunications network. This is a response to the perceived lack of local competitive networks to those of France Télécom on which the necessary access and services are held to be overly expensive (Gwenaelle Lalaux, Sipperec, interview with author, June 2000). The dark fibre of the Sipperec network will subsequently be made available for rental to all operators under the 'objective, transparent and non-discriminatory conditions' outlined by French law, and should potentially create a high level of competition in the services provided to previously underequipped areas.

---

[70] In general, it is suggested that the recent development of telecommunications infrastructures in France is still strongly linked to the development of cable networks in the 1980s (IAURIF, 1998b), and this might be borne out by the fact that IAURIF published a quite substantial document on regional cable networks only in 1998.

In February 2001, the company LD Câble was chosen to build and run the 250 kilometre network on an 18 year concession, the first of its kind in France. This will require investment of nearly 200 million francs from LD Câble. Each of the 80 communes in three *départements* (91, 92, 93) involved in Sipperec will have an entry point to this network, thus creating a potential broadband market of 3 million inhabitants and 200,000 companies (Sipperec website). There is the eventual aim in a second phase of network deployment to extend the infrastructure to a further 48 communes, thus ensuring full coverage of *départements* 92, 93 and 94 (Sipperec promotional brochure).

The necessary process of a 'publicity procedure' was started in December 1999, with three main objectives: to evaluate with the relevant actors the levels of access to infrastructures in Ile-de-France; to identify further needs in infrastructure; and to gather together the expectations of operators in the domain (Sipperec, 2000, p11). A panel of independent experts produced their findings in April 2000 following consultation with the operators of the Ile-de-France region (Sipperec, 2000). For this study, one of the most important elements to be underlined by this 'publicity procedure' report was the absolutely crucial factor of telecommunications infrastructure access to the shaping of the telecommunications markets of Ile-de-France. We will discuss this in more detail in the next chapter. Here, we can observe that the report detailed its consultations with regional operators as being largely characterised by dissatisfaction with the type and extent of telecommunications infrastructure in Ile-de-France, and with the obdurate strategy of France Télécom in relation to access to its infrastructures and the cost of leased lines. Most of the operators suggested that these factors altered their own strategies for network deployment in the region, and therefore, with regard to the proposed Sipperec network, prevented them from instigating deployment in the inner *couronne* because of the cost and unprofitability of doing so. The infrastructure 'deficiency' in these areas was therefore noted by the report on both counts outlined in the Loi Voynet: that it is not provided by market actors at reasonable price; and that it does not meet expected technical and quality requirements. The 'publicity procedure' validated, then, the plans for the Sipperec network:

> Consequently the Commission recommends that Sipperec should act on the deficiency situation of broadband infrastructures within its territory, which has and will have serious implications in terms of responding to needs, and should take the necessary measures for territorial competitiveness, with the deployment of a dark fibre infrastructure, in order not to compromise the economic, social and cultural future of its member communes (Sipperec, 2000, p.42, author's translation).

This brief look at the Sipperec initiative has demonstrated the influence of national urban and administrative policy on a regional – local strategy. This strategy could not go ahead without the approval of an independent commission agreeing with Sipperec's observations on the lack of competitive infrastructure serving its member communes, and its desire to resolve this situation by deploying its own dark fibre network. The dominant intercommunality of the Sipperec project is therefore resisting the inertia and rigidity of the (continuing) territorial monopoly in telecommunications infrastructures.

This April 2000 report and situation has since altered substantially, as, following pressure from many groups involved in the telecommunications field throughout France including Sipperec itself, the French government agreed to modify the problematic mechanisms within the 1999 Loi Voynet. Given the limitations and difficulties observed by local and regional authorities regarding their restrictive possibilities for intervention in telecommunications developments, the French government proposed a modification to the law in 2001 which aimed to make this intervention more favourable to the authorities. This occurred within the government's own *projet de loi sur la société de l'information*.

The regulation that any proposal for the construction of telecommunications infrastructure in a commune had to be preceded by a publicity procedure demonstrating a deficiency in the existing territorial offer of operators was replaced by one which decreed a simple public consultation. The other main problematic regulation concerning the maximum eight year period for the amortisation of the costs of investment was suppressed. These legislative changes have met with positive responses, but nevertheless, for Jean-Philippe Walryck of AFOPT, 'these issues are not the answer to everything. The changes are not going to automatically solve all the problems for *collectivités* in the domain of telecommunications' (Jean-Philippe Walryck, AFOPT, interview with author, November 2000).

Telecommunications and notions of territoriality We have already discussed the differing views and ideas of territoriality in France and the UK in the previous chapter. As was suggested there, the French notion and practice of *aménagement du territoire* has no clear equivalent in the UK, either as simple notion or as actual practice. It is usually set against 'urban and regional planning', which may partly recognise the multiply scaled connotation of *aménagement du territoire*, but not its important parallel notion of balancing the national space economy.

Daniel Thépin of IAURIF summarises the interaction between telecommunications and *aménagement du territoire*:

> Generally since the start of competition the notion of infrastructure has evolved considerably (in terms of *aménagement du territoire*), as in 'dense sectors' like Ile-de-France there is now the superimposition of operators' distinct networks and the 'convergence' of services supported by technically different networks (eg cable and telecoms). In addition, public power is not supposed to substitute itself for private initiatives (except in the case of 'recognised deficiency' although this is not easy to identify because it is poorly defined in French law) (Daniel Thépin, IAURIF, personal communication, August 2000).

In this way, the relationship between telecommunications developments and *aménagement du territoire* were, as we have seen, strengthened by the Loi Voynet of 1999, which clearly brought an increased spatial equality focus to the former. In an emerging 'digital *aménagement du territoire*' (Bourdier, 2000), there is now recognition of the opportunity that telecommunications infrastructures and services create for territorial development: 'The development of the information society is an unprecedented chance to rethink the planning and modern development of territory' (Veyret, 2000, author's translation). As a document by Sipperec, AVICAM and two territorial authorities makes clear: 'For territorial authorities, *aménagement du territoire* must include from now on the development of so-called 'dark fibre' infrastructures, because telecommunications are now as important a stake as railways were in the nineteenth century and motorways in the 1950s' (Sipperec et al, 2000, p.9, author's translation). Yet, this link is not strictly a recent one, given that DATAR was considering the territorial implications of telecommunications policies as early as the 1970s (Thatcher, 1999, p.125). As Briole et al argue:

> The role of the DATAR in the launching of and the support for regional communication plans, in the 1980s, must be considered as one of the primordial factors in the intellectual genesis of bringing together decentralisation and new communications technologies… The DATAR found in telecommunications, in particular, the material for a possible renewal of its missions and its alliances with territorial authorities (Briole et al, 1993, p.68, author's translation).

In early recent French government thinking about telecommunications developments and territoriality, though, infrastructures were not so much the key, as uses and users, or teleservices. More recently, however, telecommunications are perhaps being expressly seen as a tool for *aménagement du territoire*, for example in the *département* of Seine-Saint-Denis (93), just to the north of Paris, where this is one of their guiding principles within a departmental telecommunications plan presented in November 2000 (OTV, 2001a). The DATAR agency have, though, inevitably been involved in promoting more interaction between telecommunications and *aménagement du territoire* (see, for example, IDATE, 1999). As its director again states:

> The economic, social and cultural importance for all actors (homes, businesses, schools, public services) to be connected to this value chain that constitutes the Internet, shows that the information society has become a territorial stake. From now on, the key factor is no longer the offer of technologies – today this is broad and available – but the rhythm of diffusion of these technologies to territories and their cost. […] From the point of view of *aménagement du territoire*, deregulation does not lead spontaneously to offer the optimal conditions for diffusion. In this context, the role of public powers is indispensable for improving territorial coverage, and in particular for enabling cooperation between all actors (Jean-Louis Guigou, DATAR, in e-territoires, 2001a, author's translation).

On this latter point, the Sipperec initiative can again perhaps be seen as an exemplary development of telecommunications territoriality, with its improvement of infrastructure coverage in the *inner couronne* of Paris, and its intercommunal foundation enabling a substantial number of communes there to benefit from competitive infrastructures and services and no longer rely on the expensive offer of the incumbent.

Once again, though, we must contrast this increasing and evolving link between telecommunications developments and explicit notions of territoriality in France with a lack of anything similar in the UK. While we might be able to suggest an emerging territorial basis to both broadband network deployment in the London region and the London Connects strategy, this has so far been a largely discursive foundation to both developments, and we would have to await more material proceedings before being able to conclude that territorial development, as well as or rather than a market-led approach, has been a major feature of either strategy.

Planning, urban policy, and operator strategies   From a private sector point of view, it is necessary to note the limited importance of planning issues in the strategies of some operators. They allegedly play little or no part in the strategic decisions of MCI WorldCom for infrastructure deployment, for example, as they are not influenced by external factors or issues in their overall European strategy (Marc Bourgeois, MCI WorldCom, interview with author, November 2000). Nevertheless, MCI WorldCom base all their developments around a 'local-global-local' premise (MCI WorldCom website). Meanwhile, for Cable & Wireless, 'planning and urban issues are important, but it's all about profitability'. One important consideration, though, is that local authorities are more likely to allow the digging up of roads for infrastructure provision, if the infrastructure is aimed at large customers who bring value to the area (Matt Cochrane, Cable & Wireless, interview with author, January 2001). Equally, Jean Nunez of Cegetel argued that planning issues only became important 'a posteriori' to network deployment rather than 'a priori' (Jean Nunez, Cegetel, interview with author, September 2000). We can question, though, whether in the light of what we have discussed in this chapter, national planning issues and urban policy are more crucial to the territorial strategies of telecommunications operators than they might think.

Telecommunications developments and Paris and London as capitals and historical centres   On a national level, telecommunications developments in Paris and London would also appear to relate closely to a seemingly paradoxical need for the state to balance the continued growth of these cities as engines of the national economy against the containment of their growth for reasons of national territorial cohesion. We saw in chapter 5 how the role of global cities on a national level as capitals and historical centres remains very important. Equally, but on a smaller scale, the state may see promoting the competitiveness of strategic local spaces within the cities as within the national interest, but again, this should not go against either overall national or regional territorial cohesion.

The development of telecommunications infrastructures and services in Paris and London is seen in some quarters as leading the way for a more widespread national roll-out of the same infrastructures and services. For those with a vested interest in the capital, there is the inevitability that without a high level of infrastructure deployment in its capital, a national level strategy becomes substantially compromised or even impossible:

> By pioneering mass broadband access and usage, by supporting innovative pilot projects and application provision, and by monitoring, evaluating and disseminating the results, London can pave the way for the rapid spread of broadband throughout the UK. Moreover, the government will be hard pressed to achieve its 2005 target in the rest of the UK if broadband is lacking in London (Jenkins, 2001a, p.1).

Market-led telecommunications developments in London, such as infrastructure and service deployment in the City and in the Docklands, dominate where profitable clients are concentrated, and the resulting infrastructure networks are crucial for connecting (strategic parts of) the city outwards to global telecommunications networks. As the UK capital, however, public sector-led telecommunications developments, such as broadband deployment and London Connects, become crucial for connecting the city inwards, ensuring a territorially cohesive level of connection both across the London region and across the UK as a whole, as a result of the example set by the capital. The Local Futures Group report summarises some of these multiple implications of telecommunications developments in London:

> Overall, it is clear that the City is a major driver of telecommunications infrastructure investment and competition and that there are major spin-off benefits to the rest of London and the UK resulting from lower telecommunications cost, greater choice and in central London, higher bandwidth availability and improved service quality. These positive externalities - working through market competition, but not entirely captured in regional and local price differentials - also spread into the residential market, where London's telecommunications are still competitive (Local Futures Group, 1999a, p.12).

In turn, the traditional debate regarding the development of Paris compared to that of French provincial areas (part of the idea of *aménagement du territoire*) has become less important in recent years (Denis Deschamps, CROCIS, interview with author, November 2000; Conseil Régional d'Ile-de-France and IAURIF, 1998; DATAR and Préfecture d'Ile-de-France, 1999). Some people have felt that this debate has overshadowed the more important issue of the relationship between Paris and other French cities and Europe, especially in the light of the logics of globalisation: 'The weakness of France is not 'Parisian overconcentration'; it is on the contrary not having metropolitan alternatives to Paris, in other words, other cities of a really international dimension' (DATAR and Préfecture d'Ile-de-France, 1999, p.13, author's translation). We can certainly see this idea in a development like the Téléport Paris – Ile-de-France, where the main overall objective of the strategy was to increase the economic competitiveness of the Ile-de-France region in a European (and global) context. It had little or nothing to do with the balance of the region compared with other French regions. While most other French cities and regions are predominantly turned towards Paris, Paris is, like London, itself turned towards Europe and beyond.

Concluding Remarks

In this chapter, we have seen how telecommunications developments in Paris and London are still very much bound up in the continuing importance of their respective national policy contexts and cultures. This might seem to be an obvious conclusion to make, but as we already saw in the early chapters of the study, processes of economic 'globalisation' and an increasingly transnational focus to the telecommunications sector have led to suggestions of diminished, or even bypassed, national level control over, and responsibility for, sectors such as telecommunications. The empirical investigations of this chapter imply that we need to play down this globalising 'rhetoric'. In the first place, looking individually at telecommunications developments in Paris and London, and their regulatory – governance contexts, we can see how they are each inherently bound up respectively with French and UK telecommunications policies, broader national governance and economic development systems, and national cultures of territoriality. In the second place, comparing these national influences on telecommunications developments in Paris and London, we must conclude that, for all the apparent similarities and convergences between the French and UK contexts and the subsequent similarities between elements of developments in the two cities, there remain crucially important cross-national differences and divergences, which are reflected in some of the differences between the developments in the two cities. We need to consider all this in more detail here.

Following our discussions of cross-national urban and telecommunications policy contexts in chapter 2, it was hypothesised that the national level might still be relevant to the development of telecommunications infrastructures and services in global cities on two broad counts. Firstly, it is still, for the most part, the level of the formulation of telecommunications policies and regulatory practices. Secondly, there still exist dominant national planning and economic development systems, urban policies, and intergovernmental relations. Whatever the (increasing) influence of subnational and supranational bodies and agendas in these two domains, the majority of decisions, changes and processes still come from the state or central government, which, as we have seen, means that there are significant cross-national differences in the influence of the two domains on telecommunications developments in global cities.

Firstly, telecommunications developments in Paris and London reflect cross-national differences in telecommunications policy, dominated respectively, as we first mentioned in chapter 2, by the more recent 'demonopolisation' and liberalisation policy processes in France, and by the lengthier period of competition and liberalisation in UK policy. The more recent liberalisation of the main voice market in France, for example, has meant that regulatory and telecommunications policy debates and discussions seem to be much more common than in the UK.

Relating developments in Paris and London to our discussions of the differing approaches and trajectories taken by France and the UK to the shift from a monopolistic to a competitive telecommunications environment demonstrates notably the broad way in which, despite these differing approaches, the developments in Paris and London illustrate the continuing influence of national telecommunications policies and regulatory practices. Telecommunications developments in London highlight the role of much earlier market liberalisation in this country and the subsequent competitive telecommunications landscape this helped to induce. In the City of London, this regulatory situation has been the key factor in creating a highly competitive market to support the Square Mile as a global financial centre, while in the Docklands, it also contributed to the broader regeneration project by, amongst other things, helping the chief executive of the LDDC attract and play off against each other the teleports of BT and Mercury. Meanwhile, telecommunications developments in Paris have been shaped instead by the relatively recent moves away from a state-dominated *dirigisme* towards a liberalised market. This is exemplified perfectly by the Téléport Paris – Ile-de-France, which, having been discursively and (partly) materially constructed in a monopolistic environment, lost much of its *raison d'être* when the opening of the French telecommunications market was proposed. Equally, however, the long period of monopoly control has clearly influenced the local authority at La Défense, who have tried to retain a single concessionary operator for the provision of telecommunications infrastructure in the zone, in spite of the coming of a competitive market. This caused great *chagrin* among other operators, as we shall see later. The opening up of the French market has also provided new opportunities for the roll-out of alternative infrastructures at lower costs for particular groups and users, as in the cases of both the broadband research infrastructures, where the research community moved away from their earlier costly France Télécom network to obtaining funding for their own LAN, and the Sipperec project, where a group of communes have come together in a partnership to deploy an alternative lower cost cable infrastructure to that proposed by France Télécom. The commune of Issy-les-Moulineaux has configured the new competitive telecommunications landscape for its own territorial needs by both promoting itself as a location for IT and multimedia companies, and by attracting a diversity of operators to build infrastructures to serve these companies.

The continuing close relationship between the French state and France Télécom, evolving from the latter being the DGT state operator to a partly privatised operator, but with the state as majority shareholder, is one of the primary elements in the French policy context, which is still influencing the landscape of telecommunications developments in Paris. The Téléport Paris – Ile-de-France, for example, suffered from the unwillingness of local, regional and national politicians to give their full support to a project which was effectively aiming to offer an early form of competition for France Télécom in the regional telecommunications market. The regulatory difficulties which prevented the development of its proposed alternative networks were the result of the state's continuing preference for a France Télécom monopoly. Later, the 'obligation' of the French government to open up telecommunications markets under European legislation was followed by initial reluctance to initiate the unbundling of France Télécom's local loop, which has delayed substantially the presence of alternative operators within the 'last kilometre' of copper networks in the Paris region, and their deployment of DSL services, which would provide cheaper access to broadband for the SMEs and residents of Paris than the fibre networks which are restricted to large companies because of prohibitive costs. However, we did see that France Télécom was permitted to roll out its own ADSL services across Paris.

Telecommunications developments in Paris illustrate, then, the continuing dominance and 'protection' of France Télécom, with the common argument of state and incumbent being that the competitive market is not mature enough yet. In contrast, telecommunications developments in London show how BT is facing much more competition and contestation in far maturer markets. The competitiveness of the City of London as a global financial centre was a key factor in the liberalisation of the UK telecommunications sector in the early 1980s. It was felt that creating a competitor for BT and privatising the incumbent would benefit the City in terms of lower tariffs and better quality services. This was particularly the case in the light of the 'Big Bang' deregulation of financial markets. We saw, for example, how the attribution of a licence to Mercury led to the company focusing its territorial strategy on the large business users of the City of London, primarily through the construction of its own fibre network, which led, in turn, to increased BT investment in, and focus on, its own infrastructures serving central London. The abolition of the duopoly in 1991, and the opening of the market for international services later in the decade have both served to vastly increase the level of competition BT faces in the telecommunications markets in London. In addition, in contrast to Paris, BT was prevented from rolling out its own ADSL services before the unbundling process at least got under way. National telecommunications policies and regulatory practices have, then, over the last twenty years, largely contributed to the different situations both in which France Télécom and BT find themselves (apparently more favourable for the former than the latter), and, relatedly, in terms of the overall development of telecommunications infrastructures in Paris and London.

The monopolistic telecommunications environment in Paris clearly obstructed some key telecommunications developments in the region, such as the Téléport and the local loop at La Défense (which we will analyse in chapter 8). At the same time, there was clearly a diminishing focus on developing particular strategies to strengthen the monopoly, as had been the case with the Zones de Télécommunication Avancées (ZTA) in the 1980s.[71] The fact that it was France Télécom which was initially approached and involved in the roll-out of broadband infrastructures for the research communities of the Ile-de-France region in the early 1990s can perhaps be viewed as one of the last 'bastions' of a monopolistic orientation to territorial telecommunications strategy in the region. Subsequently, as we shall see in more detail in the next chapter, France Télécom became increasingly bypassed by the research communities and the Conseil Régional in favour of more competitive alternative offers. Liberalisation though was not all about removing the barriers to widespread infrastructure deployment. Some elements of regional government focus on telecommunications infrastructures in Paris were withdrawn, particularly on the financial side. The relevance of, and

---

[71] See Négrier (1995), for a discussion of the ZTA concept with particular reference to the example of Montpellier.

need for, a public sector-led Téléport was questioned, and the project went into decline and faded from view.

Against this situation in the Paris region, it is surely possible to suggest that the early opening up of markets in the UK goes some way to explaining the seemingly long-held lack of public sector interest in telecommunications infrastructures in London. Liberalisation was meant to privilege the free market and the development of competition, rather than continuing state or public sector regulation of the sector.

Secondly, and in contrast to the first point, a study of telecommunications developments in Paris and London also brings out elements of variation from these broad cross-national policy differences. In other words, we can identify some similarities in national influences on these developments. These similarities are related to the liberalised market environments of both France and the UK, which, since the former undertook to promote competition in 1998, appear, at least on the surface, to have developed some elements of convergence. For example, the French and UK national policies relating to local loop unbundling and broadband deployment, and their territorial tendencies in telecommunications developments in Paris and London, have strong similarities in chronology and nature. Despite initial signs of inertia from the French government in particular, the unbundling process in both Paris and London is now meant to firmly contest the last stranglehold of national incumbent operators on the telecommunications market, and through promoting the deployment of DSL connections, create the potential for more choice in local markets in both cities.

Telecommunications developments in Paris seem to reflect little now of the French policy orientation of the 1980s towards a national industrial strategy based on national champions, which was obviously the opposite to the UK government's neo-liberal stance of opening parts of the market early to promote competition. It is clear that the strategies of Issy-les-Moulineaux, Sipperec, and the Ile-de-France research community would not have been possible in their present form in a monopolistic marketplace, or in the context of French government policy of the 1980s and early 1990s. Within this, it is perhaps surprising that the shadow of the national Minitel system, which was, even a couple of years ago, held to be slowing French take-up of the Internet, has all but disappeared. Fibre networks in La Défense and Issy-les-Moulineaux, broadband infrastructures for research communities, and the more general process of Digital Subscriber Line (DSL) roll-out throughout the region all demonstrate that telecommunications infrastructures and services, including fast access to the Internet, are being constructed and deployed throughout the Paris region with virtually the same level of quality, even if in widely differing ways in response to widely differing contexts and specificities, as throughout the London region, which rapidly embraced the Internet. The 'information society' discussions and policies of the French government have been formulated at roughly the same time as those of the UK government as well.

With regard to some policies and processes, however, it is difficult to judge (yet) whether they reflect or differ from the theoretical assertions mentioned above. The more widespread development of wireless local loop technologies in Paris and

France compared to London and the UK might, on the one hand, suggest a more mature and receptive French telecommunications market, or, on the other hand, illustrate how the overwhelming maturity of the UK telecommunications market as a whole, following a lengthy period of competition, makes it difficult for new technological networks and infrastructures to gain a stranglehold on an already bulging market.

These cross-national policy similarities would seem to suggest a degree of convergence between France and the UK in telecommunications policy. However, given the continuing importance of the cross-national policy differences outlined earlier, added to the overall strength and lasting persistence of national traditions and cultures in this domain (as we saw from the beginning of chapter 5), it would appear to be quite likely that any such convergence is more of a superficial than of a permanent nature. It is inevitable that the different types of telecommunications technologies and networks on the market today are going to be deployed in the principal territorial markets around the world, that are global cities such as Paris and London. It is far less inevitable that French and UK telecommunications policy, which has developed along diverging pathways for a couple of centuries, and adapted in differing ways to external pressures and influences, is converging in a way which would suggest that these individual national traditions have been completely forgotten. The contrast between French and UK cultures of 'technological territoriality' is too deep-set, and is too closely related to broader national political and territorial cultures to disappear quickly.

Thirdly, telecommunications developments in Paris and London reflect cross-national differences in systems of governance, and urban and economic development policy, dominated respectively, as we first mentioned in chapter 2, by political decentralisation, strong intergovernmental relations, and increasing consideration of telecommunications in urban policy in France, and by political centralisation, weaker intergovernmental relations, and limited consideration of telecommunications in urban policy in the UK.

Whilst the French government instigated a policy of decentralisation measures in the early 1980s, in the UK central government was taking back powers from local and regional authorities. The crucial involvement of the Conseil Régional d'Ile-de-France in constructing and packaging the Téléport is perhaps an illustration of the devolvement of powers in economic development activity in France around ten years ago. The more recent strategies seem to have built on this, as not only is the Conseil Régional one of the main actors in the roll-out of broadband infrastructures in the Paris region, but the Sipperec initiative and developments at Issy-les-Moulineaux demonstrate a little of the ways in which local communes are taking responsibility and control for shaping their own telecommunications markets. In contrast, the telecommunications developments in London can be argued to have their roots in central government control or policy. In the cases of the broadband deployment initiative and London Connects, these are regional responses to the central government policy related to UK Online, while the competitive, *laissez-faire* telecommunications markets being shaped in the City and the Docklands can be viewed as the territorial descendants of the traditional and long established stance of government towards market-led territorial development.

These cross-national differences can also be seen in the way in which French telecommunications policy and regulatory processes seem to involve local authorities in one way or another. The problems encountered by local authorities and intercommunal organisations such as Sipperec with the original government legislation concerning their interventions in telecommunications developments (the need to undertake a 'publicity procedure', the need to show a deficiency in the offer from private operators, and the need to amortise their investments in the construction of dark fibre infrastructure in eight years) were diluted because the government eventually came round to an understanding of the problems and came up with alternative proposals, which favour local authority intervention in telecommunications more. This was the result of close relations and negotiations between local and central government. In the UK, by contrast, the way in which the shaping of the telecommunications sector has been politically centralised and, thus, focused on the promotion of mature competition has meant that telecommunications developments are overwhelmingly dominated by market decisions and forces. Local authority intervention in telecommunications infrastructure construction and management has been extremely rare (Kingston-upon-Hull is one example, but even here, it was the national regulatory environment which shaped local strategy).

The greater power of French local authorities compared to UK ones can be illustrated by the way in which André Santini, a former central government minister, was able to develop his ideas on the importance of telecommunications developments in the commune of Issy-les-Moulineaux. Similar examples of local politicians in the London region who have used their positions to develop territorial telecommunications strategies are much harder to find, for the simple reason that British local leaders have traditionally been given far less power and kudos than their French counterparts. In this way, local leadership in Paris and France has certainly been more creative than in London and the UK.

In contrast, telecommunications developments in London have, until recently with the London Connects project, been undoubtedly the domain of central government, and the generally weak relations between central and local government in the UK has meant that the latter has had very little of a role to play in telecommunications developments. The capacity of London 'to respond on a strategic level to economic, social, cultural and technological changes which cross narrowly defined local authority administrative boundaries' (Parkinson, 1998, pp.413-414) was very limited until recently with regard to telecommunications developments.

Fourthly, and in contrast to the third point, a study of telecommunications developments in Paris and London also brings out elements of variation from these broad cross-national policy differences. In other words, we can identify some similarities in national influences on these developments. We observed, for example, how the London Connects strategy is now firmly relying on close relations and partnership between, amongst others, central, regional and local government. This is a type of strategy that might have appeared impossible even ten years ago in the light of government centralisation in the UK and the abolition of regional government, but it now perhaps symbolises the beginnings of a renewed negotiatory and less top-down approach to governance and economic development policy, thus bringing it nearer to the political structure of strategies in Paris and France.

One would also have to argue that the decentralised form of governance associated with the French national system since the reforms of the 1980s do not always show themselves very much in relation to telecommunications developments in Paris. The state here still has a major part to fulfill in defining policy and regulatory mechanisms which are bound up in these developments in the capital. The Sipperec initiative, for example, despite being based around an intercommunal partnership in the inner *couronne* of Paris, could only proceed following the approval of a commission, set up according to the former national regulatory framework of demonstrating territorial infrastructure 'deficiency'.

Equally, on the one hand, while telecommunications developments like the research infrastructures of Ile-de-France are outlined and budgeted in the *contrats de plan* between the state and the Ile-de-France Région, on the other hand, these intergovernmental economic contracts do not seem to detail much intervention or spending in the telecommunications sector.

These cross-national policy similarities would seem to suggest a degree of convergence between France and the UK in governance and economic development systems, and urban policy. However, as with the case of national telecommunications policy discussed above, given the continuing importance of the cross-national policy differences outlined earlier, added to the overall strength and lasting persistence of national traditions and cultures in this domain (as we saw in chapter 5), it would appear to be quite likely that any such convergence is more of a superficial than of a permanent nature. The contrast between French and UK political cultures and traditions of territoriality is too deep-set to disappear quickly.

It is not too difficult to draw some parallels between recent telecommunications strategies and particular elements within the emergence of communications technologies in the nineteenth and early twentieth centuries. The territorial management implications of the development of the visual telegraph in France, which we discussed in the previous chapter, are surely the roots for the strong contemporary links between telecommunications networks and *aménagement du territoire* in France. We saw this in the regional territorial development objectives of the Téléport Paris – Ile-de-France, and also in the way the Sipperec project has embarked on an alternative territorial coverage of the communes of the inner *couronne* of Paris.

We can suggest, following all this, that the focus on the national contexts and specificities of telecommunications developments in Paris and London provides further broad evidence, through a differently scaled focus (urban developments rather than national policy), of Thatcher's conclusions to his study of the roles of national institutions in the shaping of telecommunications policy in France and the UK:

The study provides strong evidence for the importance of national institutions. In the face of powerful, common external pressures for change, Britain and France were able to maintain differing, and indeed diverging, institutional frameworks for more than two decades. Distinct patterns of policy making existed in the two countries in terms of decision-making processes and the nature of policy. These patterns and differences between them can be linked to the institutional contrasts between Britain and France (Thatcher, 1999, p.310).

This set of national policy contexts and specificities has made up the first layer of our exploration of the parallel and incremental multiply scaled territorialities of telecommunications developments in Paris and London. However, as we have seen and underlined throughout the chapter, particularly in the way that elements from the next two chapters have 'infiltrated' our discussions here, these contexts and specificities are not bounded, independent and unilateral influences on these developments, as they cannot and should not be considered independently of the other layers of territorialities that we investigate in the next two chapters. In the following chapter, therefore, we build on the discussions and analysis of this last chapter on the national influences on telecommunications developments in Paris and London, and move the focus inwards to explore the relationships between these telecommunications developments and processes, practices and actions present or constructed at the urban regional level. This focus is suggested to be incrementally made up of the wider 'place' of telecommunications within discursive and material actions on embellishing the competitiveness of Paris and London as global cities; a changing form of territorial regulation of telecommunications on the part of regional authorities, particularly with regard to the intertwined processes of competitiveness and underpinning the context of free market infrastructure provision; the territorial basis of operator strategies on a strategic regional level; and the necessary characteristics of regional public sector actors in their intervention in the telecommunications domain, both individually and within partnerships.

Chapter 7

Telecommunications Infrastructures and Services in the Global City 'Package'

Introduction

> London is a quirky and funny old place for telecoms (Charles Barbor, Cable & Wireless, telephone conversation, December 2000).

In the previous chapter, we saw how telecommunications developments in Paris and London still relate crucially to policy contexts and specificities at the national level. In this chapter, the intention is to move our discussions and analysis in to the level of territorial contexts and specificities at the urban regional level, and explore how telecommunications developments in Paris and London are bound up in practices and dynamics related to urban competitiveness and economic development, and the construction of partnerships between different actors in the global city. The discussions and analysis in this chapter, therefore, broadly fit in with the first part of what has been identified as the 'dual competitiveness challenge' for the London region, in relation to telecommunications and economic development, but which we can extend to the Paris region as well, namely 'to improve the *external* competitiveness in a national, European and even global context of inter-regional competition' (Local Futures Group, 1999b, p.14).

Telecommunications are widely seen by urban and regional bodies and authorities as being of great importance to global cities and their competitiveness (Pete Large and Pauline Irwin, Corporation of London Economic Development Unit, interview with author, August 2000; Jean-Baptiste Hennequin, Conseil Régional d'Ile-de-France, interview with author, October 2000), ranking only behind market access, availability of skills and transport connections in a survey of business location factors (Healey and Baker, 2000). As representatives from France Télécom and the Institut d'Aménagement et d'Urbanisme de la Région Ile-de-France (IAURIF) respectively explain:

> The availability of high-performance telecommunications in Ile-de-France is a strategic factor for the competitiveness of national and international businesses. It contributes to the Region's attraction, which local government authorities, planners and investors have been working to enhance in order to draw foreign companies into Ile-de-France (Fargette, 1994, p.6).

> Just as the TGV trains created a new sense of geography by changing the travel time between cities, so will teleports and new communications technologies revolutionise

inter-regional planning - where proximity and access to services are the key - by promoting instantaneous linkups and new partnerships. To take this new dimension into account and indeed to anticipate related developments, will provide new means to achieve more balanced regional development (Dufay, 1994, p.8).

Indeed, London and Paris rank highly in surveys of the most 'wired' global cities, and are seen as two of the top cities in Europe for the quality of their telecommunications infrastructures and services (Finnie, 1998; Lynch, 2000; Healey and Baker, 2000).[72] Quantitative surveys ranking cities in this way are not scarce, but they obviously fail to give us any detailed qualitative information or analysis about the particular infrastructures and services of individual cities, or how cities compare against each other in this domain, particularly when competitive advantage is supposedly diminishing as major centres are now equipped with similar high quality networks. By developing a qualitatively based comparative case study of the territorialities of telecommunications developments in Paris and London, we are able to explore and compare the deployment of particular infrastructures and services in both cities.

The Role of Telecommunications Developments in the Urban Competitiveness of Paris and London

Telecommunications Developments and Global Cities as Financial Centres

The development of telecommunications infrastructures and services has proliferated in Europe since the widespread liberalisation of EU telecommunications markets. The major urban centres of Europe have inevitably been at the heart of this development. London, Paris, Frankfurt and Amsterdam are the main European financial centres, and it is the financial sector, above all others, which requires the highest quality communications networks, and these cities have a wealth of national and multinational companies willing to pay for them. The increasing numbers of fibre optic networks connecting these four cities have led to a so-called 'Golden Square' of telecommunications infrastructures and services (Financial Times, 2000), as 'seamless 'end to end' performance' requirements in voice and data communications connections become more and more crucial (DTZ, 2000a). These cities and the connections between them became the main action scene for innumerable competing telecommunications operators to such an extent that there was a combined glut of fibre optic capacity, which sent prices tumbling and forced smaller operators out of business, as in the case of Iaxis (Financial Times, 2000), and of players – 'too many companies with too little to distinguish them and their business plans from each other' (The Yankee Group Europe, 2001, p.6). Malecki suggests that many of the new entrants in the fibre backbone market

---

[72] Although in another earlier study (Lecomte and Gollain, 1992), the cost of international telecommunications was noted to be one of the three major weaknesses of the competitiveness of Ile-de-France.

were driven by an overspeculative 'build it and they will come' philosophy (Malecki, 2003).

London is widely recognised as a more important financial centre than Paris, being part of the triumvarate of truly global centres of financial activity along with New York and Tokyo (Sassen, 1991; see also Llewelyn-Davies et al, 1996; Froment and Karlin, 1999). Although the Paris region is classed second in Europe for the quality of its telecommunications infrastructures and services, it remains well behind other cities such as Frankfurt as a financial centre (Healey and Baker, 2000). The Healey & Baker European Cities Monitor survey of 500 European business executives placed London as the top location in Europe for business. It was named best location in six of the eleven criteria, including quality of telecommunications, which in turn was named as the third most important factor for business location behind access to markets and international transport links (Healey and Baker, 2000). This means that telecommunications are obviously of vital importance to major firms, both *per se* and as a location factor. As Ireland described a number of years ago:

> The typical financial institution needs good access to head office; voice and data connections to its corporate units; the ability to centralise data flows in selected sites; and to be able to communicate with its customers and other financial institutions whether by telephone, fax, telex or data network be it public or private. These services need to be delivered within sensible timescales and at an acceptable price. The services need to be reliable and any faults repaired promptly (Ireland, 1994, p.12).

If we add the diversity of new telecommunications infrastructures and services which operators are now offering their business clients, the complexity of communications requirements becomes even more intense. Financial firms in the City use high bandwidth services and wide area networks as large amounts of data are sent between geographically dispersed places. Financial centres require high quality telecommunications links between firms, and between firms and clients (London Business School, 1995). Evidently, the financial sector of London, notably in the City, with its highly demanding requirements for quality communications technologies, has a substantial role to play in the development of telecommunications infrastructures in London as a whole.

The backdrop for the territorial regulation of telecommunications developments: the discursive nature of urban competitiveness   In the telecommunications and IT domain, urban competitiveness is the prime concern for the public sector. One of the expressed purposes of public sector intervention in telecommunications developments in Paris and London is to increase the territorial attractiveness of the city or part of the city as a location for business and investment compared to other competing cities or even parts of the same city. It can also be suggested that this reasoning does not leave the private sector (and especially telecommunications operators) impassive, as they are also likely to benefit greatly from presence in a highly competitive and attractive urban region. There are differing reasons, then, behind urban competitiveness objectives, but this means that, in terms of telecommunications developments, there is a common initial aim of increasing infrastructure and service provision and quality.

The public sector concern for the role of telecommunications and IT in urban competitiveness in Paris and London is illustrated primarily in the discursive and promotional material of city organisations and institutions. In London, the perceived role for telecommunications developments in urban competitiveness is perhaps a response to the way in which London, and the City of London in particular, is becoming increasingly conscious of continental competition (Pete Large and Pauline Irwin, Corporation of London Economic Development Unit, interview with author, August 2000). King argues, for example, that 'in the Europe of the twenty-first century, London will be on the Western periphery' (King, 1993, p.96).

Consequently, over the last ten years or so, many agencies and organisations have been trying to set out how telecommunications and IT can play a part in keeping London ahead of the competition, especially in the light of recent changing governance arrangements. In this way, Haywood suggested that 'As London modernises its administrative system the challenge will be to attract sufficient public / private sector investment to ensure that the infrastructure and services of the city will be able to meet the needs of its citizens in the 21st century' (Haywood, 1998, p.381). The new Mayor of London was compelled to produce both an Economic Development Strategy and a Spatial Development Strategy, which was envisaged to take an integrated approach including all aspects of physical planning and infrastructure development. Before the reintroduction of regional government in London, the prospectus of the London Pride partnership highlighted the need for continuing investment 'to complete a broadband telecommunications network and to build on London's current strengths in this field' (The London Pride Partnership, 1994, p.81). Indeed, the Partnership suggested that IT and telecommunications could be one of the city's lead sectors in the new millennium 'if pro-active strategies are developed to take advantage of the opportunities they present' (The London Pride Partnership, 1994, p.41). For the promotional agency, the London First Centre, 'London has the perfect mix of a highly sophisticated core telecoms provision with numerous niche operators offering customised, competitive and innovative solutions' (London First Centre, 1998). We shall look in more detail at this provision later in the chapter.

Urban competitiveness is therefore clearly of considerable importance to local and regional authorities in London, although the way in which this is framed and expressed does not always suggest a profound level of analysis (and can reveal rather patronising attitudes vis-à-vis other cities):

London is the leading European 'World City'. This means a considerable amount of inward investment, jobs etc for London, South East England, and the UK as a whole. We are out in front and the issue is how to maintain the position. If Newcastle fails, it is very sad for Newcastle and the local area. If London fails it is a disaster for the UK as a whole. Therefore there is a great deal of responsibility in getting it right. The position has been achieved over a number of years (even centuries) and is supported by a number of international hubs. For example, international travel (Heathrow Airport and

Eurostar), financial and business services hubs, and telecommunications and Internet hubs. London has more international internet capacity than any other city; 3 of the 5 largest internet city to city links; and more Internet Data Centre space than any other European city. This is vital to support Business in London (Colin Jenkins, GLA, personal communication, June 2001).

Colin Jenkins is able to identify the main European challengers to London as well:

As competition in telecommunications and internet services intensifies throughout Europe, and the driver of regulation shifts to the EC and UK (recently) slips behind the pace, cities such as Amsterdam, Frankfurt and Paris will emerge as strong challengers to London. 'New' city environments such as Berlin are also poised to make a quantum leap ahead (Jenkins, 2001a).

Subsequently, it is felt that the Mayor's office is competing with the likes of New York, Tokyo, Frankfurt and Paris, much more than Manchester and Birmingham. To this end, they would like to see, for example, back offices located within London, and not Swindon (Alex Bax, GLA, interview with author, February 2001). With these kinds of arguments, it is clear that authorities in London consider the that the UK is not big enough for the capital (cf. Taylor, 1997).

In Paris meanwhile, economic development and the competitiveness of the city are also firmly on the agenda at the current time, following the recent creation of a new regional economic development agency, the ARD (Agence Régionale de Développement). Within its role of promoting Ile-de-France in the international inter-urban competition stakes, the ARD has been allotted four domains of competency, the fourth of which revolves around the promotion of innovation and new technologies, especially for small and medium enterprises and industries (Conseil Régional d'Ile-de-France, 2000b). The creation of this agency aims to provide a more cohesive and targetted approach to developing the competitiveness of the Paris region, which was felt, in recent times, to be overly distributed among different organisations (Jean-Baptiste Hennequin, Conseil Régional d'Ile-de-France, interview with author, October 2000).

ICTs and Business Location

As we have already begun to see, one of the key elements within the relationship between telecommunications developments and global cities as financial and business centres is the influence of IT and telecommunications infrastructure on business location. The relatively high quality offer of Paris and London in this regard is a factor in their attractiveness to large multinational companies. More specifically, however, ICT firms may be looking for other elements as well, as part of their locational decision-making. In this way, a study by the Observatoire du Développement Économique Parisien (ODEP), for example, of ICT companies in the Paris region highlighted three main criteria in their choice of location: the prestige of having an address in Paris; the quality of transport networks; and the technical assets of buildings, which need to be highly functional, and with easy access to telecommunications and IT networks (Courtois-Martignoni, 1999, p.10;

Françoise Courtois-Martignoni, Mairie de Paris, interview with author, October 2000). AFOPT draws the link between these networks and broader 'new economy' benefits:

> Thanks to fibre optic local loops in Paris, as in many French cities, new start-ups are being born every day, and they already have very high numbers of employees [...] The deployment of networks and their constant maintenance lead operators to invest massively each year on a long term basis throughout the country (AFOPT, 2000, p.25, author's translation).

We can briefly consider here a specific example of an ICT firm looking to locate itself in the Ile-de-France region, namely Cisco Systems (IAURIF, 2000b, pp.9-10). An important factor in the locational choice of Cisco in Ile-de-France was where their employees lived. Cisco preferred to alter the working schedules of their employees rather than force them to move:

> We told them: you'll adjust the times you work, it matters little to us where you're based, seeing that you're doing your job and you've got access to the relevant information. We equipped the people who didn't have routers at home, we gave them an ISDN line and they can telework. Many only go to the office from 10 or 11am (they avoid the traffic jams); they all work as efficiently at home and come to the office for meetings and social interaction (Jean-Pascal Goninet, quoted in IAURIF, 2000b, pp.9-10, author's translation).

Finding suitable property space in global cities is a major problem for companies. Françoise Courtois-Martignoni of the Mairie de Paris noted how because Paris had no more office spaces of more than 5000 square metres, companies requiring such spaces were leaving central Paris and looking to set up in the suburbs somewhere (Françoise Courtois-Martignoni, Mairie de Paris, interview with author, October 2000). Cisco, for example, were looking for extra space in the Paris region, particularly in the near suburbs:

> not really Paris because of the difficulties in getting in there when you come from outside... we're still looking, we don't want to go too far and we've got problems finding square metres which are well served, near public transport and telecommunications infrastructures (Jean-Pascal Goninet, quoted in IAURIF, 2000b, p.11, author's translation).

They later found office space on land owned by Renault at Meudon, to the south of central Paris.

Thus, the relationship between ICTs and the availability and quality of office space is extremely important, especially from the point of view of these companies (see ODEP, 2000, pp.6-10; BH2, 2000). Start-up Internet and multimedia companies can clearly only afford office space at the lower end of the market when they are setting up. Despite the currently 'buoyant' nature of the property market in London, with the lowest vacancy rate since 1987 of just 4.4% (Karen Sieracki, BH2, interview with author, October 2000), for ICT companies in the City of

London, property location is more closely characterised by 'rent threshold pressure combined with a diminishing supply' (BH2, 2000, p.17). This has meant that in the City and the Fringe:

> the smaller computer / IT telecoms businesses found it difficult to articulate their business plans. The need for space is not thought out in advance. This was due to the fluidity of the business... This lack of planning combined with higher rates of growth and rental levels can cause space shortages quickly in the Grade C and Grade B stock (BH2, 2000, p.18).

In terms of property, the Docklands area has been viewed as having a significant locational advantage in that 'it is easier to build there, so clients go there because they're sure of finding space etc' (Karen Sieracki, BH2, interview with author, October 2000). In this way, then, business location, and the factors bound up in this, become important elements in the competitiveness of cities such as Paris and London, and these relate to the ICT domain, both as one of the key location factors for companies, and when it is firms in this domain deciding on locations.

The Changing Territorial Regulation of Regional Government Intervention in Telecommunications Developments

As we began to see in the last chapter, the ways in which, and the extent to which, the regional level of government in Paris and London has been able or has needed to intervene in telecommunications developments has changed over the years, according to a number of territorial specificities relating to regulatory and governance contexts, and the steadily modifying nature of the market. In this section, therefore, we explore these changes, and examine, in particular, the deeper and more complex links between specific regional telecommunications strategies and the regional competitiveness contexts we have discussed in this chapter so far.

The Conseil Régional d'Ile-de-France: from the vast and multisite Téléport to targetted research infrastructures There have been at least two different telecommunications strategies in the Paris region, in which the Conseil Régional d'Ile-de-France has been intrinsically involved, at least partly for regional competitiveness reasons, namely the Téléport (which we explored in more detail in relation to national policy contexts in chapter 6) and the development of broadband infrastructure networks for research communities and institutes. The way in which the decline of the former strategy led on a greater importance of the latter can be viewed as illustrating quite well the changing focus of territorial regulation of the Conseil Régional in the telecommunications domain over the last few years. Thus, while the Téléport Paris – Ile-de-France was an early attempt, in a monopolistic market context, at stimulating the role of telecommunications in regional competitiveness, the more recent regulatory intervention of the Région in the telecommunications domain has had to take into account the full liberalisation of the French telecommunications market.

As we saw in the previous chapter, between the late 1980s and the mid 1990s, the clearest and most important strategy in the Paris region of telecommunications infrastructures and services being deployed to add value to Paris as a global city was the proposed multi-site Téléport Paris – Ile-de-France. The proposal for the Téléport Paris – Ile-de-France was bound up with the need of the Ile-de-France region to increase its competitiveness and attractiveness to large companies and external investment sources: 'IAURIF saw a number of European teleport projects... how teleports were becoming ports for information... so it was thought that Ile-de-France should have a teleport which would be at the scale of the importance of the region...' (Daniel Thépin, IAURIF, interview with author, September 2000). The Téléport was therefore planned to be developed on five interconnected sites around the region, each with a different focus or function:

> For the Plateau de Saclay it was a service centre. At La Défense there was to be a teletranslation centre and a videoconferencing centre. At Roissy it was an EDI platform. Marne-la-Vallée had an antenna site and a mediacentre or mediabuilding – we changed the name because 'médiabuilding' was too English, and 'médiacentre' worked in both languages. This development was to attract businesses working in the domain of the image which needed applications, IT platforms, technical platforms of buildings, with networks and synergy... (Daniel Thépin, IAURIF, interview with author, September 2000).

As IAURIF noted at the time: 'This dual approach – both sectoral and geographical – aims to favour the development of key sectors in regional planning, both for major corporations and for small businesses' (Henry and Thépin, 1992, p.78). One of the principal aims of all the Téléport sites was to ensure the creation or maintenance of, and interaction between 'international access', 'regional integration', and 'local identity' (Henry and Thépin, 1994, p.17). The sites were conceived as 'places where regional development plans and local policy converge and are applied' (Henry and Thépin, 1994, p.21). Equally, these sites were supposedly to 'constitute a critical mass for the development of high-tech services which may then be extended to other zones with fewer facilities in the Ile-de-France Region and greater Paris' (Henry and Thépin, 1994, p.17). The difference was alleged to be that, historically, towns used to concentrate wealth for their own purpose, whereas the new towns of the 21st century (such as those in the Téléport) would aim to redistribute this wealth (Ricono, 1994).

As we saw in the previous chapter, however, in the last few years, the Téléport fell into decline as it lost its *raison d'être* of trying to improve and extend regional telecommunications infrastructures with the prospect of liberalisation, and the Conseil Régional has no longer financed regional infrastructures such as the Francilienne des Télécoms (Daniel Thépin, IAURIF, personal communication, August 2000). It is interesting that this particular regional telecommunications infrastructure was seen in the mid 1990s as the major network linking up various sites in the Paris region, and demonstrating the importance given to telecommunications by regional bodies, yet now it is suggested that it has 'never been a specific infrastructure but was a prod to France Télécom for them to speed up the networking of the *grande couronne* (outer suburbs) with fibre optics… It wasn't a highway. It was 2 kilometres of fibre are missing there or there etc. We need to improve the principal commutators…' (Daniel Thépin, IAURIF, personal communication, August 2000; interview with author, September 2000). In sum therefore, while 'teleports were created by the development of competition, and at the same time, they were caught up by the activity of large operators' (Daniel Thépin, IAURIF, interview with author, September 2000), it appears more as though this Téléport was created by the need to improve the regional telecommunications infrastructure, to 'substitute' for the lack of competition to France Télécom, in order to improve regional competitiveness more generally, before it was 'caught up' by the prospect of competitive markets.

Subsequently, then, in the light of the liberalisation of the French telecommunications market from 1998 onwards, the Conseil Régional has had to reconfigure its role in relation to telecommunications developments. The key element now is to promote local and regional competition in markets, which would hopefully improve and extend the quality and quantity of telecommunications infrastructures and services in the Ile-de-France region for the benefit of business location, urban competitiveness and territorial development. In this respect then, the objectives of regional territorial regulation of telecommunications developments have not changed from the objectives the Conseil Régional had for the Téléport, but the means with which to achieve this have. The Téléport can be seen as an attempt to make up for the 'deficiency' of competition in the regional telecommunications offer, and a way of encouraging the monopoly operator, France Télécom, to improve its infrastructure and service offer. The Conseil Régional is now working within a competitive telecommunications environment, and looking for ways to territorially configure this competition to the benefit of the region as a whole, local areas within the region, and particular user groups. As the OTV have summarised:

> Faced with the accelerated public deployment of these technologies and the emergence of more visible inequalities, the region seems determined to define a new policy from autumn 2001. Four axes could be promoted, such as increasing territorial access to broadband, the promotion of content (using broadband; the creation of digital poles), the progressive development of 'blind zones', and finally, access for all to existing networks (OTV, 2001a, author's translation).

Against this broad background, it has focused in particular in the last couple of years on broadband networks for the research communities of the region. This is the major element in what Jean-Baptiste Hennequin calls 'a principal role' for the Conseil Régional in the telecommunications domain, of which 'stimulating competition' is perhaps the key aspect, although as he says:

> I'm not sure that we can display that politically, because politically it's not always well viewed to say 'let's stimulate competition in Ile-de-France', because it's interpreted as an attack on France Télécom, which it isn't, as France Télécom's interest, as they well know, is to delay the process, but the more competition they have, the more the market will become important, so it's in their interest, like it was in the interest of the DGT

(Jean-Baptiste Hennequin, Conseil Régional d'Ile-de-France, interview with author, October 2000).

The co-financing of telecommunications networks by the Conseil Régional for the research institutes of Ile-de-France dates back to the early 1990s, when they were obliged to work in partnership with France Télécom: 'this is a tradition for the Region for more than ten years now' (Jean-Baptiste Hennequin, Conseil Régional d'Ile-de-France, interview with author, October 2000). As Yannick Landais of ARTESI also says: 'The state has always supported the research community. There's important implications there for economic development, and for the image of the Ile-de-France region generally' (Yannick Landais, ARTESI, interview with author, October 2000).

In 1990, the Region supported the development of the *réseau pour la recherche* (RERIF), which was connected to the national research infrastructure, Renater. France Télécom, then obviously the sole operator, was enrolled into the building and running of this Internet Protocol (IP) switching network, which offered capacity of between 64 kbs and 2 Mbs, and connected up the principal research institutes in Ile-de-France.[73] This arrangement lasted eight years, and saw the Conseil Régional invest a total of 60 million francs. Eventually however, technological evolution and the prospect of the liberalisation of the telecommunications market made the research institutes realise that they were paying over the odds for these 'obsolete' broadband infrastructures and services (Jean-Baptiste Hennequin, Conseil Régional d'Ile-de-France, interview with author, October 2000; Hennequin, 2000).

Some of the research institutes recognised that the newly liberalised telecommunications market was an opportunity to get the Région back on board to help finance the broadband infrastructures they urgently required. They decided to construct their own Local Area Network (thereby avoiding dependence on France Télécom) and looked to the Région for financial assistance. Jean-Baptiste Hennequin, *chargé de télécommunications* at the Conseil Régional, continues:

> The research institutes did not have control of their infrastructure... So in order to stop this happening again, they came once more to the Region, but this time not for it to involve France Télécom, but to build their own network, a Local Area Network connected with fibre optic... They said that there were a certain number of sites in Ile-de-France where there is a high concentration of university establishments, so it's pertinent for these closely linked establishments to create a LAN. Except that the fibre is still not there. So when these establishments had the possibility of renting fibre, they did so on long contracts. When there isn't this possibility, there has to be public works which are very expensive because you have to buy the infrastructure and all that. ATM infrastructures are expensive. So, on a regional level, they said to us 'we're going to do these plaques, these broadband optical local loops which will connect up the

---

[73] There are 10 universities in Paris alone and others elsewhere in the region, as well as research centres such as the Institut National de la Santé et de la rEcherche Médicale (INSEM), the Commissariat à l'Énergie Atomique (CEA), and the Institut National de Recherche en Informatique et en Automatique (INRIA).

establishments, and we'd like the Region to pay half the cost of the investment'. We accepted and in March 1999 financed two small optical local loops, one at Evry-Génopôle because there we'd really like to have the equivalent of Oxford – Cambridge – London in biotechnology. This sector has massive needs in terms of data exchange between sites. Hence, the significance of a local loop which allows the investment costs to be shared, and which offers broadband of 150 MB at FF 8 a month or even less. So, a great importance for these networks. It started with the Génopôle, then we continued at Marne-la-Vallée... The two last plaques to see daylight were the Parisian academic network (RAP) in 1999 which concerns more than 100 sites in Paris intra muros, 10 universities and all the establishments of the Ministère de la Recherche. We put FF 15.5 million into this project. The second in May 2000 was the project for the network in the Bièvre valley, because the Bièvre valley is in two départements of Ile-de-France to the south, or south east, Val de Marne and Hauts de Seine. We put FF 20 million into this project, so another important investment (Jean-Baptiste Hennequin, Conseil Régional d'Ile-de-France, interview with author, October 2000).

In 2000-2001, the Conseil Régional was looking at the possibility of constructing an overall regional research network linking over 400 sites, at an anticipated cost of around 300 million francs over three years. It was suggested that the cost and the number of sites might favour a sale by lots, thus allowing the Region 'to exercise a real leverage effect on the competition between telecommunications operators. This could then allow coverage of peripheral areas with competitive infrastructure networks' (Jean-Baptiste Hennequin, Conseil Régional d'Ile-de-France, interview with author, October 2000; Hennequin, 2000). In addition, this would shift financial responsibility off the public sector to the market, as the Conseil Régional has begun to question its need to be as directly involved in the financing and organisation of the deployment of these research infrastructures, which are closed networks and maintained by the research community themselves:

> For me, it's not really the job of researchers to be only looking after the running of active routing infrastructure etc. It would perhaps be more in their interest to appeal more directly to operators... We said to ourselves that it would be more intelligent to make a list of all the research sites in Ile-de-France, to see where there is actual competition between operators in Ile-de-France, and on this basis to do a market auction by 'lots', which would be launched at the regional level and which would allow us to take advantage of the competition between operators, because the more the competition is active, rich and dense, the better it will be for all territories which are connected, and the cost of telecoms will fall. Hence the idea of a call for offers, that I explain in my document, which would be launched at the level of the Conseil Régional (Jean-Baptiste Hennequin, Conseil Régional d'Ile-de-France, interview with author, October 2000).

In this way, then, the territorial regulation approach of the Conseil Régional d'Ile-de-France in the telecommunications domain has quite substantially evolved over the past few years – from supporting a large-scale strategy to develop a more 'competitive' regional network infrastructure to increase regional competitiveness, to a more territorially focused support for the deployment of infrastructures for a specific user group, and now the prospect of a combination of the two, via the large-scale promotion of competition across the region, but divided into particular territorial 'lots' for the same user group, with regional competitiveness as a continuing, but more tacit, backdrop.

The London region: from total laissez-faire to a strategic territorial approach to broadband and the 'information society' In the Paris region, then, the territorial regulation of regional government intervention in telecommunications developments has been shaped and influenced above all by the changing nature of the telecommunications market, and specifically the shift from the relative domination of monopoly to competition. In the London region, the situation is inevitably different given that, on the one hand, competition in telecommunications in the UK has existed for far longer, and on the other hand, between 1986 and 2000, London did not have any form of regional government or strategic authority. We saw in the previous chapter how both these elements of national policy context have been crucially important in helping to shape the landscape of telecommunications developments in London. We can consider these elements further here in comparing the nature of regional territorial regulation of telecommunications developments in London to that discussed above in the Paris region.

Given the lack of strategic regional authority before 2000, the territorial regulation of telecommunications developments in London as a whole sat between central government and certain (partnerships of) local authorities and groups. Perhaps inevitably, in neither case did telecommunications feature high up on their agendas. The former had been responsible for liberalising the telecommunications markets in the 1980s, thereby encouraging competition between operators, and was obviously not going to contradict this neo-liberal agenda by too much intervention. The latter perhaps did not have sufficient power, control, or simple understanding of telecommunications to intervene in and try to shape local markets. Consequently, the market dominated. This also seemed to meet the needs of most groups, particularly business, who benefitted from increasingly competitive tariffs and offers. One exception to completely market-led developments was the construction of a Metropolitan Area Network for higher education establishments across the London region.

The development of telecommunications networks and services in London is undoubtedly still very much driven by the market. The question remains, however, of the extent to which the public sector in general still needs to be involved in infrastructure deployment to ensure that all areas and businesses in London have access to the networks and services they need. In other words, the extent to which there is still a need for the territorial regulation of the London telecommunications market. Colin Jenkins, e-business advisor at the GLA, is somewhat reserved on this point:

> I back away from over regulation. If we cannot get the balance right, I prefer to be arguing for more regulation that being constrained and arguing for less. We are regulated by a European regulator with national implementers. Industry wants a level

playing field across Europe, with no local barriers to entry. The free market has served central London well in the last few years. A highly competitive market has evolved based on high quality, low cost offerings. However, we now face a significant step change to contend with affordable broadband access being the key. Given the current level of funding available and existing market conditions, industry cannot afford to drive this forward on its own. Intervention of some form is inevitable if London / UK is to maintain its competitiveness in world terms, particularly as the DTI has just taken £22 billion out of the UK industry for 3G mobile licences. If there is intervention, then it should be restricted to areas that will stimulate the market and innovation. Additionally, with the strategic planning oversight, regional authorities can make sure that the basic infrastructure (duct nests) is included in every major building project and that buildings are e-enabled by, say, CAT5 cabling (Colin Jenkins, GLA, personal communication, June 2001).

The emergence and development of new technological solutions and possibilities such as broadband, the increasing (realisation of the) importance of maximising the potential benefits of the 'information society', and the creation of a new regional government authority and mayor have together brought the question of the territorial regulation of telecommunications developments in London back to the fore. The development of broadband networks and services in London has become bound up in a national regulatory context following the publication in February 2001 of 'an action plan to facilitate roll-out of higher bandwidth and broadband services' by the e-Minister and the e-Envoy (Office of the e-Envoy, 2001) as a first step towards the formulation of a strategy for Broadband Britain. The aim is for the UK to be the G7 leader in broadband networks by 2005. The move to a broadband economy, and its importance to national competitiveness in the future were the justifications for this action plan. This document followed on from a broader 1998 government white paper on national competitiveness, which aimed to promote a 'knowledge-driven information economy' of which ICT infrastructure was one of the key foundations (Department of Trade and Industry, 1998).

It is held that the telecommunications market is likely to be dominated by Asynchronous Digital Subscriber Lines (ADSL), cable modem and broadband fixed wireless access, and against this background the report suggests that the Government should take a lead in a number of ways:

- Providing leadership.
- Driving forward competition in the supply of broadband infrastructure and services.
- Tackling barriers to the growth of the broadband market.
- Undertaking urgent research on the costs and benefits of pump-priming the market to extend services to rural areas and lower income groups.

With regard to London, broadband development is seen as crucial on two levels. Firstly, for the capital itself, as availability of affordable broadband access to the Internet is argued to be one of the three most significant challenges for London in the 'new economy'.[74] Secondly, for the whole national strategy of the government, particularly given that the UK is shown to be falling behind its western counterparts on this technological development:

> If the government is to achieve its objective for 2005, London must deliver. At the same time, broadband needs to become a mass-market phenomenon – widespread use of broadband in London will create the commercial basis for its deployment throughout the UK (Jenkins, 2001b, p.6).

The suggested governmental actions can, however, perhaps be viewed on the one hand as less critical for London than for other parts of the UK due to the unquestionable intensity of competition in the telecommunications market of the capital. London is, for example, already 100% covered by the rollout of ADSL (Office of the e-Envoy, 2001). However, the report takes into account both economic development and social inclusion aspects, so for the former, the input of the government is allegedly still needed in ensuring that the broadband market in London aids national economic competitiveness as a whole (particularly with the ongoing local loop unbundling process being overseen by Oftel), although yet again, the inevitability of a 'trickle down' effect to help other parts of the country is expressed. For the latter, there is the issue of how to increase broadband availability for citizens, especially given that universal broadband access has been argued to be too costly and too restrictive on operator entry to liberalised markets (Colin Jenkins, GLA, personal communication, June 2001).

As Colin Jenkins suggests, the market has generally served London well in terms of improving the quality and quantity of infrastructures and services on offer, but this is perhaps finally coming to be seen as no longer enough to meet the needs of the whole London region, never mind just the big business customers of the City and the Docklands: 'Whilst central London is well supported for fibre access, and this is likely to be extended to adjacent development areas, the vast majority of London has little availability of broadband access. What is available is expensive' (Jenkins, 2001a). As a result of these gaps and limitations, the e-business advisor to the GLA argues straightforwardly that "all strategies developed by the authority should support the development of 'e' in the capital, and take into consideration the possible impact of any information technology developments" (Jenkins, 2001a, p.1).

There is clearly a refocusing of public sector thinking and policy going on here, as central London and its adjacent areas have long been the 'premium network spaces' of London, with access to higher quality telecommunications infrastructures and services than elsewhere in the capital. Before broadband and the rhetoric of the information society, however, the fact that the vast majority of London had little availability of similar levels of infrastructures and services (indeed, some borough communities had low levels of telephone ownership) was either relatively ignored or simply not a topic meriting much focus. With the dual

---

[74] The other two are the development of ICT skills and of e-government services (Colin Jenkins, GLA, personal communication; Jenkins, 2001a).

goals of promoting e-business and universal broadband access, territorial development now shows some signs of appearing on central and regional government agendas, even if it remains to be seen whether this concern will amount to more than token lip service.

In addition to broadband infrastructure deployment in London, the London Connects initiative represents another major element in the 're-territorialisation' of public sector regulation and intervention in the domain of telecommunications developments in London. For example, among the principles of the strategy outlined in the declaration signed by participant groups (London Connects, 2000) are the beliefs:

- That cities are essential to the process of building a fair Information Society because they are the geographical, political, socio-economic and cultural entities where millions live, work and directly exercise their rights as citizens and consumers.
- That there is the need for conscious steps to improve social inclusion and help avoid the division between information-rich and information-poor citizens and communities.
- That the convergence and gradual globalisation of information society technologies and services need new forms of governance and cooperation.
- That there needs to be cooperation across London and across sectors to build network infrastructure and develop applications required to deliver the promise of electronic government.

Here then, we have an explicit recognition of the multiple territorial processes and implications bound up in an ICT strategy such as London Connects. Within national and global contexts, urban level territorialities are identified in the strategic importance of the city *per se*, the need for reconfigured urban governance, and the advantage of cohesive and coordinated urban infrastructure and services, while more local level territorialities are present in the recognition of the potential social and spatial inequalities or disconnections, often tied in with ICT developments.

The level of public sector intervention in the telecommunications and IT domain in Paris and London depends, first of all, as we have seen, on how this domain is viewed by the city and regional authorities, especially in terms of how it might benefit their economic development objectives. One feature which seems to be common in both cases is the setting up of working groups to discuss this domain. London Connects can be seen as a more widespread and formal type of working group in this regard. For the Mairie de Paris, ICTs are a priority for economic development (Françoise Courtois-Martignoni, Mairie de Paris, interview with author, October 2000), hence the creation of a working group in February 1999 within the Observatoire du Développement Économique Parisien (ODEP), bringing together some of the main public and private sector actors in Paris in this domain. Their position was summed up as follows: 'It matters that Paris offers privileged access to telecommunications infrastructures at the best cost, facilitates the interconnection of different networks, and promotes information and training in order to allow and encourage innovation and especially the location of innovative companies' (Courtois-Martignoni, 1999, p.4, author's translation). The second point here on the interconnection of networks was backed up by an IAURIF report on regional cable networks. This working group has, then, for example, been able to highlight certain limitations to telecommunications developments in the region, and therefore the areas in which the public sector can intervene. Firstly, the cost of broadband is a current hindrance:

> With regard to the deployment of infrastructures and the use of ICTs, Ile-de-France largely holds its own in the competition between European centres, but is a little behind for some types of network and modes of communication. The development of networks (duct, dark fibre, shared independent networks etc) must be supported. In effect, their extension to many business and commercial zones will allow more pressure to be put on lowering the costs of broadband connections which still seem high (Courtois-Martignoni, 1999, p.13, author's translation).

Secondly, although the Paris region seems to be very well equipped in terms of telecommunications infrastructures, the city authority has expressed a slight worry that the lack of complete territorial coverage can reduce the numbers of businesses locating in the city (ODEP, 2000, p.5). The director of the economic development bureau of the city authority highlighted specifically this point at the ODEP round table meeting in September 2000. Thirdly, the time it takes for operators to install connections for businesses (ADSL, leased lines) is 'a very important factor in the competitiveness of the Paris region' (ODEP, 2000, p.6, author's translation).

The Paris city authority has, nevertheless, also adapted to the 1998 legislation which changed the rules concerning telecommunications networks in the public domain. It has developed precise objectives to ensure the city, its citizens and businesses benefit from the deployment of telecommunications infrastuctures: to guarantee access to the municipal public domain under non-discriminatory conditions; to give operators the means to create their own local networks in Paris through a partnership with the Ville de Paris (Filliatre, 2000).

Negotiating the place of telecommunications in the Contrat de Plan État – Région
The development of the Contrat de Plan État – Région, an investment contract between central and regional government, illustrates the importance of intergovernmental relations in economic and territorial development in France. Some telecommunications developments in Ile-de-France have already been characterised by significant regional intervention, and the future nature of this intervention can partly depend on how this domain is placed within this intergovernmental agreement.

The joint view of the Chambres de Commerce et d'Industrie (CCI) de Paris – Ile-de-France of the relationship between telecommunications and territoriality is that the former are capable of 'shaping territories and of defining their performances'. As a result, they developed a series of propositions in an action programme called 'Organising territory to develop the economy', which aimed for three regional objectives in the telecommunications domain, within the scope of the preparation of the most recent Contrat de Plan État – Région in 1999 (Chambres de Commerce et d'Industrie de Paris – Ile-de-France, 1999, p.35, author's translation):

- To allow the emergence of poles with a strong specialisation in telecommunications, such as call centres and teleworking centres;
- To equip the whole of the regional territory with telecommunications infrastructures with a high level of performance, 'even if the market is already particularly dynamic, notably at the level of operators of private local loops';
- To promote the use of ICTs by businesses, through programmes of information and training, in a perspective of increasing competitiveness and productivity.

This was a relatively detailed view of how the CCIs can together territorially regulate the telecommunications domain in the Paris region for economic development purposes. Nevertheless, they felt obliged to clarify that 'the formulated propositions do not aim at substituting once again 'the visible hand of the state' for 'the invisible hand of the market'.' They continued: 'The CCI de Paris/Ile-de-France, as well as almost all the telecommunications operators, remain convinced of the virtues of competition' (Chambres de Commerce et d'Industrie de Paris – Ile-de-France, 1999, p.358, author's translation). In this way, they seemed to be cannily trying to justify their need to intervene in the market, without upsetting or creating tensions with the operators. This is a fairly similar position to those taken by regional bodies and authorities in London – territorial intervention, whilst keeping off the toes of the operators.

Despite the work put in by the regional Chambers of Commerce on the importance of incorporating these objectives in the telecommunications domain into the Contrat de Plan État – Région 2000-2006, the final version of the Contrat de Plan contains relatively little about this domain. In speaking about increasing the 'influence' of Ile-de-France, the actual plan states: 'If its position remains favourable at the heart of international communication networks (road, rail and air networks), it must improve its specific advanced telecommunications networks' (Conseil Régional d'Ile-de-France, 2000a, p.5, author's translation). However, from a total of twenty two articles discussing the domains in which the State and the Région should intervene, ICTs are only included in one of these articles alongside higher education and research. As Irène Le Roch of France Télécom observed: 'In Ile-de-France we would be happy to discover what the strategy of the Conseil Régional is, because it's not very visible at the current time. We received the Contrat de Plan État-Région, and apart from research...' (Irène Le Roch, France Télécom, interview with author, September 2000). The Plan seems to demonstrate that the State and the Région feel that they do not have too much of a role to play in this domain, and that their small interventionist role should be focused on the relations between ICTs and research. The total cost of this intervention is stated as FF131 million, compared to the suggested FF525 million outlined in the preparatory document of the regional Chambers of Commerce. Jean-Baptiste Hennequin of the Conseil Régional offers some reasons:

> I calculated how much we put in the Contrat de Plan in relation to the total investment, and it's 0.2%. What happened? We simply didn't have any ideas or any projects. The Région is after all a small administration, which is organised in a traditional manner with a directorate. Each directorate has a sector of intervention… New technologies don't mean much to a lot of people. We see that it's to do with the Internet. What can we do? Well, maybe cybercafés, but what else? So, in effect, we didn't have a vision. With this sectoral organisation, there weren't many people in the services capable of developing ideas. In particular, we weren't organised. There wasn't a person delegated to that or several people capable of reflection. I had some ideas, and I expressed them, but they weren't able to 'climb' the administrative ladder. Today, it's different. I have ideas, and I'm doing this internal document because we've noted that the Région needs to announce something (Jean-Baptiste Hennequin, Conseil Régional d'Ile-de-France, interview with author, October 2000).

The likely implications of this lack of proposed regional intervention are uncertain as yet, but seem to shift the relations between the public sector in the Paris region and telecommunications networks towards a market-led approach, recently favoured by the public sector in the London region.

The place of telecommunications in the Economic Development Strategy for London The place of the telecommunications and IT domain in the recent Draft Economic Development Strategy for London (London Development Agency, 2000) appears to mirror that of the Contrat de Plan for Ile-de-France. There are fairly token statements about the capital's 'high quality and capacity telecommunications' underpinning the global reach of the financial sector. However, in addition, the brief section on telecommunications encouragingly focuses more on the domain's current limitations:

> The quality, capacity, coverage and cost of London's telecommunication infrastructure currently inhibit some Londoners from participating fully in its economy. As technologies change, it is imperative that access to these is made available to all Londoners at reasonable cost. The LDA will work with its partners in business as well as the Boroughs and the GLA in making a case for London to maintain its competitiveness in telecommunications technologies and arguing for balanced provision throughout the capital (London Development Agency, 2000, p.32).

The e-business advisor to the GLA suggested that a later draft of the Economic Development Strategy would contain 'more on 'e' things' (Colin Jenkins, GLA, personal communication, June 2001), but this is so far unclear.

One would have to conclude that the Draft Economic Development Strategy is a positive step forward for London on the whole. There are signs that a territorial perspective on the capital is seen as advantageous, something which would perhaps not have been stated so explicitly even just a few years ago: 'London can also be seen spatially – as a region; in relation to the adjacent regions; and as a series of sub-regional local economies, each with its own characteristics and each interdependent' (London Development Agency, 2000, p.4).

The Extent of Market-Led Telecommunications Developments in Paris and London

In the previous section, we explored how the public sector is playing an important part in territorially regulating telecommunications developments in Paris and London, essentially as an element within a wider role of improving urban or regional competitiveness. What this also illustrated, however, was that this role seems to be undertaken frequently (and perhaps increasingly) in the shadow of market-led operator developments, hence perhaps the relative lack of strategic proposals in this domain in both regions. The vast majority of telecommunications infrastructures and services in both Paris and London have been developed by private operators, so what we need to explore now is the extent to which telecommunications developments in both cities are being left to the market, the ways in which different operators are deploying infrastructures in the two cities, and whether this negates, redirects or reaffirms the role of the public sector in this domain. In light of this, we can further question what form the regulation of the market takes and how this varies spatially, if the increasing maturity of competition in telecommunications markets means that operator developments dominate key zones.

Trust in Competitive Urban Telecommunications Markets

To say that operators are the key players now in the landscapes of telecommunications developments in Paris and London is perhaps stating the obvious. What is most interesting about these landscapes is how the public sector fits in with or reacts to operator developments. This is all a question of the level of trust that central, regional and / or local government can have in the capacity of the market to deliver. If they can trust the telecommunications market in the city to deliver the required level of infrastructures and services, to meet all demand, then they may adopt a *laissez-faire* approach. If, on the other hand, they are able to identify sectors (both geographical and user) whose demand is unlikely to be met by the market, then they clearly have a role to intervene and try to shape the territorial regulation of the market to include those sectors.

In these sections, we build on what we have already seen, regarding the ways in which the former situation has been more the case in the London region, and the latter situation more characteristic of the Paris region. This can be closely corrolated to the early liberalised markets in London and the relatively recent liberalisation of the market in Paris, where the public sector has been used to playing an interventionist or regulatory role. Nevertheless, while demand is such in the City and the Docklands that the market probably does deal with most requirements, cases of businesses in central London having problems gaining access to the necessary infrastructure are not unheard of:

There's also a problem with getting that sort of infrastructure for small companies. There's a company recently who wanted to set up an office in Fleet Street, and were taking a small amount of space, and they needed access to a fibre optic cable or something, and BT told them it would take them two years before they got round to doing them, which we couldn't believe. We thought there'd been some mistake, but there hadn't been because it would just be too expensive for the amount they were, you know, getting. So there is a bit of a problem... And the provision of electricity as well is a bit of a stumbling block for the large telecommunications / IT companies. It's likely to involve putting in a new substation, and that can be, you know, six months, a year... (Pete Large and Pauline Irwin, Corporation of London Economic Development Unit, interview with author, August 2000).

We need, therefore, to explore in more detail the territorial specificities of the differing levels of market dominance between Paris and London. For example, Colin Jenkins of the GLA refutes the statement that the lack of intervention in the London telecommunications market from the public sector demonstrates a complete trust in the ability of the market to deliver the required networks and services. Instead, 'it demonstrates a historic lack of London focus, and absence of need – prior to broadband' (Colin Jenkins, GLA, personal communication, June 2001). This 'historic lack of London focus' can perhaps be attributed to the early liberalisation in the 1980s, which has meant that for around twenty years, the telecommunications market has been the principal force in developing infrastructures and services in the city, and therefore the public sector has not needed or has not wished to greatly intervene in this market. It is surely a little harder to argue that there has been an 'absence of need' to intervene and to territorially regulate the market provision, given that through the progressive stages of liberalisation, the operators coming into the market have chosen to focus principally on the key parts of the market, namely the business sector. This has led to increasing geographical and user sector variations in the choice and quality of infrastructures and services available. To argue otherwise is to suggest that broadband is the first telecommunications development of the last twenty years which merits public sector intervention in the London region to try to ensure complete territorial coverage.

The Territorial Concentration of Market Delivery

The London telecommunications market is estimated to be worth around £1300 million, roughly the same as for Paris (Colt Telecommunications; quoted in Graham, 1999, p.940).[75] Dozens of operators, either with their own networks or as resellers, are now present in both regions. There are, for example, around 200 licensed operators in the London region alone (Rob Tapping, BT, interview with author, January 2001). Tables 7.1 and 7.2 offer an idea of the different operators with their own fixed infrastructure in Paris and London.

---

[75] The ART estimated the overall French telecommunications market to be worth nearly 25 billion euros (around £16 billion) in 1998, and growing at 7-8% per year (AFOPT, 2000, p.23).

Table 7.1 Operators with fixed infrastructure in central Paris

OPERATOR	TOTAL INFRASTRUCTURE (KM)
Cegetel	400
Bouygues Télécom	15
Siris	20
Kertel	6
WorldCom	400
Belgacom	2
Eurotunnel Télécom	1
COLT	300
GTS Omnicom	37
RSL COM	6
Telcité	5
Completel	160
Viatel Operations SA	150
Tele2 France SA	10
Level 3 Communications	100
France Télécom	200
Global Crossing	30
9 Télécom	6
Global Telesystems	40
Linx Telecommunications	50
Carrier 1 France	70
Metromedia Fiber Networks	120
Louis Dreyfus COM	30
AXS Télécom	2
21st Century COM France	40
Global Metro Network France	
TOTAL (September 2000)	**2200 km**

Source: Filliatre (2000).

However, these figures do not hide the advantage that London is still held to have over Paris and other rival cities in its telecommunications markets:

> Today, there is more internet data centre floor space in service and available for service in London than in any other European city. Almost the entire UK international telecommunications and internet capacity, as well as much of the European connectivity with North America and the Far East, is focused on London, giving it economies of scale over and above those of other cities in the UK and elsewhere in Europe (Jenkins, 2001b, p.4).

More specifically, as Colin Jenkins highlights: 'large businesses located in the central areas of London are well provided for through a number of competitive fibre networks. Central London offers the business community a wider choice of high-quality, low-cost options than any other world city' (Jenkins, 2001b, p.6). The inward investment agency London First Centre (LFC) view telecommunications networks in the capital as not so much a factor in themselves to attract companies and investment, but certainly a locational quality to be flagged up – 'we make sure we stress that you can never be more than 25 metres from optical fibre in central London' (LFC representative, telephone conversation, May 2000).

Table 7.2   Operators offering telecommunications services to end users in the City of London

OPERATOR
British Telecommunications PLC
Cable & Wireless / NTL
COLT
Level 3
MCI WorldCom
Storm Telecommunications
Energis Communications PLC
UUNet
AT&T
Global Crossing
Thus PLC
Global One

Source: Kerbes, 2001, p. 12.

London and Paris dominate, then, their respective national telecommunications markets. Colin Jenkins of the GLA, for example, suggests the territorial advantages of London over the rest of the UK:

London has certain advantages. It is better placed than other UK regions from an infrastructural viewpoint. It has a wider range of competitive options for core network connectivity to other key UK cities and to other countries, as well as relatively high population densities. Local access to the core networks should also be better in London than in the rest of the UK. Most BT local exchanges have been enabled for digital subscriber lines and the potential for high-speed internet access via cable TV networks using cable modems is greater than in the rest of the UK (Jenkins, 2001b, p.6).

This advantage has been seen to have built up over a number of years, however, relating back to the national policy context we looked at in the previous chapter. As Kennedy observed ten years ago:

Since 1981, London has enjoyed a comparative advantage over its European rivals in telecommunications as a consequence of liberalisation of the regulatory regime. It appears not so much to be the cost of telecommunications as the basic quality of provision and the innovative potential in supply which influences locational decisions. [...] The telecommunications restructuring process in London has so far only been matched in New York and Tokyo (Kennedy, 1991, p.141).

London now has many fibre optic infrastructure networks underlaying its core business areas. Many more telecommunications network operators are present in London and the UK than other European countries (Local Futures Group, 1999a, p.10). According to the Corporation of London, 'more countries – 230 to be exact – can be dialled directly from London than from anywhere else on earth; the country's largest internet exchange and the world's largest satellite earth station are both in London' (Corporation of London, 2000, p.29). One in ten of the UK's ISDN lines is located in the City (Corporation of London, 2000, p.29), while:

> Low costs are the reason why London is rapidly becoming Europe's telecoms hub, the centre through which European and multinational companies route their internal communications. In consequence, London is the leading destination for inward investment in telecoms services, helping create an unrivalled infrastructure for the knowledge-driven economy (Corporation of London, 2000, p.29).

Such concentrations of communications-intensive businesses provide the justification for operators to spend large amounts on infrastructure there: 'The City still represents a major source of income for both BT and Mercury' (Ireland, 1994, p.39). This is clearly still the case even now when there are at least a dozen main operators, and no longer just the two. Equally, financial firms have substantial purchasing power in the domain to demand high quality telecommunications. Each of the main networks is part of a wider pan-European, or in some cases, global infrastructure being constructed by many of the operators, of which BT, Cable & Wireless, Colt, MCI WorldCom and Energis are just the major ones in London.

A further measure of the importance of London within the UK telecommunications market is the fact that Rob Tapping, who is BT Access Planning Manager for London, is in charge of one of seven such regional sections for BT UK South, but his region actually generates a quarter of the demand in the South (Rob Tapping, BT, interview with author, January 2001). It has also been estimated that something like 80% of all telecommunications traffic in France starts and / or ends in Paris (Finnie, 1998). What these figures and the alleged 'glut' of fibre, alluded to earlier in the chapter, suggest for telecommunications developments in Paris and London is that the size of the market in the two cities has brought intense competition, which has, at least on the surface, become the dominant shaping force in infrastructure provision. As we said before, it would seem that most regional planning agencies and local authorities need not worry initially over attracting infrastructure and service provision to their territories, as a host of operators are fighting to provide this for seemingly any company that wants it.

However, within the two cities, this multitudinous provision is somewhat limited to the key financial zones, where the biggest profits are to be made:

> Putting a fibre optic local loop in place is a heavy investment, notably in civil engineering terms. Operators opt therefore for a 'cream skimming' strategy and choose zones with high concentrations of large companies, where, they hope, the return on their investment will be quickest. [...]The time when it is no longer necessary to be large and Parisian for operators to be knocking at your door has not yet arrived (01 Informatique, 1999, author's translation).

Indeed, only around 3500 buildings in business zones in the whole of France are connected to fibre optic local loops (AFOPT, 2000, p.30). The deployment of competitive fibre infrastructures is, then, a practice with an inherently territorial basis and implications.

The 'publicity procedure' for the Sipperec network and access to telecommunications infrastructures in Ile-de-France We saw in the previous chapter how, due to the original French national policy context, the Sipperec agency had to instigate a 'publicity procedure' involving a panel of independent experts in order to demonstrate that their project for a broadband network in the inner *couronne* was a response to the 'deficiency' of infrastructure in these areas. The subsequent report vindicated the position of Sipperec and permitted the deployment of their network. This procedure and the report also outlined broader findings relating to the quality and extent of telecommunications infrastructure in the Ile-de-France region, which we can consider here in more detail, because they reveal important elements in both the territorial basis of operator strategies and the overall territorial concentration of the telecommunications market in the region.

One of the most important findings of the report for this study is its overwhelming underlining of the crucial importance of access to infrastructures in the telecommunications domain in Ile-de-France. As the report notes:

> The market for access to infrastructures conditions very clearly the deployment policies of operators, as much in terms of investments as in levels of penetration of the market, of the perimeter of intervention or of the nature of their offer to clients. The desire of operators to control their service fully, in terms of support network and applications, leads them to favour ownership of infrastructures (civil engineering, trench sharing, medium or long term dark fibre rental) rather than purchase of bandwidth (leased lines), which remains for them a temporary solution, and besides which is not always possible under acceptable economic conditions (Sipperec, 2000, pp.17-18, author's translation).

However, the availability of, and access to, telecommunications infrastructures in Ile-de-France are held to be unsatisfactory by many of the operators consulted within the context of the Sipperec 'publicity procedure'. 81.5% of the operators had a negative view of regional infrastructure availability. As one operator stated:

We don't enter these markets on equal terms with France Télécom: it's as if Renault owned the motorways, the toll booths and the cars. We're suffering from the lack of departmental telecommunications directorates which might resemble the existing departmental infrastructure directorates (*Directions Départementales de l'Equipement*) (quoted in Sipperec, 2000, p.18, author's translation).

This all means that within Ile-de-France, there are substantial territorial disparities in the level of infrastructure deployment:

> In total, 92.6% of the operators see a deficiency in the offers of broadband infrastructures in terms both of capacity and geographical presence. Only France Télécom considers these infrastructures to be geographically well distributed and adequately available. The business zones of central Paris and part of western Paris (La Défense) are the only areas judged to have capacity. The needs of companies in other geographical areas cannot, therefore, be satisfied (Sipperec, 2000, p.19, author's translation).

Bound up in the territorial extent of these infrastructures are the territorial strategies of the operators. According to the report, 66.6% of the operators questioned considered that the unsatisfactory availability of infrastructures constituted an obstacle to their regional deployment strategies:

> The majority of operators, 72.6%, indicate not being able to serve all the zones of the periphery of Paris and the Sipperec communes as a result of 'lack of infrastructures'. This is less linked to economic potential itself as to the associated criteria of the cost of traffic collection with regard to this potential. For the operator, as soon as the client is located beyond the perimeter of a fibre optic infrastructure, the cost of the service to the client will increase with the cost of the leased line allowing access to the client. Only clients with substantial telecommunications budgets will assure the operator of a return on such an investment. The choice of geographical coverage derives, then, primarily from these two economic parameters: telecommunications consumption threshold and cost of access (Sipperec, 2000, p.20, author's translation).

The territorial concentration of telecommunications infrastructures and services in Ile-de-France is therefore quite substantial. One operator told the panel that: 'We will never come to the *petite couronne* with the current cost of network construction' (quoted in Sipperec, 2000, p.21, author's translation). Another operator highlighted the territorial differentiations in pricing of services dependent on location in relation to infrastructure: 'Our standard and inclusive broadband offers, valid within our loops, cannot be reproduced outside if there are no alternative fibre infrastructures. Today, on this principle, we only serve La Défense, Marne la Vallée, Vélizy, Cergy and Saint Quentin' (quoted in Sipperec, 2000, p.21, author's translation).

The high cost of leasing lines from France Télécom is highlighted as another important factor in the territorial provision of telecommunications infrastructures and services. One operator said that: 'Today we pay more for a connection between Paris and Marne la Vallée than for Paris – Frankfurt or Paris – Zurich' (quoted in Sipperec, 2000, p.27, author's translation). This is bound up in the territorial basis of the strategy of France Télécom regarding the pricing of their leased lines. They have started to distinguish between 'dense zones' and 'non-dense zones' when calculating the tariff for broadband leased lines at 2 Mbp/s: 'The level of concentration of their clientèle has to permit them to qualify the density of the zone. The operator observes that these zones assure it of scale economies which justify the proposed tariff differences' (Sipperec, 2000, p.30, author's translation). Thus, if both the end points of a connection are in a 'dense zone', the tariff is reduced by 30%, while if only one of the end points is located in a 'dense zone', the reduction is only 10% (Sipperec, 2000, p.30).

This brief look at the 'publicity procedure' report for the Sipperec network has revealed a more general set of findings than those permitting the construction of this network. It is evident that territoriality is a key element in the telecommunications markets of Ile-de-France, both in the quality and extent of infrastructures in the region, and related to this, decisions regarding the nature of operator strategies. The emerging territorial differentiations in service pricing within the region, according to location in relation to infrastructures, seems crucial in this regard, as it highlights the importance of territorial infrastructure provision in the first place, and throws up questions about how under-equipped local areas can compete with the 'dense zones' where operators are guaranteed to make a profit. Strategies such as the Sipperec network seem, therefore, to be crucial in France for ensuring that territorial differences in telecommunications provision do not augment.

The Differing Urban Territorial Strategies of Traditional and New Operators in Paris and London

What we have been working towards in the last couple of sections in this chapter is an understanding of the nature of territoriality of telecommunications infrastructures and services in Paris and London. At this point, we are, in some ways, at the heart of the subject of this comparative study. We are looking to explore here how telecommunications operators deal with the development of networks in Paris and London differently, in response to the wider differences between the two cities in politics, regulation, local government and cultures of territoriality. As Eric Guerquin of the Conseil Économique et Social Régional (CESR) in Ile-de-France suggests: 'It is desirable that we move towards a regionalisation of actors in order to better take into account the specificities of our region and the aspirations of its inhabitants' (Guerquin, 2001, p.11, author's translation). This is as much the case for private operators as public sector actors.

Our emphasis here on operator strategy also responds to the distinct and very surprising lack of focus in previous research on telecommunications and cities on the spatial aspects of actual corporate strategies (although see Warf, 1998; Musso, 1997; Jeannot, Valeyre and Zarifian, 1999; Rimmer, 1997). Warf has indeed argued:

> that the processes internal to individual firms are important, that not all companies within the same industry behave in precisely identical ways, and that the decision-

making processes and firm-specific dynamics of such institutions are critical to comprehending how geographic space is produced and restructured (Warf, 1998, p.256).

Studying the strategies and dynamics of firms can be, then, 'a window into the manifold ways in which decisionmakers respond to the political and economic imperatives of late twentieth century capitalism and create new geographies in the process' (Warf, 1998, p.266). In this way, we explore and compare in the following sections the territorial basis of the strategies and activities in the Paris and London regions of both two of the largest competitive operators in Europe, Colt and MCI WorldCom, and of the respective incumbent operators, France Télécom and BT.

Colt in Paris and in London Colt is a British telecommunications company which started out by developing infrastructures and services in the City of London from 1993 onwards (Colt – City of London Telecommunications), following the end of the duopoly in the UK market. They were obviously able to benefit here from the precursory liberal regulatory environment that allowed new entrants into the UK telecommunications market several years earlier than the markets of other European countries. By March 2000, Colt's SDH fibre optic network in London covered 257 kilometres, including the City, the Docklands, the West End, Westminster, Hammersmith and Camden (Colt Telecommunications website). This meant that by the time they were able to set their operations up in France (and elsewhere in Europe[76]) and in Paris in 1997-98, Colt already had several years experience in the UK and in London. We might, therefore, suppose that their operations in Paris were based on those in London, and this is, to a large extent, the case: 'At the start, Colt completely followed the way it had set up in Great Britain, but with differences in terms of price, process etc...' (Emmanuel Tricaud, Colt, interview with author, November 2000). Their strategy for France from 1997-98 onwards was developed according to experience of the liberalised UK market. By 1999, Colt in Paris had a fibre optic network of 164 kilometres, serving more than 120 clients in 143 different buildings (Colt Telecommunications website).

It is possible to see telecommunications developments in France more generally following those in the UK with its earlier liberalisation, and operators such as Colt present in the UK market before the French market (Emmanuel Tricaud, Colt, interview with author, November 2000). Emmanuel Tricaud said that there is 'not much difference' now in the infrastructures and services of Colt in Paris and in London, and the reasons behind its strategy in the two cities. Nevertheless, Colt has a policy of classing regions according to their competitiveness and attractiveness on a European level, with Frankfurt currently ahead of Paris, so this would suggest the company configures its strategy, at least to some extent, according to specific territorial considerations. M. Tricaud agreed, however, that the role of telecommunications in the competitiveness of European

---

[76] The Colt EuroLan, its pan-European fibre optic network, now covers 12 countries and includes 32 metropolitan local loops and 15,000 kilometres of fibre (Colt Telecommunications website).

regions is balancing out because the main operators such as Colt are deploying their infrastructures and services in all the principal regions in a largely homogeneous manner. Nevertheless, DSL infrastructure seems to be problematising the territorial homogeneity of telecommunications deployment. For Colt, their roll-out of DSL across Europe 'will be expanded according to regulatory aspects distinctive to each country' (Colt Telecommunications website). We can already see evidence of the influence of national regulatory contexts in the territorial strategies of Colt, as it has been able to deploy its own services on unbundled lines in the whole of central Paris since the end of 2001 (Colt Telecommunications website), whereas in London, the opening up of BT exchanges to competitors such as Colt is proving to be a more stubborn regulatory obstacle to be overcome. Colt in the Paris region envisages eventually extending their networks into more peripheral, less competitive areas, but using ADSL, rather than fibre (Emmanuel Tricaud, Colt, interview with author, November 2000).

The territorial basis of the deployment of the networks of Colt is extremely well illustrated by Jean-Raphael Hardy, project director at Colt France, when he describes how they decided on the routes for their network in Paris:

> It was very simple. We took a plan of Paris and we located on it the buildings which housed our principal clients, the banks, the insurers, the 'courtiers' etc. To know their addresses, we simply used the Yellow Pages. Then we explored the terrain on foot, and determined in this way, as near as possible, the ideal future itinary for our fibre optic loop. The deployment of a network always results from a compromise between technical and economic constraints – here, possibility of passage against potential revenues! (Jean-Raphael Hardy, Colt France, quoted in Géomatique Expert, 2002, p.30, author's translation).

The territorial basis of the network deployment strategies of Colt is therefore strong. As M. Hardy again states: 'We're always looking to adapt ourselves to local constraints, by negotiating with the *mairies*. It's in our interest' (Jean-Raphael Hardy, Colt France, quoted in Géomatique Expert, 2002, p.31, author's translation).

Perhaps the major difference between strategy in London and in Paris is in terms of type of growth. Colt in London is growing 'internally', while the growth of Colt in Paris is more 'externally oriented' as demonstrated by the fact that it acquired the Internet Provider Imaginet during 2000 (Emmanuel Tricaud, Colt, interview with author, November 2000).

Another important difference between London and Paris lies in the locations they use for laying cables. In London, they run them in ducts under the pavement. In Paris, they have to use the sewers, which has the advantage of being less expensive in terms of works costs, but the disadvantage of being more expensive in terms of rent: 'We could probably eventually get the right to dig up the pavement in Paris but they'd probably make us wait 5 years to do it' (Emmanuel Tricaud, Colt, interview with author, November 2000). As Jean-Raphael Hardy elaborates, with reference to the territorial specificities of Paris:

It's not always necessary [to dig underground]. In Paris, which is a unique city in many ways, practically every street is 'doubled' underground by a sewer gallery. There are 2,500 kilometres of them![77] This network has pretty much a larger capacity than needed to simply collect used water. Its galleries, which we could qualify as multi-technical, make life easier for all those wanting to lay networks. They are visitable, secure, and have easy access. Besides, EDF, France Télécom and others have already resorted to them. It's just as well these undergrounds exist. Imagine if it was necessary to dig up, every three months, the pavements of the Champs-Elysées! The sewers constitute the priority passageway of all networks, because the urban policy of the City of Paris authorises civil engineering works only under the absolute condition of no other solution being possible. It's stipulated in black and white in the contract we signed with the City. But, this facility is very costly. Rights of passage in the Paris sewers, for four ducts of 40mm, currently amounts to around 15,000 euros per year and per kilometre, which is enormous! (Jean-Raphael Hardy, Colt France, quoted in Géomatique Expert, 2002, p.30, author's translation).

Meanwhile, in London, Colt entered a market previously dominated by BT and Mercury. It might be suggested that the situation was slightly easier in Paris where there was only France Télécom which had significant advantage. For example, Colt was able to be the first to offer alternative services at Issy-les-Moulineaux (Emmanuel Tricaud, Colt, interview with author, November 2000).

Differences between the UK and French telecommunications markets clearly influence the strategy for infrastructure and services development in Paris and London. Emmanuel Tricaud therefore noted the more powerful role of Oftel, compared to the ART, as a competition adviser and regulator, and the higher bandwidth possibilities in London. In France, a key issue was the way in which businesses had been traditionally encouraged for a long time by France Télécom to use the Transpac system rather than an IP (Emmanuel Tricaud, Colt, interview with author, November 2000). Equally, when questioned about the role of planning and urban policy on the strategy of Colt, M. Tricaud mentioned the problem of interconnection which is much greater in France than the UK. In effect, network interconnection in France requires multiple points of presence:

This is greatly linked to the administrative complexity of telecoms in France. In Britain, for example, to be present everywhere, you connect at one point to British Telecom and the network is done. In France, the interconnection catalogue of France Télécom necessitates being present in each PRO (operator relation points: regional collection centres). There are 18 of these in France. […] The logic of the interurban network is therefore not necessarily economic, but it is subordinated to the interests of France Télécom – and it probably won't change very soon, as all operators need the services of the historical operator (Jean-Raphael Hardy, Colt France, quoted in Géomatique Expert, 2002, p.32, author's translation).

---

[77] In comparison, the strategy of Colt for network deployment in other large French cities such as Lyon and Marseille is necessarily 'completely different [as] these cities were not as fortunate to have their own visionary engineer or prefect, like Belgrand and Haussmann in Paris' (Jean-Raphael Hardy, Colt France, quoted in Géomatique Expert, 2002, p.31, author's translation).

In this way, the national regulatory contexts again strongly influence operator strategies differently in Paris and London, and in the French case, it appears to be still configured in favour of the incumbent.

There are probably far fewer differences between Colt in London and Colt in Paris when it comes to general elements of strategy such as promotion and marketing. They use 'all the usual means, especially presentations to businesses', and frequently promote their offer building by building focusing on all the companies in a building at the same time (Emmanuel Tricaud, Colt, interview with author, November 2000).

The Aristote report suggests how 'one of the main characteristics of Colt is its capacity for constructing networks in buildings to serve its clients' (Aristote, 1996, p.20, author's translation). Colt was serving 600 such buildings in London by the end of 1996. Its commutation sector grew tremendously at this time too – from 47 million minutes of traffic in 1995 to more than 400 million minutes in 1996. Colt was the first operator to bring SDH to the level of its clients. As the Aristote report clarifies:

> SDH allows the offer of completely certain and reliable services to two types of clients: on the one hand, the consumer, meaning the banks, the insurance companies and the law firms; on the other hand, telecommunications companies. Colt works in close collaboration with telecommunications companies and extends its networks to allow these companies to serve their clients. For example, Colt serves, amongst others, AT&T, Global One (formerly Sprint) and France Télécom (Aristote, 1996, p.20, author's translation).

The London network of Colt is, therefore, understandably complex:

> at the regional level, the two main commutation centres are those of the West End and Bishopsgate, as well as the transmission nodes of the Docklands and Canary Wharf. This network is connected by STM-4 at 622 Mb/s. In each zone, STM-1 loops are installed at 155 Mb/s. Today, Colt has more than 70 STM-1 loops serving 400 buildings. In all, Colt administers 400 SDH multiplexers 24 hours a day, 7 days a week, to offer its clients a superior service. Nobody in Europe has such experience of SDH development at the client level or the local scale (Aristote, 1996, p.20, author's translation).

Colt in London, then, has major expertise in SDH services. However, they also offer ATM services to their clients. The main users of ATM have been the insurance sector for switching information images between the main London insurance brokers, and the Sohonet media and advertising project in the West End which requires video communication between agencies.

An example of a major firm in London which has taken advantage of the infrastructure and services offered by Colt is the Hong Kong and Shanghai Banking Corporation (HSBC). Their dealing rooms occupy 9000 m² with 700 dealers, who use 1500 telephone circuits and 100 Ethernet segments. When the firm was largely located on proximate sites (and before they had made important acquisitions), they used simple BT circuits at 16 Mb/s without routers. When different parts of the Bank moved elsewhere, the two buildings were connected by two 10 Mb/s circuits from Colt routed on different paths: 'For a network administrator, it is no longer a question of managing virtual networks, but virtual buildings' (Aristote, 1996, p.24, author's translation). The network has been improved through ATM services at 155 Mb/s which join up the Token Ring networks, illustrating that '[t]he motivation of bandwidth is no longer the load of the network, but the response times' (Aristote, 1996, p.24, author's translation).

Table 7.3   **The inherent scalar focus of Colt's various data transmission services**

Scalar focus	Services on dedicated bandwidth	Services on commutated bandwidth
Local		COLTCell via ATM COLTFrame via Frame Relay COLTFrame²Cell with ATM and Frame Relay on one network
Urban	COLT*Link* and COLT*DataLink* on SDH networks	
National	COLT*CityLink*	
International	COLT*Eurolink* and COLT*CarrierLink*	COLTEuroCell, COLTEuroFrame and COLTEuroFrame²Cell

Source: Adapted from Colt Telecommunications (2000).

Colt entered into a cost-sharing agreement with the operator Level 3 in May 1999 for the construction of networks in northern Europe. Level 3 will contribute to the financing of Colt's German intercity network, in exchange for Colt co-financing Level 3's network linking Europe's major financial centres. This illustrates the new alliance-dominated environment in European telecommunications, but more importantly perhaps, an increased focus on local networks. In other words, Colt is happy to share the cost of building pan-European infrastructure, in order to reach local urban markets, within which it can develop and deploy its own range of infrastructures and services according to territorial specificities such as particular market requirements in individual cities or parts of cities. Another emerging example of this is the leading role Colt has played so far in Paris in driving forward a competitive ADSL offer for businesses to that of France Télécom over unbundled local lines. 'Local' level services are thus becoming an important element within their strategy, which is increasingly dominated by a logic of global localisation. For example, the whole spectrum of Colt's services for data transmission is 'divided up' in terms of territorial reach (see table 7.3), but as this range of services relies on a coherent, end-to-end infrastructure (the EuroLAN), they and their scalar focus must be seen as flexible and overlapping.

MCI WorldCom in Paris and in London   The focus in this research on the strategy of MCI WorldCom and their deployment of infrastructures and services in Paris and London predates the well-publicised financial scandal which beset the American division of the company in the second half of 2002.[78] The analysis in this section should, therefore, inevitably be considered with this subsequent context in mind, although what follows remains interesting as a brief study of the territorial strategy of one of the main 'alternative' telecommunications operators in Europe at the height of the telecommunications 'boom', particularly as a comparison with the strategies of the other operators we focus on, such as Colt. Equally, during these problems, the company has tended to stress the limited effect on its European operations, suggesting their relative independence from its North American operations. Indeed, the website of the restructured MCI company now promotes a similar full range of voice, Internet and data solutions to companies in the UK and France as before.

MCI WorldCom is an American telecommunications company which moved across the Atlantic and set up a European end to its operations as European telecommunications markets were gradually opened up. In this way, as for Colt, they started out in Europe in the UK, due to its early liberalisation measures. Being interested primarily in business users, London became the heart of their UK network, with a metropolitan infrastructure of 180 kilometres in place by the start of 1998 (MCI WorldCom website). They now have four or five infrastructure loops in the UK, such as the Triland backbone (Marc Bourgeois, MCI WorldCom, interview with author, November 2000). At the back end of 2000, this implicit distinction in corporate strategy between business and residential customers became somewhat more explicit, as the company decided to split its operations into two businesses – WorldCom for corporate clients and MCI for residential ones (Yankee Group Europe, 2001, p.7). Nevertheless, they promote their extensive local, continental and transatlantic networks as exemplifying 'truly global reach allied with industry leading products and services plus a proven commitment to

---

[78] MCI WorldCom offers, therefore, a particularly good example of the impact of the downturn in the global telecommunications market, discussed briefly in the preface, but which took shape following the research for this study. By October 2002, the company had been forced to file for Chapter 11 bankruptcy protection in the United States as a result of financial mismanagement and fraud at a senior management level. Widespread over-investment in infrastructure deployment had not been fully accounted for in the balance sheets of the company, partly in order to obtain further funds from investment banks, meaning that the apparently healthy finances of the operator were in effect covering over a financial crisis. In April 2003, WorldCom filed a plan of reorganisation as detailed in its Chapter 11 restructuring schedule, and changed brand name to MCI. The wholesale business of the company was sub-branded as UUNET. A three year business plan has been drawn up 'to position the company as a leader in the move toward convergence of local, long-distance and data services' (http://www.mci.com).

local support', and a corporate 'local-global-local' philosophy (MCI WorldCom UK website).

There seems to be slightly more investment from and presence of telecommunications companies in London than Paris. For example, American operators such as MCI WorldCom have designated London as the base for their European operations, often citing language and culture as part of the reason for their choice. So while this may not necessarily mean more infrastructure and services, it is still an example of the differing territorial strategies of an operator like WorldCom in London and Paris.

In Ile-de-France, MCI WorldCom has a cable infrastructure of some 200 kilometres. When they started out in France and in Paris, they decided to concentrate on the axis between La Défense and the Bourse in central Paris, obviously the focal location for the large companies they are most interested in. The first 20 kilometres of their regional network was initiated in May 1996 on the axis running from Étoile – Bourse – Opéra, using the extensive 19th century sewers. Since then, however, they have seven other loops in the region covering such zones as the western crescent and Marne-la-Vallée (Marc Bourgeois, MCI WorldCom, interview with author, November 2000).

Where the territorial strategy of MCI WorldCom seems to diverge slightly from that of Colt is in its explicit embracing of a cross-national logic of 'having the same offer everywhere in Europe'. While other operators might provide similar infrastructures and services in different cities and regions, they do not highlight their territorial homogeneity as clearly as WorldCom: 'this is a very specific strategy among all operators... The idea is to keep a level of homogeneity everywhere, at a pan-European scale. This makes management of networks a lot easier' (Marc Bourgeois, MCI WorldCom, interview with author, November 2000). The idea is to have the same basic technical level and the same types of products available and accessible everywhere, with a centralised supervision office in Amsterdam. This means that a company in London can expect to be able to use the same type and standard of network and subscribe to the same services as a company in Paris. With a completed pan-European network already in place, the infrastructure for this territorially homogeneous strategy is present. The best example of this strategy at work is when WorldCom decided to bid for a Universal Mobile Telecommunication Standard (UMTS) licence in the UK and for a wireless local loop licence in Germany, with the intention of starting out in these new markets in the two respective countries before widening them to their operations in the rest of Europe. In both cases though, their bids were turned down, so WorldCom decided not to concentrate on either UMTS or wireless networks in the whole of Europe, as they would not have been able to develop a territorially homogenous offer in either market (Marc Bourgeois, MCI WorldCom, interview with author, November 2000). Here then, national telecommunications regulatory processes influenced the global strategy of one of the leading private operators in Europe.

It is interesting that many operators always run infrastructure networks which are partly owned by themselves and partly rented from other operators: 'The important thing is to have control over the networks. Having full ownership of the infrastructure is not so important' (Marc Bourgeois, MCI WorldCom, interview with author, November 2000). This is best seen at La Défense, which is obviously one of the most important zones for WorldCom's business, but where they do not have their own cable. WorldCom themselves have a wholesale section to their business which deals in renting or selling their infrastructures and services to other operators.

The strategy of WorldCom, then, is an international one with clients everywhere, as might be expected of an American operator in Europe. Other operators have more of a focus on the national level, and clients also at a national level: 'they're only starting now to realise that clients really want international communications' (Marc Bourgeois, MCI WorldCom, interview with author, November 2000). M. Bourgeois seemed very keen to emphasise how truly global or international WorldCom is compared to other operators, distinguishing between this and the strategies of other operators who use joint ventures to penetrate the markets of other countries *à la* BT in their French joint venture with Cegetel.

In terms of differences between France and the UK (and therefore Paris and London), the roles of France Télécom and BT were compared. It was argued that the former was a much stronger and more appreciated historical operator than the latter. The fact that France is the only country where operators have to make a financial contribution to the continuation of universal service was also noted. WorldCom would prefer not to have to pay France Télécom in this regard (Marc Bourgeois, MCI WorldCom, interview with author, November 2000).

WorldCom are not at all influenced by urban policy or developments. Their deployment strategy is not influenced by anything external to the company: 'WorldCom has its strategy. We've know what we wanted to do in Ile-de-France and elsewhere for several years' (Marc Bourgeois, MCI WorldCom, interview with author, November 2000). Whether this completely internal focus is actually the case, however, must surely be open to question to some extent.

Local loop unbundling and the development of ADSL were felt to be really important, with France Télécom having a major advantage with the latter. The wireless local loop might eventually become important. These new developments illustrate important opportunities in a rapidly changing telecommunications landscape (Marc Bourgeois, MCI WorldCom, interview with author, November 2000).

In London, WorldCom offers local and international telephone services, as well as broadband data services via an SDH network, onto which they also transplanted a permanent ATM virtual circuits network. The Aristote report discusses an example of a WorldCom ATM-based service specifically developed for London, called Media-Express, which was introduced early in 1997:

> Traders, who already have 200 television channels at their disposition including several financial information channels (CNN-FN, Bloomberg,…), no longer want to wait for the piece which interests them to come round on the channel again. Instead, a web page proposes a list of pieces diffused on all these channels in the previous 24 hours. Double clicking on the title of a piece starts a video stream transported on the ATM network (Aristote, 1996, p.21, author's translation).

An example of a major firm in London which has taken advantage of the infrastructure and services offered by WorldCom is the French bank Paribas. Paribas has its headquarters in Paris, but its technology centre is in London: 'In effect, the telecommunications offer, and the demand of the professionals of the bank for this, are infinitely stronger in London than in Paris' (Aristote, 1996, p.25, author's translation). They turned to WorldCom for high bandwidth connections between their Marylebone Gate and Kinsley House sites – an ATM link on SDH at 622 Mb/s:

> The dealers have PCs equipped with three 21 inch screens. Their applications frequently demand more than 10 Mb/s. Video demand is great: 200 television channels, including 9 financial channels, are received in each building. The users frequently carry out file transfers of more than 25 MO. When they are using electronic mail, spreadsheet (with files on the network) and video channels at the same time, it is important to allocate well priorities on bandwidth... (Aristote, 1996, p.25, author's translation).

Overall, then, while MCI WorldCom seems keen to stress the 'homogeneity' of its networks and services across Europe, their parallel 'local-global-local' philosophy must be seen as indicating more of a territorially specific element to their strategies in the Paris and London regions than M. Bourgeois, for example, emphasised. It is likely, for instance, that the deployment of their infrastructure in Paris is subject to the same possibilities and constraints as we noted for Colt.

The territorial strategies of former monopoly operators: France Télécom in Paris and BT in London   The principal general difference between BT in the UK and France Télécom in France, and therefore one of the main influences on their territorial strategies in London and Paris respectively, is the length of time they have been competing in liberalised national telecommunications markets. As we have already seen, BT has had a competitor since Mercury was given a licence in the early 1980s, whereas France Télécom has been sheltered from competitors right up to the 1998 EU imposed limit for opening up markets. BT was also privatised early in the 1980s, whereas France Télécom still remains in the hands of the French state to the tune of 51%. The privatisation of BT was aimed at improving its standard of service and economic performance. As Hulsink has written:

> BT's corporate culture successfully developed from a former state bureaucracy to a market-led, vertically integrated multi-product firm. The centralised functional form was replaced by a more decentralised divisional form, that has moved to neatly defined geographical markets and moreover to BT's various customer groups (Hulsink, 1998, p.152).

Nevertheless, it is not strictly the case that the former monopoly operators have either suffered from, or have been totally against, the introduction of competition in telecommunications markets. As AFOPT has argued with regard to France Télécom:

Far from suffering from increased competition, it is benefitting fully from the growth in the telecommunications market, within which the volume effect makes up for the reduction of the unit price of call minutes on national and international communications. We only have to see its stock exchange capitalisation of some 224 billion euros (in March 2000), which today represents a sixth of the French GDP, to understand how much competition is generating wealth and stimulating creativity (AFOPT, 2000, p.24, author's translation).

For David Ellis of BT too, 'the growth in the market has meant that now everyone has a slice of the pie... London has benefitted from having other competitors in telecommunications' (David Ellis, BT, interview with author, January 2001).

Tensions between France Télécom and the telecommunications developments of the Ile-de-France Région We saw earlier in the chapter how the Conseil Régional d'Ile-de-France has been involved in two main strategies in recent years to expand regional telecommunications infrastructures, the Téléport and then broadband research networks. In both cases, France Télécom was involved initially, before being somewhat bypassed, which created certain tensions between the operator and the development of the strategies.

France Télécom was part of the early discussions regarding the setting up of a teleport in the Ile-de-France region. This was a little inevitable, given its status as the public monopoly. However, in later lists of members of the Téléport association, it is France Télécom which is conspicuously absent. The Téléport, by this stage, appeared to have a primary objective of improving the quality of telecommunications infrastructure in the region, largely to make up for the absence of competition. Daniel Thépin of IAURIF describes the increasingly obdurate nature of relations between the incumbent and the strategy:

There were a certain number of difficulties regarding telecommunications in Ile-de-France. We had trouble in making the operator (France Télécom) accept projects. It was a very political situation as well... For twenty years, the operator had worked to put everything everywhere and at a very high standard... Then needs started to appear which were much more important in terms of transmission and bandwidth, and in specific zones in Ile-de-France. As there was a single tariff everywhere, it was very very expensive, much more expensive than what was happening in countries which had already opened their telecommunications markets to competition. Paris – New York was much dearer than New York – Paris... So there was a regulatory delay, but there was also less demand in France than over there. The Teleport began, in a way, from these observations. We wanted to concentrate on certain axes, to create the conditions for the development of activities based around the theme of information and communications technologies, and to do this it would be necessary to put in place a certain number of capacities which would offer exceptional conditions at that particular place... After this, we were talking of synergy at the Institut. One can't abstract oneself from one's era... At the same time, it wasn't clear because France Télécom, the national operator, didn't at all agree. With the monopoly and the Institut there were political pressures... So we were trying to do this project with a thousand year old,

monopolistic telecommunications operator which didn't want the project to go ahead...
(Daniel Thépin, IAURIF, interview with author, September 2000).

Ultimately, as we saw, the Téléport never really fully got off the ground, due to a combination of the lack of cooperation with France Télécom, the inability to run alternative networks before the mid 1990s because of regulatory restrictions, and the eventual irrelevance of the project with upcoming liberalisation.

In addition, we have seen how, in the ten years or so that the Conseil Régional d'Ile-de-France has been involved in the development of telecommunications infrastructures for research institutes, its role has had to constantly change in order to meet, in parallel, an increasingly complex regulatory environment, the needs (financial and technical) of the research institutes, its central intervening role between these users and France Télécom, and its own overall requirements and strategy (for economic development) for the whole region in the domain of telecommunications networks. Even in the two years since they agreed to finance a new set of *plaques* in and between various sites around Ile-de-France, these factors seem, for the most part, to have intensified:

> What's happened is that in two years, the deal for telecoms has changed. Competition is intensifying, competitive operators to France Télécom are becoming more and more aggressive. They're investing, they're present in a number of growth sectors of Ile-de-France. We started to notice that the local *plaque* was good for all sorts of reasons. We have lower costs, much lower than those of the market. It allows researchers to do experiments on protocols etc. (Jean-Baptiste Hennequin, Conseil Régional d'Ile-de-France, interview with author, October 2000).

Unlike the original research networks of the early 1990s, however, these new infrastructures had been initiated without the involvement of France Télécom, in order to benefit from the newly liberalised French telecommunications environment. The incumbent did not remain quiet about this snub:

> The concession for the local *plaque* provoked proactive steps, even aggressive ones, from France Télécom, who, seeing that a clientele was in the process of getting away, came to see the Conseil Régional to say 'Stop doing these local *plaques*, it's not the best solution, it's not their job, it's a job for an operator. Whatsmore, these are closed networks etc, so stop financing these networks which are closed and proprietary' (Jean-Baptiste Hennequin, Conseil Régional d'Ile-de-France, interview with author, October 2000).

This contestation did illustrate a regulatory problem for the Conseil Régional, and demonstrates the increasing complexity of the nature of territorial regulation of telecommunications developments in an environment of liberalisation, increasingly demanding actors, and increasingly taut relations and opposing needs between actors:

> It's true that in the case of an extended local network, when you deploy fibre, this fibre is designated for a specific use. But, if you constitute in French law, a closed user

group, as for the RAP and the Génopôle, you cannot share these networks with users who were not specified at the time of the deployment of the network. That assumes that there is a closed user group... This is seen as a problem, not so much at the level of France Télécom, but especially at the level of the Région, where we were saying 'why are we doing closed networks?' In actual fact, these networks are not closed on the Internet, because that was understood like that. They're connected to the Internet. There's an access provider called Renater, but which isn't really an access provider because Renater doesn't manage the IP system etc... In my opinion, amongst all these militant arguments against the *plaques*, there's only one which is valid, which is that operators today know their jobs better and better, and researchers reckon that they've taught IP and the Internet to France Télécom, which isn't untrue, but nowadays it's not really the job of researchers to be only looking after the running of active routing infrastructure etc. It would perhaps be more in their interest to appeal more directly to operators... (Jean-Baptiste Hennequin, Conseil Régional d'Ile-de-France, interview with author, October 2000).

The more widespread development of broadband infrastructures for research groups in Ile-de-France has been proposed to take place via a kind of auction by lots, which the Conseil Régional hopes can lead to competitive offers for the construction of each network. France Télécom may, therefore, be able to get back on board in this way.

The strategies of both France Télécom in the Paris region and BT in the London region are evidently related to their former national monopoly status. Having been the sole telecommunications providers up to liberalisation, their basic infrastructures and services are inevitably territorially dense and widespread, covering the vast majority of both countries and therefore virtually the whole of the Paris and London regions. Nevertheless, an important distinction between the two can be made in relation to the territorial density of their networks, which is strongly linked to the different periods of monopoly and liberalisation in the two countries. As BT was subject to much earlier competition and privatisation, they had to be concerned with ensuring profitability and did not necessarily have the investment potential to deploy a vast fibre infrastructure across the London region including its farthest corners. The much longer monopoly of France Télécom meant that they could develop a substantial fibre network in the 'public' interest. They are thus able to highlight how their fibre optic network in Ile-de-France of more than 570,000 kilometres makes it 'among the densest in the world'. They emphasise the near complete territorial coverage in the region:

Each commune in Ile-de-France is either crossed or directly served by fibre optics, or very near a commune which is... Whatsmore, the copper network of France Télécom, almost entirely underground (more than 20 million km of copper pairs), serves all businesses, buildings and homes in Ile-de-France, thereby guaranteeing universal service (France Télécom, 2000b, author's translation).

For Daniel Thépin of IAURIF, this dense fibre network constitutes:

perhaps a specificity of the Ile-de-France region. Other European regions don't have such coverage, which is a result of the old regime. We are certainly one of the few European regions to have such an important network deployed by the incumbent operator, and that's a specificity (Daniel Thépin, IAURIF, interview with author, September 2000).

In attempts at assuring the social and territorial cohesion of regions like Paris and London, then, incumbent operators still have a key role to play, 'especially those who have dense territorial networks. They enjoy a positive image of Republican egalitarianism linked to their history. In France, where the public service has never fallen from grace, this is a heritage to yield a profit' (Guillaume, 2001, p.5, author's translation).

In the cases of both France Télécom and BT, however, it is now necessary to distinguish between their traditional copper networks and their new wideband or fibre optic networks. The access network (the PSTN or old copper network) planning and wideband (above 2 mb) planning of BT in the London region is thus split in two according to volume drivers. This means that each section is allotted a relatively even quantity of demand (Rob Tapping and David Ellis, BT, interviews with author, January 2001).

'Streaming' the City   The territorially-based focus of BT's main frame relay service, FrameStream, and its tariffs became the subject of contestation from Energis, Cable & Wireless and MCI WorldCom in the second half of 2000. These operators alleged that the initiation of new tariffs for access circuits for FrameStream in the City Zone and Central London Zone were 'anti-competitive and will have the effect of squeezing the complainants profit margins to an extent that competition in frame relay services will become uneconomical. This may have harmful consequences in the development of the UK data services market as a whole' (Oftel, 2000). This brief vignette reveals not only the inherently territorial focus of deployment of one of BT's main business products, and the potentially differing territorially-based tariffs applicable to businesses dependent on their location in London, but also the overall importance of telecommunications developments in London for national territorial development, in terms of the alleged possible effects of a lack of *local* competition in central London on *national* competition in data services. In this way, competition in telecommunications markets seems to be bound up in scalar imbrication or parallel multiscalar territoriality.

In April 1993, BT introduced a new service aimed at the financial institutions of the City of London, DealerStream. This service aimed at homogenising tariffs for point to point circuits, and was run over the City Fibre Network. Previously the tariff for a point to point circuit depended on the distance between both user points and their closest respective local exchanges, and the distance between these two exchanges (Ireland, 1994, pp.42-43), so therefore, this service can be seen as taking the geography out of telecommunications tariffs as the two calculated distances no longer influence the cost of the service.

Shaping infrastructures and services for Ile-de-France – the incumbent way
Within Paris, the network of France Télécom, under roads and along the sewers, follows the main west-east and north-south arteries, and focuses on serving government ministries, embassies, the Bourse and company head offices (Matheron, 2000, p.117). The dense territorial coverage of the networks of France Télécom in Ile-de-France is illustrated by the major infrastructure project of the Stade de France. It was eventually constructed in the Saint Denis area, although for some time Melun to the south east of Paris was the preferred location. For France Télécom though, 'we would have provided the telecommunications for the Stade de France anywhere in Ile-de-France' (Claude Martin, France Télécom, interview with author, October 2000). It is thus served by a secured double local loop offering capacity of 2.5 Gbits/s (France Télécom, 2000b). Because France Télécom has infrastructures everywhere in Ile-de-France then, and other operators have infrastructures in certain key sectors too, M. Martin argued that competition was taking hold more in actual services than in infrastructures.

One element in the strategy of France Télécom specific to the Paris region, and its national capital and international status, is the offer of a *Point d'Accueil International* (PAI – international welcome point) from the Business Branch of the company, which aims to give help to and make it easier for foreign businesses and companies to locate in the Ile-de-France region (France Télécom, 2000a).

The territorially-specific strategies of telecommunications operators can also be made clear from the ways in which they deploy infrastructure and deliver specific services to valued individual clients. Take the example of France Télécom and Disneyland Paris, one of the biggest tourist poles in Europe. The latter clearly requires suitable, dedicated and efficient telecommunications services, and chose the Expertel Services & FM division of France Télécom to deliver these for a site which hosts its head office, the theme park, and several large hotels. These services are meant to be characterised by 'availability, reactivity and quality' (France Télécom, 2001b). As a result, the developed telecommunications strategy for Disneyland Paris includes a dedicated team of technicians to look after a unified network for the administrative centre, which is made up of 3 PABXs and 3,700 lines, a unified network for the hotels, which has 8 PABXs and 8,000 lines, and the infrastructure for the reservation call centre. The needs of the site in terms of telecommunications are discussed and assessed every month (France Télécom, 2001b). This brief example demonstrates how operators develop particular territorial strategies for particular clients, whose loyalty needs to be safeguarded at all costs. Disneyland Paris clearly has specific and unique telecommunications network and service requirements, which necessitates a specific and unique response from an operator such as France Télécom. Consequently, then, this is a further 'sub-local' element in the territoriality of telecommunications developments, within the already territorially-based deployment of infrastructures and services by operators at a regional and local level.

France Télécom has a similar type of strategy for *collectivités locales* clients as well. As their promotional material states:

The sites of a *collectivité* turn out quite often to be varied and scattered locally. Ensuring the smooth running and cohesion of these different places becomes complex. In addition, the needs of *collectivités locales* in terms of networks means an increase in bandwidth. To respond to the expectations of the *collectivité*, France Télécom proposes a full range of solutions for which the implementation is adapted to the local context: the *Réseau Intra-cité*, which links the technical services, the schools, the library, and the theatre around the town hall; the *Réseaux Mixtes Dédiés*, which take into account the existing infrastructure... (France Télécom, 2001a, author's translation).

Many of France Télécom's network services are particularly aimed at large businesses, which are often concentrated in the Ile-de-France region. For example, their promotional material lists the *Service Multisite Haut Débit* as 'an offer for a broadband inter-site network, specific to Ile-de-France which uses secured fibre optic loops', and the *Offre d'Accès Fiabilisé* which 'allows to meet head on the growth of network IT (Internet, Intranet...). This offer particularly adapted to Ile-de-France proposes exceptional security levels' (France Télécom, 2000a, author's translation). In both cases, it is evident that the territorial context of the Ile-de-France region (the capital and major urban region of France; the density of large businesses located there; the subsequent density of the network infrastructure of France Télécom) has shaped the development of these service products of France Télécom. Equally, they go on to discuss the specificity of their local development relations in the region:

France Télécom has a network of agencies which is particularly dense in Ile-de-France. Traditionally rooted in the local fabric, thanks to its long history on the ground and to a desire to base its organisation on proximity, France Télécom knows the needs of local actors. It brings the communes, the departments and the SEMs the necessary adapted solutions... (France Télécom, 2000a, author's translation).

Once again, then, territoriality is utterly bound up in the regional strategy of telecommunications operators.

Both France Télécom and BT are organised along a combination of functional and geographical bases. In terms of the latter, both have recently restructured their regional directorates. BT wanted to try to match the recent redrawing of some regional government boundaries in Britain, so that their delimited territories paralleled those of the new RDAs (Sue Davidson, BT, foreword to Local Futures Group, 1999b). Before the restructuring of France Télécom, the internal organisation of the French incumbent was divided more according to function than geography, with geographical divisions only coming into play below the level of the operational sub-divisions (Jeannot, Valeyre and Zarifian, 1999, p.39). The subsequent restructuring (which aimed to maximise internal coordination) was primarily focused on market changes: 'The segmentation of activities by markets aims to better meet an increasingly specific and diversified demand with the multiplication of telecommunications products and services' (Jeannot, Valeyre and Zarifian, 1999, p.40, author's translation). Separate sections were therefore created to work in the residential, professional and large company markets respectively.

Conflict in France Télécom strategy between operational markets and territorial coverage seems, however, to exist:

> Through the composition of its networks and its public service assignment, France Télécom has both a strong and a diffuse territorial inscription. However, whilst the requirements for geographical proximity with users and infrastructures remain important, a major tendency towards the concentration of activities is emerging and constitutes a source of tension with which the organisational changes find themselves confronted (Jeannot, Valeyre and Zarifian, 1999, p.41, author's translation).

The full transformation of France Télécom's market from being public service *users* to company *clients* is still relatively recent, and involving as it does the operator to remain in the first instance territorially diffuse, and in the second instance territorially concentrated or proximate, means that this is necessarily one of the key elements in its overall strategy. This tension is unlikely to be resolved either, while the operator must both fulfill universal and public service obligations and maximise revenue from competitive markets. In this way, the territorial strategies of incumbents and new entrants differ enormously. Hulsink summarised the two-pronged concerns of France Télécom:

> The current strategy of France Télécom is to become a strong overseas competitor, while remaining integrated at home in order to provide high-quality universal services. The company is free to distribute video signals and use radio in the local loop (unlike BT in the UK). To prepare for international and domestic competition, two burning issues needed to be addressed in the early 1990s: the still centralised and functional structure of the organisation (hampering quick and effective decision making and its strategic and operational activities), and the vague legal status of the state handicapping its international acquisitions, alliances and access to financial markets (Hulsink, 1998, p.267).

Both former incumbents have prioritised cutting their debts in recent corporate strategy. The difference is that BT has taken the corporate restructuring route, while France Télécom has been confident of shifting its debt load within its current organisational framework based around the full variety of networks and services. Within its internal restructuring, BT developed plans to float parts of its separated businesses such as BT Wireless (mobile) and Ignite (broadband), and sell other worldwide assets.

Both BT and France Télécom are also involved in joint ventures in each other's countries – BT with Cegetel in France, and France Télécom with Global One and MetroHoldings in the UK. BT has also developed the 'Farland' network which connects its joint venture operations in France, Germany and Spain (Financial Times, 2000).

This section, like our analysis of Colt and MCI WorldCom previously, has illustrated the ways in which BT and France Télécom shape and develop differently their strategies for the London and Paris regions respectively, according to key territorial contexts and specificities. Whereas with Colt and MCI WorldCom being present in both cities, we were able to see how the same operator is developing a different strategy for each city according to particular territorial contexts, as well as comparing how the two different operators are developing their strategies, here we have only been able to see how similar (incumbent) operators are developing different strategies for London and Paris. Put together, however, and it can be argued that these sections have begun to demonstrate, perhaps for the first time, the variety of ways in which telecommunications infrastructures, operator strategies and territoriality are all inherently intertwined in global cities.

The Aterritoriality of the Pan-European Carrier Market?

In telecommunications, both Paris and London are now part of ever more extensive global fibre and satellite networks. Being important 'nodes' on these networks means that both traditional and new operators have to reconcile decisions about deployment of their infrastructures and services in individual cities with their national and increasingly pan-European (and global) strategies.

Consequently, another increasingly important element in the global city telecommunications 'package' of Paris and London is the pan-European carrier market. All the major operators in Europe have either constructed their own continental backbone networks linking the main cities and regions or have purchased or leased fibre from other operators in order to be able to provide services between these cities and regions (Flanigan, 2000). Following in the footsteps of the EU's Trans-European Networks (TEN) projects, the Hermes Ring was the first pan-European telecommunications network, which was substantially built along the tracks of the European rail system (Financial Times, 2000). Newer operators seem to have been ahead of the game in network construction, with traditional historical operators like BT and France Télécom often left to lease the fibre they require (Andersson, 2000). However, fears of a glut in supply of fibre, as mentioned before, seem to be well founded: 'Carriers invested in new network infrastructure with the expectation that Internet-related services would fill pipes and more than make up for declining voice revenue, but data service revenues don't yet make up for voice shortfalls' (Yankee Group Europe, 2001, p.1). Indeed, they have often yet to provide sufficient revenue to allow carriers to break even on the cost of building their infrastructure, as 'competition for business customers has also intensified, and more and more frequently carriers are providing networks at prices that are below cost, just to win business' (Yankee Group Europe, 2001, p.6).

This kind of development exemplifies the flexible, permeable boundedness of the 'global city' in terms of telecommunications, as the definition of high quality telecommunications infrastructures and services in global cities such as Paris and London must now necessarily take into account high capacity, trans-continental, cross-border networks, which are a scale up from the local or urban level networks specific to each city. Telecommunications developments in both cities, then, still need to be framed within an important cross-national context. For example, the suggestion of a convergence in infrastructure provision in cities should mean that operators need to concentrate on providing niche market networks and services, or value-added types of services, which might well imply differing territorial influences, specificities and implications.

'Global end-to-end provision': the changing strategy of Cable & Wireless   In 1998, Hulsink described the corporate strategy of Cable & Wireless as:

> aimed at strengthening its position as a major force in worldwide telecommunications through the establishment of a federation of telecommunications companies consisting of a global super-carrier and local niche operators. This worldwide strategy is made up of three related objectives: the provision of global end-to-end services for business customers, the expanding of basic telecommunications services (particularly in less developed countries) and the building up of a solid base in mobile communications around the world (Hulsink, 1998, p.162).

This has all changed in the last couple of years, with a refocusing of strategy firmly on the first of these objectives (Matt Cochrane, Cable & Wireless, interview with author, January 2001), as Cable & Wireless 'has transformed itself from a holding company with investments in a patchwork of carriers around the world, to a streamlined carrier focused on the provision of data services to corporations around the world' (Yankee Group Europe, 2001, p.9). This change in strategy has been 'a real right angle turn' (Matt Cochrane, Cable & Wireless, interview with author, January 2001). This corporate restructuring included selling its mobile interest, One2One, and its residential cable and telecommunications arm in the UK to NTL (Matt Cochrane, Cable & Wireless, interview with author, January 2001). All this is a much different strategy from BT or France Télécom, and has led to one big advantage: 'Rather than struggling to cope with the reduction of a high debt load, Cable & Wireless now must decide what to do with the more than $6 billion in cash it has gained from these deals' (Yankee Group Europe, 2001, p.9).

In turn, Matt Cochrane identified the liberalisation of European markets from 1998, and the related massive increase in telecommunications requirements in Europe as tied up with the corporate refocusing on providing seamless end-to-end services for business clients (Matt Cochrane, Cable & Wireless, interview with author, January 2001). With its aim of becoming the European and world leader in IP service provision for businesses, some of this money has gone on the acquisition in the last two or three years of 15 or so European-based Internet companies, and much more (around £1 billion) is planned to be spent on extending its European IP network to cover 200 cities by 2002 with a single architecture design (Costa, 2000, p.29). As Matt Cochrane stated: 'we needed to improve our backbone on the continent'. They also purchased several web design companies, which is another element in their 'end-to-end' objective (Matt Cochrane, Cable & Wireless, interview with author, January 2001), and there are plans to build their own ADSL network in collaboration with France Télécom. The European network, however, is just part of C&W's £4 billion plans for an extensive global IP network (Costa, 2000, p.29). This illustrates clearly the continual telescoping of scales at work in the telecommunications sector. Every network is just a small part of a much bigger network.

Cable & Wireless do not have a specific strategy for telecommunications in London, but it was stressed that 'London is never the place to try things out, because it's so different to the rest of the country' (Charles Barbor, Cable &

Wireless, telephone conversation, December 2000). Nevertheless, the operations of Cable & Wireless in London benefit greatly from the advantage of a huge old network running throughout the city (Matt Cochrane, Cable & Wireless, interview with author, January 2001). This highlights a continuing importance of both the territorial specificities of London itself, and a continuing national focus to telecommunications developments, within the strategy of C&W focused on 'global end-to-end provision'. This observation was re-emphasised by BT who carry out trials in other parts of the country, but have to be wary of differences when products and services are subsequently implemented in London. Often, the services and networks rolled out in London are 'firsts' for BT, in keeping with the high profile of London in the telecommunications sector (David Ellis, BT, interview with author, January 2001).

Territorial Authorities and the Construction of Telecommunications Developments in Paris and London: Configuring the Space of Flows

So far in this chapter, we have looked at how telecommunications developments are increasingly seen as an important part of the wider processes and practices of urban competitiveness, and how it is the market which is playing a crucial role in different ways in the shaping of telecommunications developments in Paris and London. This section builds on what we have seen so far and explores how the various elements of the public sector in both cities are involved in the construction and packaging of these developments, and the extent to which this territorial authority intervention is both tied up in their competitiveness agendas and a response to perceived benefits and failings of market delivery, particularly in terms of the territorial concentration of infrastructure provision. As Colin Jenkins remarks in his report for the GLA: 'to ensure the capital's competitiveness, business expansion and inward investment are maintained and further developed through the impact of e-business, central, regional and local authorities will need to work closely together, as well as with industry' (Jenkins, 2001b, p.1). Here then, intergovernmental relations are highlighted as a key factor for telecommunications developments in global cities.

While telecommunications developments in Paris and London are perhaps inevitably dominated by operators, the public sector has often identified roles for itself in both increasing local and regional coverage of infrastructures and services, and regulating local markets to improve choice, quality and cost of telecommunications. As the French OTV advises and lays out to local government:

> For the planner and the developer who has become a Mayor, it is impossible from now on to neglect ICTs. But, will they aim for promotional communication of the municipality or of its socio-economic actors? Or do they believe more in the development of the new economy for local SMEs? Will they want to use ICTs to assist businesses in their creation, their development, and their export expansion? Or will they bank on the leverage effect of dynamic communications infrastructures to ensure

their strike force? A wide spectrum of initiatives is within their capability (OTV, 2000, author's translation).

Territorial authorities can, thus, be viewed as the 'catalysts' for attracting operators and ensuring access to networks and services for citizens and businesses (Jacques Douffiagues, ART; quoted in Moreau, 2001).[79] Indeed, it is possible to go so far as to suggest that telecommunications developments depend a great deal on the will, capability and knowledge of the public sector as a whole, and local authorities more specifically (Jean-Philippe Walryck, AFOPT, interview with author, November 2000; Emmanuel Tricaud, Colt, interview with author, November 2000). Nevertheless, 'we have to accept that telecommunications operators cannot be bypassed. This is our *métier*. It's too complicated for *collectivités locales* to deal with by themselves. Telecommunications is, after all, not a child's game' (Jean Nunez, Cegetel, interview with author, September 2000).

Equally, however, pressure from interest groups such as local research and education establishments can lead to local and regional authorities looking to develop telecommunications infrastructures and services, especially broadband (OTV, 2001b). We have already seen this in the discussions surrounding the role of the Conseil Régional d'Ile-de-France in the construction and funding of broadband infrastructures for research institutes in the Paris region.

However, generally, in order to carve a role for themselves in the telecommunications domain (and therefore to avoid leaving developments totally in the realm of the market), territorial authorities at both the local and the regional level still require at least three traits: the ability (know-how and a lack of internal or external constraints), the willingness (desire) and the resources (financial or otherwise) to recognise the importance, potential benefits, and restrictions of telecommunications infrastructures and services. It is the need to maximise these elements in the construction and packaging of telecommunications developments which, in many cases, leads to the creation of partnerships of different types, which include groups and agencies who might possess one of more of these elements in particular. Shaping the space of flows, and its constitutive layers of physical supports – circuits of electronic impulses such as telecommunications; a network of place-based nodes and hubs; and the spatial manifestations of dominant groups and interests – is a highly complex undertaking. We look here at how this is being attempted by local and regional authorities in Paris and London in two key stages of the territorialisation of telecommunications developments.

Understanding the Territorialities of the Space of Flows

The ability of local and regional public sector actors to intervene and promote telecommunications and IT may very well be continually influenced by internal and external actors, events and forces altering or negating this wish. This would

---

[79] With 37% of GDP in London coming from the public sector, it is very important to remember that for business, the public sector is a key client (Alex Bax, GLA, interview with author, February 2001).

suggest that the urban telecommunications domain, like most other domains in urban development, is strongly characterised by negotiations of power between different actors.

Equally, the willingness of local and regional public sector actors to play a role in the telecommunications agenda of Paris and London can sometimes be questioned, given that they do not necessarily see telecommunications and IT as the most important area of priority. Employment and wider economic development issues are more likely to be favoured, because of the perception that those are the issues which the public wants to see tackled. The results of intervening in the telecommunications and IT domain are less visible, and therefore may offer less *kudos* to local and regional authorities, especially when it comes to being re-elected.

The resources of public sector actors in Paris and London clearly also have great influence on their subsequent role in the telecommunications domain. If they do not have either the finance or the manpower for intervention, then intervention is unlikely to go ahead. With the number of areas and projects which have a concerted grip on the public purse, it is inevitable that telecommunications and IT are not very high up on their list of priorities. Nevertheless, for the Mayor of London, Ken Livingstone:

> It is vital for London to stay at the forefront of both private and public sector investment in new information and communication technologies. […] It is important that private and public policy fully support an e-London, as its development is crucial not only for the capital, but also for the whole of the UK economy (Ken Livingstone, foreword in Jenkins, 2001b, i).

In this way, the ability of regional government to recognise the importance of, and the stakes involved in, the telecommunications domain, and its willingness to act or intervene, creates a policy justification for resource investment, which would not exist without this ability and willingness.

'A regional network of networks': the LondonNet project The capacity of regional bodies like the GLA to obtain funding for projects in the IT and telecommunications domain is quite small. 85-90% of funding is already allocated to other existing projects, and there is not much for new projects. The LDA does have some money, and there are bits of European funding to be had as well (Alex Bax, GLA, interview with author, February 2001). This funding problem was already illustrated a couple of years previously when Alex Bax was working at the London Research Centre. In October 1999, with the support of the boroughs and other groups, they produced an Invest to Save Bid (ISB) proposal to the government for a project called LondonNet, looking for £5.9 million (Alex Bax, GLA, interview with author, February 2001; London Research Centre, 1999).

The LondonNet ISB proposal was an initial design for a strategic regional infrastructure founded on ICTs, which would later be reformulated as part of the London Connects initiative: 'LondonNet will establish a regional network of networks, linking 33 boroughs, the new London Authority and its agencies,

Government agencies and other cross-London bodies – eventually providing a round-the-clock 'one-stop shop' for public services' (London Research Centre, 1999, p.1). Its three main objectives were to deliver better services, to improve access, and to construct a coordinated technical framework.

The government rejected the proposal in February 2000, although the run-up to the GLA elections at the time was felt to have something to do with this decision. The transformation of London governance clearly meant that it was probably preferable to wait for the establishment of the GLA later in 2000, before making a firm decision on the creation of a London-wide ICT project (Alex Bax, GLA, interview with author, February 2001).

Telecommunications and IT is a vast and complex area, and local and regional actors probably need a finely worked-out strategic plan of their action in order to successfully make things happen for them. We began to discuss earlier how the Conseil Régional d'Ile-de-France was considering the possibility of increasing competition in the telecommunications market regarding research infrastructures by dividing this territorial market up into lots. This initiative would thus combine two of the main 'roles' identified by M. Hennequin for the Conseil Régional in the telecommunications domain, namely a 'principal role' largely based around promoting competition, and a 'territorial planning role' which aims to promote offers for more peripheral and disconnected areas of the region (Hennequin, 2000):

> Plus, effectively, if we are able to undertake an intelligent sale by lots of the market, through geographical lots or especially technical lots, in this way we could make competition work… This is a project which is being studied, it's not settled. The president [of the Conseil Régional] knows about it, and they're saying 'should we do it or not?'. It's true that in this domain, things move so quickly, we don't have control over what the region is doing, we struggle a little to follow technically what's going on, because I'm all alone in dealing with telecoms and there's so many things that I don't know, or I don't have the time… One person is not enough. Therefore, we have to be modest in our interventions. It's not yet settled, and I have only one worry, that in this domain, either we launch things rapidly, or we take too much time to reflect, and in six months or a year, when we come to launch the project, it's already out of date or been overtaken (Jean-Baptiste Hennequin, Conseil Régional d'Ile-de-France, interview with author, October 2000).

The lack of time and personnel resources and technical comprehension are, in this way, slightly hampering the intervention of the Conseil Régional in the telecommunications domain. As Jean-Baptiste Hennequin of the Conseil Régional further observes:

> Information technologies mobilise all sectors of regional intervention: economic development, research and training, culture, and *aménagement du territoire*. The difficulty for a regional authority consists of having a global vision, then of elaborating a strategy for the whole, without necessarily being able to count on a structure of expertise, coordination, and even instructions for dossiers. This is a complex field, requiring extremely broad and diverse technical knowledge which runs from network and IT technologies to Internet use in rapidly growing sectors (economic development, culture, education)… [Furthermore] in this domain, it seems that quickness of action

matters, as much as the slow elaboration of long term programmes, which causes potentially static plans of action (Hennequin, 2000, p.2, author's translation).

For a rich capital region like Ile-de-France, the frequent lack of intervention, action, or even interest in the domain of telecommunications infrastructures and services on the part of various public actors can perhaps be at least partly explained by a feeling that there is no need to intervene and that everything will necessarily be sorted out almost naturally (by the market) to the benefit of everybody. This might be the case to such an extent that some public actors have never thought that they could have something to do with regard to telecommunications (Jean-Philippe Walryck, AFOPT, interview with author, November 2000). This seems comparable to the situation in London.

In spite of the proclamations that ICTs represent a priority for economic development in Paris, the Mairie de Paris also recognises its own limits: 'There is not really an urban policy for ICTs. We don't have sufficient ability to step back to get an overview of the situation. We don't even know if they are a good thing or not...' (Françoise Courtois-Martignoni, Mairie de Paris, interview with author, October 2000). In London too, DEAL admitted to not knowing too much about the ICT sector (Gill Marshall, DEAL, interview with author, October 2000). 'The dotcom and telecom opportunity' paper aimed at the Docklands East London (DEAL) board recognises the need for the agency both 'to improve its knowledge of the dotcom and telecoms sector in order to develop projects to attract additional investment to the area' and 'through linkages with local universities and other interested parties consider and seek to take forward joint initiatives to gain benefit from this growing sector' (DEAL, 2000, p.4). The Economic Development Unit of the Corporation of London admitted too that they 'didn't deal with IT / telecommunications much' (Pete Large and Pauline Irwin, Corporation of London Economic Development Unit, interview with author, August 2000). This distinct lack of action or involvement is reinforced by the Local Futures Group regional report:

> In the absence of an economic development strategy for London – a vacuum to be filled by the London Development Agency – there is no strategic region-wide approach to telecommunications. Presently, it appears that 'no organisation is taking responsibility for London's telecommunications agenda'. Telecommunications is not on the current LDP agenda. The only real points of reference are within discussions on sectors of the economy. However, 'there is a vague idea that it is tremendously important' but it is not really understood (Local Futures Group, 1999b, p.25-26).

At this stage, then, the local authorities of the City of London and in the Docklands have limited understanding of the territorial specificities of telecommunications developments in the space of flows. This seems to follow our earlier identification of the overwhelming dominance of market-led developments in these areas. A number of competitive offers in telecommunications provision are known to be available, which businesses take advantage of, but beyond this, authorities have tended not to be overly interested in any form of direct intervention. This local context played out on to the regional level in London, where, prior to the creation of the GLA, there were no strategic bodies involved fully in the telecommunications domain.

In the last year or so, however, this situation in London has changed. The Mayor of London, Ken Livingstone, has taken a particular interest in ICT developments for the capital, and the GLA nominated its own e-envoy to match those of central government. Perhaps the main initiative, however, came under the business secondment programme of the GLA, 'under which company executives work within the new authority, contributing their expertise and experience' (Ken Livingstone, foreword in Jenkins, 2001b, i). The GLA decided to use this programme in the domain of ICTs, and, following the recommendation of London First, appointed Colin Jenkins, a Director of Special Projects at the telecommunications operator Energis, to the post of e-business advisor for an initial period of six months. His mission was 'to ensure London's competitiveness, business expansion, and inward investment is maintained and further developed through the impact of e-business' (Colin Jenkins, GLA, personal communication, June 2001). In turn, this role 'significantly increased the authority's understanding of e-technology and the policy responses required in London' (Ken Livingstone, foreword in Jenkins, 2001b, i). This secondment illustrates, to some extent, a reliance of the newly created regional government body of London on the knowledge and expertise of the private sector, both *per se*, and in shaping actual public policy in this domain, thereby increasing the ability of the GLA to reflect positively, and to intervene when and where necessary. For example, the e-business advisor has put in a great deal of effort in highlighting the importance of broadband infrastructure for London, and has forged and promoted a key role for the GLA in developing broadband:

> The GLA can help to reduce the price of this high-speed technology. Working with application providers, service providers and planning bodies to develop a mass-market for broadband, the authority can take steps to redress and pre-empt a new e-divide between households that can or cannot afford access to broadband services. The GLA can also address various sustainability issues, while supporting training initiatives (Jenkins, 2001b, p.1).

Here, on a regional level, we can suggest that this identification of the territorial importance of developing broadband infrastructures is an initial step for the GLA in shaping the space of flows in the London region. The next step is to formulate and then implement a strategy for the 'reterritorialising' of the broadband networks which are the first layer of the space of flows, taking into consideration the spatial manifestations of key groups which are the third layer, in order to promote the second layer of a network of place-based nodes and hubs. A recommended strategy has certainly been formulated for the GLA, as we shall see in the next section, but the implementation step has not quite yet reached full fruition.

Implementing a Reterritorialisation of the Space of Flows

Overcoming the obstacles to affordable broadband in London   The desired regulation of the broadband market by regional government in London to achieve lower prices has both economic development and social cohesion stakes, and is needed to avert 'potentially a major crisis for UK companies and for London' (Jenkins, 2001b, p.9). However, there are several interwoven restrictions currently in place making this difficult to promote. These include the lack of actual broadband services, the lack of a large market necessary for the development of these services, and the extremely high cost of broadband access[80] which is preventing the development of a large market (Jenkins, 2001b). As Colin Jenkins summarises: 'As there is no mass market for broadband as yet, businesses have little incentive to develop broadband services for the mass market or to offer affordable broadband access' (Jenkins, 2001b, p.7). The implications for London of a lack of intervention or regulation on the part of regional government are suggested to be quite stark:

> The high price of broadband slows the rollout and developments of large markets, for example. It also hinders the take-up of productivity-enhancing innovations by small and medium-sized enterprises, denying these companies access to leading-edge services, such as application service providers, reducing their competitiveness and slowing the regenerative effects of SME growth. In addition, high-priced broadband leads to excessive business clustering in areas with high-speed fibre connectivity. It risks creating an 'e-divide' between households that can afford broadband and households that cannot, which may re-enforce and accelerate other sources of inequality (Jenkins, 2001b, p.7).

While there may appear, on one level, to be a somewhat oversimplified cause and effect reading of the situation here (costly broadband leads to 'e-divide' leads to wider problems for London), the e-business advisor's suggested way forward takes into account the undoubted complexity of shaping a feasible regulatory response:

> The GLA fully understands the commercial obstacles to lowering the price of access. A rapid fall in price jeopardises existing investment in broadband's predecessors, such as ISDN (Integrated Services Digital Network) for example and a heavy capital commitment to an intermediate technology (ISDN or cable-modem) carries a commercial risk because these technologies will be by-passed in the short to medium term. The GLA should therefore foster a climate for technological advance in which risk is distributed in such a way as to ensure the most rapid and flexible development. This will involve appropriate partnerships of providers, users and government. The authority should attempt to identify and evaluate effective partnership models and disseminate best practice (Jenkins, 2001b, p.8).

It is not just regional government which is concerned about the development of broadband. Central government has created a broadband stakeholder group,

---

[80] Currently around £30 to £40 per month for households, but even more expensive for businesses (Jenkins, 2001b, p.7).

made up of representatives from all levels of government and business, to discuss and negotiate possible ways forward. Partnership between the public and private sectors is, then, evidently held to be a crucial factor, with the former being shown to have a more significant role to play than has perhaps been the case up to now. Indeed, on this latter point, the e-business advisor postulates about the eventual demand for even superior access speeds in the next decade: 'This will need considerable investment, either in putting fibre into BT's local access network, or in building substitute technologies' (Jenkins, 2001b, p.9). This raises the problem of obtaining the necessary investment costs in the current climate, which, in turn, delineates a role for the public sector: 'given the current non-availability of debt financing in the information and communications technology sector, some intervention by government will be required' (Jenkins, 2001b, p.9).

Local and regional authorities in Paris are widely considered to be only just getting round to thinking about how telecommunications and IT might benefit their territories (Jean-Philippe Walryck, AFOPT, interview with author, November 2000; Jean-Baptiste Hennequin, Conseil Régional d'Ile-de-France, interview with author, October 2000; Valerie Aillaud, CCIP, interview with author, March 2001). This seems to be largely the case for London authorities as well in their views and knowledge about the telecommunications and IT domain:

> The Government has set a target of 2005 for all services to be accessible electronically. This will have a huge impact on local government. There is some outstanding ICT expertise in some of the local and regional government authorities in London, but it is patchy and nowhere near high enough to meet the challenges. To overcome this, we have formed London Connects, an umbrella organisation of all local and regional government entities in London. The objective is to pool expertise and experience and create projects of suitable scale. The objective is to become a lead adopter of 'e' to match London's standing and to encourage and support development outside the central area (Colin Jenkins, GLA, personal communication, June 2001).

London should continue to benefit from having its own regional authority to deal with issues and participate in strategies in the telecommunications and IT domain, especially through:

> joined up thinking on planning and the strategic development of London; a focus and co-ordination role. For example, it is vital to ensure that power capabilities are included early in any development plan, particularly as data centres are high consumers of power. London Connects should be a driving force to ensure a granular e-government service role out (Colin Jenkins, GLA, personal communication, June 2001).

Packaging telecommunications developments through partnerships We have seen that there are a number of factors bound up in the interventions of the public sector in the telecommunications and IT domain. This is also the case with regard to private sector actors and how their interventions in this domain concord with the objectives of the public sector. This situation is by no means specific to the telecommunications and IT domain. It is the case with most areas of urban economic development practices. In these other areas, we have seen the creation of partnerships to solve and / or share some of the problems which come up when there is just one authority or organisation to intervene in a process, and this idea of partnerships has become quite common in telecommunications and IT projects and strategies as well.

For Colin Jenkins, the e-business advisor at the GLA, partnerships of differing groups and bodies (public and private; local and central government) are perhaps the key to success for the telecommunications and IT domain in London (as in London Connects): 'Although I am a market forces supporter, I cannot see how the 'market' can cover all elements unaided. There are also issues around the 'digital divide' and 'social exclusion'' (Colin Jenkins, GLA, personal communication, June 2001).

The Sipperec initiative in Ile-de-France which involves a partnership between the Sipperec public agency, local authorities and the private operator LdCâble has been held up as exemplary in this regard: 'it offers all the necessary guarantees for the development of competition and the concessionary mechanism put in place allows the authority to limit to the maximum the financial risk' (Jacques Douffiagues, ART; quoted in Moreau, 2001, author's translation).

Partnerships in the telecommunications and IT domain can also involve new actors who would not have participated in the domain before, or without the interaction with other agencies and organisations. As FEDIA suggests:

> During the last three years, the *gestionnaires* of infrastructures (roads, motorways, railways, waterways, tunnels....) have contributed greatly in France and in Europe to the emergence of new operators. They have allowed the development of a competitive offer on long distance services (national and international). [...] They are also key actors for the development of competition in the local loop. In effect, metropolitan fibre optic networks are deployed by new operators entering the market by using notably infrastructures developed by the actors of the urban world (metro, tramway, public highway, water supply...). [...] Infrastructure actors present in the large urban centres and on the main inter-urban axes have therefore greatly contributed to the development of a much stronger competitive dynamic (FEDIA, 1999, author's translation).

In the Paris region, one of the most important actors in telecommunications developments since the opening of the French telecommunications market has been the RATP, the public transport authority which runs the métro system. The RATP first got involved in telecommunications in the Téléport initiative. The relative failure of the Téléport Paris – Ile-de-France does not mean that it has had negligible influence on subsequent telecommunications developments in the region:

> Certain members of the Téléport have been able to create and develop an activity as they envisaged from the start of the Téléport. For example, the RATP has become an operator of telecommunications networks, and we weren't able to do the network. But the RATP now has had the right for a number of years to use its métro and RER network to be a telecommunications operator, to rent fibre to other operators, and so there is effectively a network... this was a terminology that we used at the start of the Téléport which has been taken up by the RATP to do... Probably then if there hadn't

been the Téléport, the RATP wouldn't necessarily have followed up what they wanted to do within it early on, even if they would certainly have constructed broadband networks... (Daniel Thépin, IAURIF, interview with author, September 2000).

The RATP created a subsidiary organisation at the start of 1997 called Telcité to work in the telecommunications domain. This company had three main objectives: to open up the RATP system to as many operators as possible through mutualising resources; to promote the development of telecommunications in Ile-de-France; and to make use of the heritage and experience of the RATP in this domain (Telcité website). Telcité constructed a homogenous fibre optic network for its clients running along metro lines in Paris and RER lines in the wider Ile-de-France region, which now covers 640 kilometres (Matheron, 2000, p.123). This network is fully interconnected at a regional (for example, to the local loop at La Défense) and extra-regional level. There are two elements to their strategy: other operators have the opportunity to rent dark fibre (this has been taken up by, amongst others, Cegetel, WorldCom and Bouygues); and Telcité also offers its own services based on an SDH network (ESIS website). The use of the Telcité network by a number of operators has surpassed expectations, and is held up as the prime example of the overwhelming needs of operators in Paris for dark fibre along which they can offer their services (Sipperec, 2000, p.34). As Jean-Raphael Hardy of Colt explains:

> The RATP represents therefore a partner and a competitor. We used them when we decided to extend our Paris intra-muros network to La Défense. We rented Telcité connections situated in the tunnels of line 1 of the métro and the RER A. Then, there, we followed this up by using the services of Fibre Optique Défense (Jean-Raphael Hardy, Colt France, quoted in Géomatique Expert, 2002, p.30, author's translation).

London Underground have taken a different and slightly less extensive approach to telecommunications networks than the RATP in Paris. Cable can be laid in the tunnels of the underground system, although the Circle Line has been equipped with dark fibre already, but also along their tramducts under the roads, which were traditionally used for tramway power supplies. For example, the deployment of a fibre network in central London by the MetroHoldings joint venture partly relied on the cables Energis had already laid in the underground system (ESIS website).

In both London and Paris, telecommunications is being seen as an area to be tackled by economic development agencies, which bring together a variety of regional actors from both the public and the private sectors. In Paris, the creation of a new regional agency for economic development (the ARD) in 2001 seems to respond to a feeling that Ile-de-France has too many actors in this domain, and that a single agency will make it easier for the region to promote its assets (such as telecommunications networks and services) and attract new companies to locate in the region (by acting as a one-stop shop which can respond to all the needs of these companies and investors). It will also act as the main point of contact with other European regions. One interesting point about this new agency is that it is using

London First as a model (Jean-Baptiste Hennequin, Conseil Régional d'Ile-de-France, interview with author, October 2000). Francis Chouat, adviser to the president of the Conseil Régional d'Ile-de-France on economic questions and international affairs, highlights the importance of partnerships when working in the telecommunications and IT domain:

> At the heart of the regional development agency, and in parallel with its main functions, it is important to create an actual strategic committee where we can regularly update progress, obstacles being faced, and demands, develop also an understanding between private and public powers of the constraints which we are each working with, and especially be able to produce, as we go along, analysis, decisions and assistance to territorial development based around new ICTs (IAURIF, 2000b, author's translation).

Concluding Remarks

In this chapter, we have seen how telecommunications developments in Paris and London are bound up in a number of territorial processes and practices at the urban regional level, in addition to those of the national level which we focused on in chapter 6. The overall 'packaging' of telecommunications infrastructures and services in these two global cities can be viewed as the outcome of the interaction between both common and discordant understandings, negotiations, procedures and operations of territoriality on the part of private operators and public agencies.

These territorialities of telecommunications developments in Paris and London were approached, highlighted and analysed from three varying parallel, but interrelated, angles, which were shown to be prominent in both cities, albeit with important differences. Firstly, the discursive and material embedding of telecommunications developments within broader mechanisms, policies and ideals of urban competitiveness was highlighted. Secondly, there was the varying nature, extent and importance of competitive market delivery in the two cities, and the regulatory, interventionist and strategic responses to this. Thirdly, following on from this, there was a further exploration of the differing roles and characteristics of territorial authorities in the development of telecommunications strategies, and their enrolling of various actors in partnerships as a means of producing more effective and successful regulation of, and alternatives to, this market delivery.

These analytical angles have drawn out crucial elements both for our comparison of telecommunications developments in Paris and London, and in relation to our theoretical discussions, in particular in chapter 3 regarding the territorialities of global cities and telecommunications. We need to consider all this in more detail now.

Firstly, the broad urban competitiveness context of telecommunications developments in Paris and London is important in both cases, but influences and is influenced by these developments in somewhat different ways in the two cities. Our discussions of global city theory and the financial centre status of these cities has proved to be highly relevant with regard to telecommunications developments in Paris and London, as the financial and business sectors dominate their telecommunications markets and ensure the highly competitive nature of these markets through the presence of many different operators. However, at the same time, we must note the more important position of London as a global financial centre compared to Paris, which derives largely from the historic tradition and finance culture of the former, particularly in the City of London. This key contextual difference between the two cities tends to be borne out in our analysis of the shaping and quality of the telecommunications infrastructures and services deployed there. Telecommunications developments in London are dominated by the City of London as a key global financial centre. The high quality and quantity of telecommunications required by the financial sector and other businesses there were a major factor in opening up the UK telecommunications market to competition. The financial markets of Paris do not quite have the global importance of those of London, although this does not necessarily mean there is less demand for high quality telecommunications networks, as we saw as evident from the fact that Paris ranks second behind London in surveys of European 'wired' cities.

Urban competitiveness objectives seem to provide a foundation for regional authorities in both Paris and London to attempt to configure their telecommunications infrastructures in relation to global networks, processes and dynamics. If telecommunications are no longer perceived as offering a competitive advantage in the European and global spatial economy per se, authorities in global cities must nonetheless ensure that their infrastructure and service offer still meets the requirements of potential multinational businesses and investors, and therefore compares favourably to competitor cities around the world. We saw how much of this urban competitiveness element is discursive in nature. It was quite difficult to pinpoint material practices which clearly constitute a territorial *mise en oeuvre* of competitiveness objectives highlighted by the likes of Colin Jenkins at the GLA and Jean-Baptiste Hennequin of the Conseil Régional d'Ile-de-France, although in the case of the former, this might be explained by the banality of some of the comments.

Following on from this, the place of telecommunications developments in the most recent regional government strategic plans or documents is comparable between the Paris and London regions. Both the Contrat de Plan État-Région for the former and the GLA's Economic Development Strategy for the latter contain only relatively limited discussions regarding the role of the ICT domain in regional development and competitiveness objectives. In the case of Paris, it was shown that the regional Chambers of Commerce had outlined a wider, more detailed role in the ICT domain in their contribution document to the preparation of the Contrat de Plan, but that this had evidently been diluted in the final plan, and the proposed investment much reduced. In the case of London, it was noted that Colin Jenkins of the GLA had stated that there was more of an 'e' focus to a subsequent draft of the Economic Development Strategy, but this remains to be seen. Thus, in both the Paris and London regions, there seems to be a great initial willingness to include and act on the potential advantages of telecommunications developments for competitiveness, yet, there appear to be substantial problems in translating this discursive element into material regional government action and investment. The potential costs of this action, together with the widespread feeling that territorial competition should dominate more than state or regional intervention and regulation, have led to this current state of affairs in both cities.

Nevertheless, this is not to oversimplify the situation in either city, nor to draw the conclusion that the territorial regulation of regional government intervention in telecommunications developments is largely comparable or similar in Paris and London. We saw how this is far from being the case. In Paris, we observed how regional intervention and action in telecommunications has had to be reconfigured in the light of a new competitive context. Whereas the Conseil Régional was previously primarily concerned with a regionally comprehensive strategy in the shape of the Téléport, such a strategy became increasingly contested and redundant as liberalisation approached – 'caught up by the activity of large operators'. They have subsequently refocused their strategy on territorially specific areas, and in particular, on the research communities of Ile-de-France, both to boost the status of the region as one of the principal research and development poles in the whole of Europe, and to promote the competitiveness of the region in general. In London, meanwhile, we observed a very different state of affairs in which the precursory liberalisation of telecommunications in the 1980s had very much engendered a limited and *laissez-faire* type of relation between territorial authorities and telecommunications developments, in which the market dominated completely. This is still the case in the current London context, although there is evidence of the beginnings of a reterritorialised approach to the ICT domain, as the emerging contexts and mechanisms of broadband and of the wider 'information society' start to focus the whole issue, and highlight particular territorial differentiations across the London region, which may have broader repercussions for regional competitiveness. In some ways, then, a comparison of the role of telecommunications developments in the competitiveness of Paris and London has brought out the wider contrasts between the two cities. On the one hand, regional authority involvement in telecommunications developments in Paris is shaped by a still relatively new context of liberalisation, and seems currently, therefore, to be characterised by an approach somewhere between relative *laissez-faire* and targetted territorial intervention – in other words, they want to let the market lead the way for the first time, except for territories or groups who are unlikely to benefit from competition. On the other hand, in the London region, the market has already lead the way for a long time, and whilst it is generally recognised to have served quite well the needs of London in the domain of economic competitiveness, regional authorities are, on one level, heading in the opposite direction to those in Paris, in reasserting a reconfigured territorial regulation / intervention role for themselves, in order to attempt to ensure the widespread roll-out of broadband across the London region, and benefits for all territories and groups within the 'information society'.

Relating back to the historical perspectives of chapter 5, telecommunications in Paris and London can also be positioned within an evolving social and symbolic rationale, which seems to be closely linked to technological and regulatory shifts over the years. Kaika and Swyngedouw (2000) have suggested that there was a substantial shift from celebrating urban technological networks in the beginning of modernity, which might be strongly linked to the emergence of the 'modern infrastructural ideal' in western nations (Graham and Marvin, 2001), to their subsequent underground burial during high-modernity, which was the era of nationally standardised PTTs in telecommunications. Our discussions of contemporary developments suggest that there has been something of a shift back towards a 'celebration' and a 'spectacle' (Debord, 1995) of telecommunications networks in their widest possible sense, which includes the material and discursive constructions of an 'information society', as well as the more specific ways in which mobile and broadband networks are now held as '*the* embodiment of progress' (Kaika and Swyngedouw, 2000, p.125) in many quarters. For example, broadband networks in Paris and London were seen to be in the process of being differentially rolled out according to the ways in which they had been discursively constructed by each region – as aiding regional competitiveness in Ile-de-France through promoting high quality research, and as bridging socio-territorial gaps in London through a dual focus on e-business and lowering costs of access for all. The powerful symbolic connotations of broadband for both territorial competitiveness and territorial cohesion should thus not be underestimated.

Secondly, the nature, extent, and importance of competitive market delivery of telecommunications infrastructures and services in Paris and London varies, and subsequently, there are differing regulatory, interventionist and strategic responses to these market specificities. This intensive deployment of infrastructure harks back in many ways to earlier times, as seen in chapter 5. If laying underwater cable between France and the UK in the 1850s was an example of Victorian hegemony, as Mattelart made out, then the global deployment of high speed fibre optic networks today can perhaps be seen as an example of capitalist or market hegemony. Their foundations though are firmly in the global infrastructure criss-crossing of the latter half of the nineteenth century, which similarly was undertaken by predominantly private capital, and in the case of the British network dominated by private operators. Little seems to have changed, then, in respect of the business areas of London, where we saw that, since early liberalisation, it is private operators constructing and running fibre optic networks throughout the City and the Docklands. In contrast, larger state control of networks in France in the late nineteenth century also finds a parallel with the larger state control of the telecommunications sector *per se* in the 1980s and first half of the 1990s, as partly seen in the Téléport, where regional authorities tried to configure an advantageous set of territorial infrastructures and services to counter the limitations of the monopoly of France Télécom and be able to compete with regions such as London.

In addition, we drew a comparison between the dominance of business users of the pneumatic tube systems of London and Paris and their similar dominance of metropolitan fibre networks today. It is hardly surprising then that the telecommunications markets of Paris and London are territorially focused on the Bourse and La Défense in the former, and the City and the Docklands in the latter. The favourable pricing systems developed for intensive users of the pneumatic tube networks have been replicated for intensive users of fibre networks (low trunk costs to the USA; internal dialling between London and Paris). We also mentioned the division in control of the London telephone system in the 1880s according to main and local lines, which was, for a time, an early competitive telecommunications market, characterised, as currently, by negotiations over co-location and the renting of underground space.

Nevertheless, we can note, in particular, the wide range of actors and agencies who have become involved in the process of the construction and packaging of telecommunications infrastructures and services in Paris and London. We can, in both cases, distinguish between telecommunications developments which are generally private initiatives, managed by operators concerned principally with profit and capital accumulation, and those which are more the result of partnerships between public and private actors, but such is the (continuing) importance of public territorial regulation in both cities that even in the former case, we cannot simply ignore or reduce the role that urban and regional authorities have in shaping all telecommunications strategies.

The deployment of many telecommunications infrastructures in Paris and London is intrinsically configured to global processes and dynamics because of the overall global strategies of many of the operators involved in deployment. We saw in detail how the territorial strategies of both incumbent and alternative new operators in Paris and London are increasingly part of wider more aterritorial global or pan-European strategies, aiming at maximum market presence and almost a homogenisation of infrastructure roll-out, that become the mainstay of subsequent 'global grids of glass' (Graham, 1999). Nevertheless, it has also become clear that territorially-based differences between operators are becoming crucial in determining which operators dominate which markets in what ways. We quoted the report of The Yankee Group Europe, which observed the very real possibility of a 'glut' of players in the European telecommunications domain: 'too many companies with too little to distinguish them and their business plans from each other' (The Yankee Group Europe, 2001, p.6). On one level, at least, these necessary distinctions between companies and company strategies are being made in terms of territoriality, as operators focus on particular (territorial and functional) markets in particular cities in particular countries, as we saw in the cases of Colt and MCI WorldCom.

As well as the interactions between territorial processes and specificities and *different* operators, we have also been looking to explore how the *same* operators deal with the development of networks in Paris and London *differently*, in response to the wider differences between the two cities in politics, regulation, local government and cultures of territoriality. Here, we looked particularly at the examples of two of the largest competitive operators in western Europe, Colt and MCI WorldCom, and tried to distinguish differences in their strategies for infrastructure and service deployment between the Paris and London regions, and the possible reasons behind these differences. This argument was more evident in the case of Colt. This operator started out in the City of London around 1993 in the liberalised markets of the UK, and has since steadily expanded across Europe, including within France and especially Paris. It was stressed, then, that their overall strategy was developed originally for the City of London, and that their experience there had influenced to some extent their more recent operations in Paris, for example, albeit with key differences in terms of regulatory contexts. One of these

territorial differences between the two regions, influencing deployment of
networks by all operators, including Colt, is in the nature of the interconnection of
telecommunications networks, whereby in London and the UK, an operator can
link its network to that of BT at one point in order to obtain full interconnection,
while in Paris and France, an operator needs to be present in each of the regional /
national connection points for interconnection to the network of France Télécom.
The case of MCI WorldCom also illustrated some of the points regarding the
construction of differing territorial strategies between Paris and London, but at the
same time, there was more of an initial emphasis on highlighting the
'homogeneity' of their networks and services in the large urban regions of Europe.
Part of the difference in this regard between Colt and MCI WorldCom would
appear to result from Colt having started out in one territory in Europe, while MCI
WorldCom is an American entrant into Europe, which has evidently regarded
presence in Paris and London as equally important from the start. We were able,
nevertheless, to begin to go into some of the detail of the strategies of MCI
WorldCom to draw out some of their territorial differences between Paris and
London, but it would have required greater depth to carry this out fully, which
would necessitate more detailed information from the operator, which they are
understandably reluctant or indeed completely unwilling to provide for commercial
reasons.

We also illustrated some of the differing territorial influences and specificities
between the Paris and London regions by analysing the strategies of the respective
incumbent operators in the two markets, France Télécom and BT. This can also be
viewed as an exploration of the ways in which similar operators (incumbents) have
developed their infrastructures and services in Paris and London respectively
differently, in response to the wider variations between the two cities in politics,
regulation, local government and cultures of territoriality. In the case of these
incumbent operators, it has particularly been national regulatory policy and
practice which has greatly influenced the development of their territorial strategies
in Paris and London, and therefore illustrated differences in these strategies
between the two cities. The varying territorial concentrations of the fibre optic
networks of France Télécom in Ile-de-France and BT in London demonstrate this
situation well, as it was suggested that the lengthier period of monopoly in France
permitted the former to develop a more comprehensive regional fibre network in
the public interest, and in keeping with the French concern for *aménagement du
territoire*, whereas BT was subjected earlier to a competitive context and had to
obey certain market and financial restrictions and necessities above any widespread
territorial deployment considerations.

Nevertheless, we saw how both France Télécom and BT have developed
important services specifically for the Paris and London regions respectively. For
the former, this includes services aimed at a variety of clients from Disneyland and
the Stade de France to the local authorities of the region. For the latter, the business
zone of central London and the City appears to concentrate much of the interest of
BT in this regard, as in its controversial low tariff Framestream product which
targetted companies in this particular territorial market. The apparently closer links
between France Télécom and local authorities than those of BT, discussed already

in the previous chapter when we highlighted how the French public operator had created the OTV, relate evidently to its lengthy public service mandate up to the late 1990s, but also to the nature of French intergovernmental politics within which local *élus* were, and generally still are, less inclined to 'rock the boat' by bypassing the national operator, particularly if they hold themselves a national political post as well as a local one.

Our explorations of the strategies of these different operators in the Paris and London regions has stressed, then, the absolutely crucial territorial basis and implications to these varying strategies. Overall, then, we can agree with the observation of Stephen Graham, when he writes that 'it is paradoxical, however, that an industry which endlessly proclaims the 'death of distance' actually remains driven by the old-fashioned geographical imperative of using networks to drive physical market access' (Graham, 1999, p.937). We can extend this further and suggest that this network deployment driving access to markets is itself driven by a series of parallel territorial contexts and specificities which ensure that this deployment consistently varies in nature and form in differing global cities such as Paris and London.

Thirdly, our attempt at exploring the differing roles and characteristics of territorial authorities in the development of telecommunications strategies, and their enrolling of various actors in partnerships as a means of producing more effective and successful regulation of, and alternatives to, market developments also illustrated some of the key differences between the territorialities of telecommunications developments in Paris and London.

Following Castells, we viewed the actions and practices of territorial authorities in telecommunications strategies as attempts at configuring a territoriality of 'the space of flows'. In order to do this, they must configure and control its constitutive layers of physical supports – circuits of electronic impulses such as telecommunications; a network of place-based nodes and hubs; and the spatial manifestations of dominant groups and interests. This is a difficult task because each of the three layers seems to intrinsically interrelate to the others, insofar as each appears to be at once a presupposition, a medium, and an outcome of the other two. Nevertheless, territorialisation of the space of flows can be viewed, at least initially, as a parallel and interrelated process of positioning a specific place within or in relation to the more global 'network of nodes and hubs', and shaping 'the spatial manifestations of dominant groups and interests', in order to construct fixed telecommunications infrastructures. From this, however, develops a reverse mechanism, whereby these territorially-based infrastructures thus reinforce the dominance of certain groups, which, in turn, may reinforce or at least transform somewhat the position of that place in relation to global networks.

We saw examples or elements of this territorialisation of the space of flows in both Paris and London, but with important differences between them, as inevitably, in the light of the above, each point in the parallel and interrelated process depends on and in turn influences territorial contexts, specificities and practices. The LondonNet ISB project was a proposal for a regional ICT infrastructure, which was dependent on financial support from a dominant group, central government, but which would have in turn become a 'spatial manifestation' of the London boroughs and other regional agencies, and better positioned both the latter as public service providers and the London region as a whole in terms of coordinated governance. The participation of the Conseil Régional d'Ile-de-France in the construction of broadband infrastructures for research groups can be viewed in a similar way. The aim was to position the Ile-de-France region as a centre for research and development, through shaping and bringing together the mutual requirements of both the Conseil Régional (regional competitiveness) and the research groups (financial support for telecommunications infrastructures), in order to construct the broadband networks. In contrast, however, we saw how the territorialisation of the space of flows in relation to broadband deployment in the London region had different objectives, different 'spatial manifestations of dominant groups', and difficult obstacles to overcome.

The objectives of public agencies and authorities in the telecommunications domain are more varied than those of private operators. We saw how telecommunications developments are being viewed as an important part of wider urban competitiveness needs and objectives. In addition, we can note their role in an *aménagement du territoire* capacity, not just in Paris but also in the London region. The London Connects strategy, for example, is attempting to formulate a strategic and territorially concordant approach and response to local and regional information society requirements.

Nevertheless, one of the overwhelming impressions left by a comparison of telecommunications developments in Paris and London is how relatively more defined the role of public sector actors in this domain seems to be in Paris compared to London. London may have the most highly regarded telecommunications infrastructure in Europe (see Healey and Baker survey; Finnie, 1998), and many parts of the city (notably the key business areas of the City of London and the Docklands) may be served by the numerous networks and services of every main operator under the sun, but the fact remains that, not only are other parts of London underserved by telecommunications networks, as we will see in the next chapter, but there is a noticeable lack of actors in the public sector, both at the regional and the local level, who deal with the telecommunications sector, and those whose remit would seem to include telecommunications do not devote much attention to it. The deployment and development of telecommunications infrastructures and services in London is virtually entirely left to market forces and to the private sector. This situation may seem to be a natural follow-on from the Thatcher-led market liberalisation policy of the 1980s, but it still comes as something of a surprise to find so few people in London, one of the biggest telecommunications markets in the world, discussing them, never mind actually intervening in the sector. It is undoubtedly taken for granted that the market will deliver every aspect of the telecommunications infrastructures and services that even the most intensive multinational business user could require. To compare telecommunications in London with telecommunications in Paris, then, highlights both the relative lack of a territorial perspective in the former case, and, perhaps associated closely with this, more trust that the market will deliver the necessary infrastructures and services.

Our exploration in this chapter of the relations between telecommunications developments in Paris and London and particular urban regional contexts and specificities has made up the second layer of our exploration of the parallel and incremental multiply scaled territorialities of these developments. However, as we underlined at the end of chapter 6, these contexts and specificities are not bounded, independent and unilateral influences on these developments, as they cannot and should not be considered independently of the other layers of territorialities that we looked at in the previous chapter and will investigate in the next chapter. In the following chapter, therefore, we build on the discussions and analysis of these last two empirical chapters, and again move the focus inwards to explore the relationships between these telecommunications developments and processes, practices and actions present or constructed at the local or intra-urban level. This focus is suggested to be incrementally made up of explorations of how local governance has shaped particular developments or the regulation of particular local markets, how intensive corridors of fibre are developing between new and old networked urban spaces, and how the construction and shaping of telecommunications developments is bound up with the parallel construction of disconnected spaces and a subsequent lack of socio-territorial cohesion in the cities as a whole.

Chapter 8

Constructing New Networked Spaces and the Territorial Cohesion Imperative in Global Cities

This society eliminates geographical distance only to reap distance internally in the form of spectacular separation (Debord, 1995, p.120).

Introduction

In the previous chapter, the focus was on how telecommunications developments in Paris and London related to territorial processes, practices and contexts at the urban regional level. In the same way as that analysis had shifted the scalar focus from the national level of the first strand to the regional level of the second strand, in this chapter we again zoom in from the regional level to the local or intra-urban spaces that are the focus of the third strand of the research. Recalling our research questions, the aim here must be threefold:

- To investigate how processes and practices of local governance, and their implications, are influencing or being territorialised in telecommunications developments in Paris and London.
- To illustrate how different types of new networked space are being constructed in Paris and London, and connected to juxtaposed old networked spaces by corridors of fibre.
- To explore whether the parallel importance of new and old networked urban spaces in Paris and London reinforces wider, previously existing problems and tensions of socio-territorial cohesion.

Given our focus in the third research strand on local governance processes and practices and territorial cohesion, we can view the second of these aims as a kind of link between the two. Local governance practices have been bound up in the shaping and packaging of particular local or intra-urban spaces as privileged locations for economic development activity, but this 'spatial selectivity' has wider territorial cohesion implications for Paris and London. Equally, following on from the previous chapter, the discussion and analysis in this chapter, therefore, broadly fit in with the second part of what has been identified as the 'dual competitiveness challenge' for the London region, in relation to telecommunications and economic development, but which we can extend to the Paris region as well, namely 'to improve the competitiveness of local and sub-regional economies *within* the Region, where cohesion and growth are both crucial to sustainable development' (Local Futures Group, 1999b, p.14).

Local Governance, Economic Development and the Territorial Regulation of Telecommunications Markets in Paris and London

We discussed in the previous chapter the ways in which, and the extent to which, the intervention of regional government in telecommunications developments in Paris and London has changed over the years, largely due to the steadily modifying nature of the market and the role of the state. In the same way, we also need to investigate the changing roles of local authorities, communes and boroughs in telecommunications developments. One of the main questions we need to explore is the extent to which the liberalisation of telecommunications has led to the 'liberation' of communes or boroughs in Paris and London to territorially regulate their local market themselves. In this respect, we can look at evidence from La Défense, the Sipperec initiative, and local boroughs in London such as Lewisham and Camden.

One of the clearest and most revealing examples of the complexities involved in the territorial regulation of local telecommunications markets in either Paris or London has been the negotiations and tensions characterising the construction of telecommunications infrastructures and services at La Défense, the main business area to the west of Paris. Here we draw out and analyse in detail the salient points of the story, as an illustration of the territorial regulation of a new networked urban space.

Shaping the Telecommunications Market of a New Networked Space: Local Conflict at La Défense

As we saw in chapter 5, the French state made the development of the La Défense area, to the west of the centre of Paris, from the 1950s onwards a prestigious *grand projet*. It has since become the French capital's primary business zone, extending to 160 hectares, with more than 2 million square metres of office space, and 3600 companies, including 13 of the top 50 companies worldwide (EPAD documents).

The local loop telecommunications network became operational at La Défense in 1997. Before this, France Télécom, in keeping with its monopoly on the French market, was the only operator with telecommunications infrastructure in the zone. The public planning agency for the area, EPAD, in consultation with the Conseil Régional and the Téléport Paris – Ile-de-France, had recognised that quality and transmission costs of telecommunications were becoming a major location factor for firms and was anticipating the deregulation of the market in 1997-98, so they decided in 1995 "to promote the construction of a high tech fibre optic network adapted to the dimensions of the main business area in Europe" (EPAD website, author's translation). EPAD's specific objectives were:

- To promote the development of competition within its policy of user services
- To satisfy the management and security needs of the technical galleries and service ways within its responsibility

In October 1995, EPAD invited bids from operators for a 10 year optical connection concession for the La Défense zone:

> Because we wanted to connect all the office blocks in an equal manner – all 80 of them – and not just according to where the best clients are located, the only constraint on the eventual provider was that they would have to have a 'site' in each building (Joel Renvoisant, EPAD, interview with author, September 2000).

Half a dozen operators responded, including France Télécom, Bouygues, Colt, MFS, and La Générale des Eaux-Unisource (Réseaux & Télécoms, 11 March 1997). The bid of the latter was accepted on 3 September 1996, although La Générale des Eaux had by then split with Unisource and held onto the latter's share in the subsidiary group Fibres Optiques Défense (FOD). FOD was to build and operate the network on the ten year contract, connect the majority of the 80 office blocks in the area, and make it open to all firms and telecommunications operators (EPAD website). The total cost of 30 million francs was entirely met by FOD,[81] apart from a subsidy of 8 million francs from the Conseil Régional d'Ile-de-France (Réseaux & Télécoms, 11 March 1997). Cohesion was the main reason for developing a single telecommunications infrastructure for the whole area (EPAD website; Joel Renvoisant, EPAD, interview with author, September 2000).

From January 1997, then, it was possible to rent dark fibre on the new local loop. Connection was offered between buildings in the zone via the LTI (*locaux techniques d'immeubles*, or building technical premises), and externally to the regional networks of France Télécom, the SNCF (Société Nationale des Chemins de Fer – the national railway system) and the RATP (Régie Autonome des Transports Parisiens – the autonomous Paris public transport agency) via the LTR (*locaux techniques de réseau*, or network technical premises) (FOD brochure; Joel Renvoisant, EPAD, interview with author, September 2000).

The total annual telecommunications consumption of the 3600 companies at La Défense has been estimated at around 1 billion francs, or 15% of the total consumption of the Ile-de-France region (Réseaux & Télécoms, 11 March 1997). The zone is promoted as 'the most important concentration of telecommunications infrastructure and services in Europe' (FOD brochure, author's translation). Growth in office space in the area was forecast to be 20% over two years, but the local loop was constructed to be able to cope with such an increase in demand (EPAD website).

---

[81] Theoretically, FOD could not really be classed as a telecommunications operator, because its activities were 'passive', insofar as it was only renting dark fibre to actual operators rather than running its own services over the network (Joel Renvoisant, EPAD, interview with author, September 2000).

However, there has been a recent series of regulatory and competitive obstacles to the development of telecommunications services at La Défense, which illustrate particularly well some of the ways in which shaping local telecommunications markets in the Paris region in a newly liberalised context have been fraught with difficulties and have been undertaken through a series of negotiations and tensions between an increasing number of institutions and groups, all looking for a degree of territorial power through the deployment of telecommunications infrastructures and services.

It was arguably the parallel developments of the continuing expansion of the business zone and the liberalisation of the French telecommunications market, which lay behind the difficulties encountered at La Défense. Prior to liberalisation, the slightly smaller number of companies located in the zone relied on the monopoly supplier France Télécom to provide all their telecommunications needs. However, with the prospect of upcoming liberalisation, the Conseil Régional and the local planning body, EPAD, wanted to reconfigure the telecommunications provision of the zone. As Jean-Baptiste Hennequin of the Conseil Régional outlines:

> In 1996 we financed the deployment of a *shared* optical infrastructure, not for the whole of La Défense but a major part of it, and we gave FF8 million to EPAD for putting in cables and creating underground galleries for the telecoms cables. This was just before the 1998 law on telecoms deregulation. The Region participated so that this infrastructure would be put at the disposition of all active operators in the market. What happened? EPAD took the money from the Region and designated a concessionary – they chose a manager for the networks. A few months later, this concessionary passed under the control of a telecoms operator, Cegetel, and, what can we do – the law cannot prevent this type of thing (Jean-Baptiste Hennequin, Conseil Régional d'Ile-de-France, interview with author, October 2000).

This proved to be to the detriment of the development of competition in this local market, which had been the original aim of the Conseil Régional and EPAD:

> So, four months after the start of this promising initiative, we find ourselves with a concessionary under the control of an operator which is applying prohibitive prices. Colt is obliged to rent ducts for fibre from FOD at a price which makes it lose money, but Colt is obliged to lose money in order to be present at La Défense near certain clients. So, we have an abnormal situation about which something has to be done (Jean-Baptiste Hennequin, Conseil Régional d'Ile-de-France, interview with author, October 2000).

Having been in conflict with EPAD since 1997, France Télécom and Colt brought the matter to the administrative tribunal of Paris the following year as they felt it unfair that operators other than FOD were not allowed to construct and run their own networks in what is, after all, the largest (and most profitable) business zone in France. Colt, in particular, had clients in nearly half the buildings at La Défense, and was therefore forced to rent dark fibre from FOD in order to meet the needs of its clients, rather than being able to deploy its own network. On 19

January 2000, then, the tribunal ruled against the decision of EPAD to award an exclusive concession for the La Défense local loop to Fibres Optiques Défense, a subsidiary of Cegetel. There was then a cassation, which left the situation far from fully resolved (Jean-Baptiste Hennequin, Conseil Régional d'Ile-de-France, interview with author, October 2000).

According to Marc Bourgeois of WorldCom: 'It was Colt's turn to attack. 9 Télécom had attacked on Internet access. WorldCom for something else...' This then demonstrates the good relations between competing operators (through associations like AFOPT), especially when all feel that certain markets or territorial zones are not fully embracing the newly liberalised telecommunications environment (Marc Bourgeois, MCI WorldCom, interview with author, November 2000).

It was suggested that EPAD did not like the idea of having technicians who were not working for them in their galleries (Emmanuel Tricaud, Colt, interview with author, November 2000). In the case of France Télécom, it was a question of them being allowed to extend their previously existing network from when they were the incumbent operator. The tribunal ruled that if France Télécom already had a network at La Défense, then there was no reason for Colt to be prevented from building their own too. Rather ironically, France Télécom suggested that the judgement was fair 'on the grounds of competition' (Réseaux & Télécoms, 24 January 2000). For Irène Le Roch of France Télécom, 'it was a classic case of the *arroseur arrosé*. Competitive operators complain about France Télécom, but when we can prove that they are also beyond the yellow line, we'll remind them, because they remind us' (Irène Le Roch, France Télécom, interview with author, September 2000). Marc Bourgeois of WorldCom reckoned that 'we should avoid such a model in the future' (Marc Bourgeois, MCI WorldCom, interview with author, November 2000). The concession awarded to FOD in 1996 was annulled by the tribunal starting from four months later. France Télécom and Colt were trying to agree a way forward with EPAD, while EPAD and FOD were also deciding on their next moves. The latter may have the right to compensation given that its exclusive concession has been withdrawn (Les Echos, 2000a; La Tribune, 2000a). Joel Renvoisant of the Direction de l'Exploitation of EPAD did not know what was going to happen following the ruling. He thought that EPAD and FOD would probably have to try and work out some kind of arrangement with the other operators at La Défense. For him, it would be difficult to blame EPAD for the problems: 'We're only the people on the ground. We don't know much about this type of thing' (Joel Renvoisant, EPAD, interview with author, September 2000). EPAD also counter that France Télécom initially ignored the concession:

> France Télécom had its own infrastructures at La Défense. It still has them. France Télécom did not come immediately to EPAD following the allocation of the concession to FOD. They had their own arrangements for being able to continue working for a few months nonetheless. Then France Télécom came to EPAD to ask if they could develop their own cable network. EPAD did not cut all France Télécom's cables after the agreement of the concessionary contract with FOD (Joel Renvoisant, EPAD, interview with author, September 2000).

Despite the tribunal decision, and the apparent prospect of the deployment of fibre by many other operators at La Défense, this was still a situation which was very much up in the air as to its resolution, according to Jean-Baptiste Hennequin of the Conseil Régional:

> The annulment of the concession won't solve the problem in the media. What the Region did, having seen what happened, was done under the old executive of M. Giraud from the right. We, the new executive, discovered this affair and we're trying to understand, and three months ago, a consultant working for Colt came to see us to say 'something has to be done, the Minister has to get involved'. We agreed to do something and we sent four letters signed by M. Huchon to Laurent Fabius, to Jospin's adviser, to Jean-Claude Guessot, and we asked the Minister to do something to resolve the situation once and for all. A judiciary decision doesn't resolve things, so we have to exercise 'médiatisation'... We wanted to show EPAD that we didn't like what happened and that we wanted to resolve the matter. This is typical of the kind of thing that we, the Region, wish to do... We haven't been contacted again by the consultant, so usually that means that the matter is being resolved... The result is that the cost of telecoms at La Défense is abhorent and abnormal. EPAD reckons that it isn't responsible, and one of its arguments is that there is not enough space in the underground galleries to put all the cables of operators. There doesn't need to be 15 or 20, as 5 or 6 would suffice... (Jean-Baptiste Hennequin, Conseil Régional d'Ile-de-France, interview with author, October 2000).

This was a complex situation inherently revolving around configurations and negotiations of territoriality. All this involved the competitive market that the Conseil Régional hoped would develop – locally at La Défense, and then regionally, as a result of the economic importance of the zone, the local territorial cohesion desired by EPAD, the profitable local concession taken up by FOD, and the complaints of Colt and France Télécom, amongst others, who were prevented both actual territorial access to key clients as well as a local 'fixing' of their regional – national – global networks, within a context of changing national regulation and legislation.

Telecommunications Developments and Intercommunality: The Sipperec Initiative

An example of a telecommunications strategy in Paris – Ile-de-France which is based on the regulation of the regional market, but which has key implications for local telecommunications markets is the Sipperec project. The joint regional – local dynamic is a good indication of the increasingly multiscalar nature of telecommunications developments in global cities. The basis of the project, as we have already seen, is an intercommunal partnership which tries to work against the weight of France Télécom.

Since the start of competition in the telecommunications market, Sipperec (which only developed an interest in the field in 1997) has been able to act as a counterbalance against the strength of France Télécom. The basis for their interest has been the seemingly continuing refusal of France Télécom to offer fair prices to local communes, which has led many of the latter to join the Sipperec group to seek more competitive tariffs from other operators: 'If France Télécom offered reasonable tariffs, there would be no problems for local communes, but this just isn't the case' (Gwenaelle Lalaux, Sipperec, interview with author, June 2000).

Intercommunality can be seen as an example of territorial cohesion. In the domain of telecommunications developments, and for broadband infrastructure in particular, this notion offers 'territorial consistency, greater simplification and visibility' and 'makes developments happen much quicker which is an important consideration for *élus*' (Jean-Philippe Walryck, AFOPT, interview with author, November 2000; see also Bourdier, 2000). Local and regional politicians are becoming more and more interested in and concerned with the deployment and management of broadband networks: 'They see in it an important element of *aménagement du territoire* to combat the 'digital divide'' (e-territoires, 2000, author's translation). The often prohibitive cost of this deployment for a single commune, and the often insufficient size of a single commune for attracting companies with their broadband infrastructure makes the development of intercommunal projects in this domain seem advantageous to all involved. As a memorandum by Sipperec and three other territorial authorities argues:

> Intercommunality and telecommunications networks are indissociable as much in terms of geographical pertinence as economic stability. The current debates on the theme of intercommunality must deal explicitly with the question of telecoms infrastructures as operating modes of intercommunality. Equally, we ask that this notion is taken into account in both *aménagement du territoire* policies and telecommunications development policies. The shortage of alternative resources in the local loop has numerous effects on territorial organisation. A genuine telecommunications strategy is imperative, integrating intercommunality, competitive markets, dark fibre networks and the effects of unbundling (Sipperec et al, 2000, p.19, author's translation).

Telecommunications Developments and Local Governance in London

Colin Jenkins of the GLA did not think a project like the Sipperec strategy, a regional infrastructure project connecting local communes which aims to equip these areas with a broadband network at a much lower cost than the offer from the incumbent, would be necessary in the London region:

> The London problem is last mile access rather than regional access. This is more applicable to the other regions. Where affordable access is not available some local London authorities, schools etc are experimenting with free space lasers. For example, the high rise council flats in Newham are now linked with this technology. The buildings are also CAT5 cabled. The local authority has run into a few problems. For example:
> there is insufficient broadband content available
> the authority is now expanding to areas where residents have brought their council property. To serve these properties, the authority have to obtain a Public Telecoms Operator licence
> We are looking at ways in which central and regional government can work with industry to step up the pace of broadband local access (Colin Jenkins, GLA, personal communication, June 2001).

The development of broadband is likely to intensify the networked corridor between the two most networked urban spaces of London, the City of London and the Docklands, although there are still important issues that will need to be resolved:

> Given that some of the development is likely to be in East London, initially between the City and Canary Wharf, [the approach to configuring business and residential accomodation for telecommunications, as discussed in chapter 7] is expected to encourage the extension of the current competitive fibre networks in these areas into the new developments. However, this can only be achieved if the cost of using the ducts laid in these areas is economical compared with the alternative of additional duct provision. Therefore, if any charge is levied for the use of ducts, it should be minimal, in keeping with the objective of supporting the development of e-London (Jenkins, 2001b, p.12).

This illustrates the importance of the territorial specificity both of duct provision, and of the control and pricing of duct use (who owns the ducts, whether they let an operator use them, and for how much), which is likely to influence the nature of telecommunications developments in key networked areas. In less networked areas, the role of government in promoting telecommunications developments becomes more important, but, as Jenkins makes clear, the way forward in these areas remains stuck at the strategic negotiation stage, rather than the stage of negotiating the use of actual ducts:

> Although this approach will address the development of the expanding central areas, there will still be areas of deficit where competitive networks do not exist or exist only in part. These areas need to be identified, and the local authorities, the construction and ICT industries brought together to consider local solutions. This approach should apply equally to commercial and residential property (Jenkins, 2001b, p.12).

Perhaps the closest, or most comparable, strategy in London in recent years to the intercommunal Sipperec initiative in Paris was the London Boroughs Information Society Network (LBISN), which was originally set up in 1994 (albeit without the name) and coordinated by the former London Research Centre (Bax, 1999). The coordination role of the LRC was initially outlined as a possibility in a report for the Centre and BT in 1991 (Yeomans, 1991). This report noted the important role for local authorities in London 'in regulating the development and use of telecommunications services' within the increasingly local focus of competition in the sector, and the potential for a strategic cross-borough approach: 'The development of telecommunications has, like other infrastructures, trans-borough implications: solutions may be similar from borough to borough, economies of scale achieved through co-operation' (Yeomans, 1991). The eventual setting-up of the LBISN a few years later can be seen as a kind of precursor to the more recent London Connects strategy, and as such, did not have the same infrastructure focus as Sipperec does. The LBISN was involved in promoting:

information exchange about emerging technologies, best practice and innovation; coordination of borough information management initiatives; development of innovative use of new forms of electronic communication for the London boroughs; information sharing and support on joint projects or bids for telematics funding; and organisation of regular meetings with presentations and demonstrations (Doulton, Harvey and Wilson, 1997).

Whereas Sipperec has been concerned with wiring communes in the inner *couronne* of the Paris region to broadband infrastructure as a more cost-effective alternative to the network of France Télécom, the LBISN was initially concerned with developing e-mail use in local government in London (Powell, Page and Bax, 1997), and then more broadly with borough websites, intranets and information access (Bax, 1999). Nevertheless, it can still be seen as having stressed the importance of partnership and shared objectives as a form of local territorial regulation in the ICT domain.

Four of the London boroughs were most involved in ICT projects – Camden, Islington, Lewisham and Newham. All four had EU-funded telematics projects which also included partners in Europe. Camden had an information database on the Internet aimed at disadvantaged groups; Islington was developing multimedia skills to further people's employment chances; Lewisham had a videophone system allowing document scanning; and Newham was working with the Metropolitan Police on kiosks for public information and police purposes respectively. These projects could be said to be illustrative of the rather 'simplistic' nature of local authority intervention in the ICT domain a few years ago, but were nevertheless evidently fairly successful in building networks of participants from various groups and organisations: 'The questions for the boroughs was whether and how far the authority could or should play a part in facilitating such a network and where economies of scale could be found' (Doulton, Harvey and Wilson, 1997). So at the workshop at which these four projects were presented in February 1997, the issue of the balance between public sector intervention and market-led development for telematics in London was already being posed:

> It was commented that it was critical for the public sector to pro-actively develop the telematics infrastructure London requires to support burgeoning multimedia and group-working developments. They cannot afford for the infrastructure development to be driven purely by commercial imperatives. Suppliers have vested interests, including the major capital they have tied up in older technology their 'cash cow'. They have a poor track record in investing in areas of deprivation which are likely to be most in need of generation, but least commercially attractive. It is in suppliers' interests to 'divide and conquer' public sector organisations; public bodies are much more powerful working together (Doulton, Harvey and Wilson, 1997).

One of the other major points to emerge during the workshop discussions was a common feeling of a lack of a London wide focus to the ICT domain, and, implicated in this, a lack of a specific coordinating London authority. The creation of the GLA and the subsequent development of the London Connects initiative seem to respond to these earlier worries about the limitations of what the boroughs were doing before. Nevertheless, London Connects has been primarily concerned with the development of e-government and the electronic delivery of services by local authorities. The configuring of local telecommunications markets to increase infrastructure and service provision, competition and choice, as with Sipperec, seems to be only a peripheral consideration in the initiative as a whole. Leaving the deployment and development of telecommunications infrastructures and services to the market in many boroughs of London far from guarantees satisfactory provision for their businesses and citizens.

This would all appear to be tied in with the lack of agencies and organisations in the ICT domain at a local or regional level. Even the London Docklands area, one of the most networked spaces in the capital, seems to be characterised by a distinct lack of territorial actors dealing with the telecommunications sector. Previously, this would have fallen under the remit of the London Docklands Development Corporation, but since the winding up of the latter, its agendas and areas of interest seem to have been passed on to various agencies. Docklands East London is now the main cross-borough economic development body, but IT and telecommunications are not really one of their main concerns (Gill Marshall, DEAL, interview with author, October 2000). The Docklands seems to exhibit exactly the way in which telecommunications developments in London are being completely left to the market.

The Territorial Regulation of IT and Telecommunications: Configuring E-Advantages in Lewisham and Camden

The borough of Lewisham was originally part of the South East London Consortium and its Telematics Forum, with neighbours Southwark and Greenwich. The Council's more recent ICT strategy focuses more on local plans within a set of parallel regional, national and global contexts – it is 'a wider strategy for Lewisham as a place' (London Borough of Lewisham, 2000, p.2). This strategy recognises that:

> The Council has a pivotal role in making sure that Lewisham is 'where it's @'. Lewisham's citizens need to be connected to new media and expert in its use; local businesses need the infra-structure and skills to exploit these developing opportunities; and public agencies locally (including the Council) need to grasp new technologies to transform the quality of services they provide to citizens locally (London Borough of Lewisham, 2000, p.2).

This ICT strategy was initiated by the chief executive of the borough, Barry Quirk, who has 'always been alert to this kind of development, possibly because he has a PhD in geography. In particular, he developed a 'digital divide' / social exclusion point of view' (Steve Pennant, Lewisham, interview with author, January 2001). Indeed, he was an advisor on the Framework programme to the European Commission (Antoinette Moussalli, Lewisham, interview with author, January 2001).

This broad strategy builds on the high level of interest within the Council over the last few years in telematics projects. Lewisham developed its own website quite early on, and this helped to encourage residents to participate in and respond to developments in the borough (Antoinette Moussalli, Lewisham, interview with author, January 2001; Steve Pennant, Lewisham, interview with author, January 2001). One of the main reasons for Lewisham becoming so involved in IT projects was the taking advantage of the European context and EU schemes in particular. The borough found it difficult to get money from Europe as it did not have Objective area status, and was restricted from joining Eurocities because of just being a borough of London rather than a city in itself. Therefore, they came to hear about the Telecities group in the EU and decided to join that (Antoinette Moussalli, Lewisham, interview with author, January 2001). Although part of the European department of Lewisham council, Antoinette Moussalli spends more time on IT developments than economic development as a whole, because the funding is there to support the former, more so than the latter (Steve Pennant, Lewisham, interview with author, January 2001). With EU funding, they developed a project called DALI (Delivery and Access to Local Information and Services), within which an innovative and highly successful video-conferencing initiative, TellyTalk, became the key element. This was at a time when Lewisham did not have any real technological infrastructure (Antoinette Moussalli, Lewisham, interview with author, January 2001). TellyTalk has offered citizens living in peripheral parts of Lewisham the opportunity to have access to council staff via a communications link set up in a local office, so that they can discuss matters relating to tax and benefits and get general advice without having to travel to the council's central offices. Further EU funding has been made available for the GALA (Global Access to Local Administrations) project, which has developed the TellyTalk initiative again, as well as created an extranet for Lewisham, Southwark and Greenwich (Antoinette Moussalli, Lewisham, interview with author, January 2001; Steve Pennant, Lewisham, interview with author, January 2001). On the whole though, Antoinette Moussalli reflected back on how 'it was a bad move to introduce IT services piecemeal. They become too many different things for different people' (Antoinette Moussalli, Lewisham, interview with author, January 2001). One of the main features of the ICT strategy of Lewisham is the way in which the telecommunications infrastructure of the borough has been completely dominated by BT. The nodes of the infrastructure are based around BT exchanges, and BT is the only operator with fibre across the whole of the borough (Steve Pennant, Lewisham, interview with author, January 2001). The network of NTL, for example, only covers half of Lewisham. Offices and schools do have access to 2 Mb HDSL connections, but 'the borough has still very much been at the mercy of the monopoly' (Steve Pennant, Lewisham, interview with author, January 2001). This is linked in though to the status of the borough as more of a commuting centre than a place of work or business centre. As Steve Pennant summarises then:

> in Lewisham, BT has been more for residents than for businesses. In any case, there is little benefit to the borough as a whole from Citibank's use of telecommunications, or from a Cable & Wireless node here. Basically, there's been no investment in ICTs in

Lewisham apart from BT, Citibank, and some small businesses. With regard to the council, even there, the actual money spent on regenerating Lewisham through IT has been very small (Steve Pennant, Lewisham, interview with author, January 2001).

The borough of Camden created a website called Camden Connects in 1996. This is run separately from a team working on e-government issues which was set up in 2000. IT and information society developments in Camden 'were partly a response to the government agenda in this domain', which acted like 'a political kick for the boroughs of London' (Andrew Stephens and Alasdair Mangham, Camden, interview with author, January 2001). While the aim is to develop a vision of e-government in Camden, it really has more to do with government than the electronic aspect.

Camden is a polarised borough. They are trying to use technology to improve this situation, through a focus on social exclusion and democratic renewal, but 'the problem is that technology is usually tacked on to the regeneration stuff, and so is sometimes the first bit to be lopped off, when resources start to dwindle, for example' (Andrew Stephens and Alasdair Mangham, Camden, interview with author, January 2001). They admitted that Camden did not have such good leadership in the IT domain as Newham or Lewisham. There is a slight lack of technical knowledge, for example, and a time lag between the 'reality on the ground' and what policymakers realise is happening. In terms of infrastructure, there is a lack of connections, a lack of knowledge about the infrastructure, and access problems: 'London is well cabled, but access to it is needed. If you don't provide this, it's not much use at all...' (Andrew Stephens and Alasdair Mangham, Camden, interview with author, January 2001).

New Networked Spaces and Corridors of Fibre Within Paris and London

Juxtaposing the Territorial Strategies of Operators and Local Authorities

In the previous chapter, we looked at how operators have been concerned with the development of networks in the Paris and London regions differently, in response to the wider differences between the two cities in politics, regulation, local government and cultures of territoriality. The local territorial strategies of operators in Paris and London will depend, to a large extent, on their relations with local authorities, communes and boroughs. In other words, there is a necessarily close link between the deployment of telecommunications networks and services and local politics, which is not always evident or desirable to all parties. As one interviewee stated off the record:

> I think telecommunications operators lack a little reactivity. They're very good at what they do, but they know little about local authorities. They find it hard to understand how all that works, and for them, like for many people, the word 'politics' has a pejorative meaning. Politics is synonymous with clients, delays, slowness, reluctance, and they're wrong. The operators which succeed most are those which use local

authorities. Why? Because they're dependent on authorities for the deployment of their networks, in sewers, in the public domain. They're dependent on authorities because authorities place a lot of money in telecoms. As soon as some operators obtain their licences, they completely lose contact with authorities. It's crucial. There's a lack of comprehension on both sides. Equally, there are many people in local authorities who deal with economic and territorial development, who don't even know what broadband wireless is. That's a problem (interview with author).

On this latter point, Irène Le Roch of the local authorities directorate of France Télécom elaborates and identifies some of the institutional and cultural tensions that telecommunications developments inherently imply:

Authorities need to have projects, as that's what they want to do with their networks. It's not just to have networks. A network is not an empty 'road'. There needs to be information passing along it, so there needs to be projects and users. Therefore, starting from the notion of network is often the first mistake. It's better to start from the notion of needs, projects, partners, and then afterwards see what type of network we can put in place. But that's an argument which is not a logic of local authorities. Local authorities are used to being in a two-dimensional world. They have their maps and look at their maps, and we're working in a dimension where telecommunications flows deform the traditional notion of distance. If you like, then, we're between two universes, between politicians who are managing a flat territory, and us who have distance capabilities which don't depend on the distance in kilometres... The 'autoroutes' metaphor is completely false, because it makes the use of telecommunications synonymous with an infrastructure which has been shaped for a certain type of flow... It's like those who think that if they don't have an autoroute or the TGV, they're disconnected from globalisation. The TGV can't stop everywhere, because then it's not the TGV. An autoroute can't have access roads everywhere, because it's not a departmental road. I think that the problem of local politicians is that they suppose all infrastructures to be departmental roads. Because they have to cover their area (Irène Le Roch, France Télécom, interview with author, September 2000).

This all means that territorial authorities need to improve their links with private operators. Jean-Baptiste Hennequin of the Conseil Régional d'Ile-de-France made an interesting point in this regard:

Through our new development agency, we'd like to develop good relations with companies, notably telecommunications operators, because today we have very few direct links with them. They're a little suspicious of *collectivités locales*, you have to say, at least of some of them, because in the past we were the clients of France Télécom, so some of them find it hard to believe it's changed, and yet it's all changed. Today we're no longer an annex of France Télécom, we work very well with France Télécom, but we're not at their disposal. They have to understand that. We need a time where we can initiate partnerships with them, which are mutually beneficial: 'if you go to a certain zone, then we can help you with the training of your engineers or your technicians, or with a commune which doesn't want to grant you a permission etc' (Jean-Baptiste Hennequin, Conseil Régional d'Ile-de-France, interview with author, October 2000).

Relations between operators and local authorities appear, nevertheless, to be perhaps more advanced in France than the UK. As Irène Le Roch of France Télécom argues:

> There is a competitive context which is beginning to be very strong. I think that France is one of the countries where competition in telecommunications is fairly active, at least as much if not more than in the UK. French territorial authorities are not the same as UK territorial authorities, so I get the impression that operators in the UK are not interested in local authorities, while in France they've understood that they are both clients and prescriptors, so they're very active (Irène Le Roch, France Télécom, interview with author, September 2000).

In both the Paris and London telecommunications markets, however, competition in the business sector is greater than in the residential sector. Many operators such as Colt and MCI WorldCom only specifically target business customers in key business areas, which means both that in many parts of the Paris and London regions, France Télécom and BT face no competition, and that competitive operators often have only token links with local authorities.

Relations between Colt and French *collectivités locales*, for example, are 'very weak, as Colt never deploys networks for *collectivités*' (Emmanuel Tricaud, Colt, interview with author, November 2000). Apparently, the richer the commune the less welcoming they are to operators looking to do engineering works to deploy or look after their infrastructure. The communes are already rich, so they think that there is no need for such infrastructure, and especially the hassle of operators digging roads and pavements up to deploy it. Communes tell Colt that 'un mètre de tranchée, un électeur perdu' – 'for every metre of trench, a voter is lost' (Emmanuel Tricaud, Colt, interview with author, November 2000). However, it is not just a matter of electoral votes: 'It is also a little bit of lethargy on the part of communes. They are barely interested in the economic impact of telecommunications, and they don't understand it either' (Emmanuel Tricaud, Colt, interview with author, November 2000).

M. Tricaud agreed that territorial telecommunications developments definitely needed the willing participation and a certain level of knowledge on the part of *collectivités locales*. With regard to the controversial notion of the *constat de carence*, he argued that this was linked closely to the delay created by the politicians in implementing liberalisation measures, which makes it something of a paradox, and for M. Tricaud 'almost like a joke' (Emmanuel Tricaud, Colt, interview with author, November 2000). Of all the *enjeux* for the future of the development of telecommunications infrastructures and services in Ile-de-France, M. Tricaud highlighted a wish 'that mayors and local politicians might finally understand the economic impact of telecommunications... but this will be a long battle' (Emmanuel Tricaud, Colt, interview with author, November 2000).

In contrast, relations between MCI WorldCom and *collectivités locales* do not encounter any particular problems (Marc Bourgeois, MCI WorldCom, interview with author, November 2000). It was suggested that this might have something to do with the way in which the L33-1 licence for operators regarding the construction and running of a network cannot be prevented by a particular commune, only delayed. However, if WorldCom did suspect they were going to have difficulties with a certain commune in their operations, they would be forced to go elsewhere. The interrelations involved are complex, as: 'before, people in the communes didn't see the importance of competition three or four years ago. Now, they do, especially for attracting businesses to locate there' (Marc Bourgeois, MCI WorldCom, interview with author, November 2000).

In contrast to M. Tricaud of Colt, M. Bourgeois felt the notion of a *constat de carence* to be undertaken by *collectivités* is 'fairly normal – it avoids a lot of civil engineering, as well as the need to spend a lot of public money'. It is more normal in any case than the eight year limit of the period of amortisation of investment, which is 'too short, a little constraining, and not very profitable' (Marc Bourgeois, MCI WorldCom, interview with author, November 2000).

Constructing New Types of Networked Space

Recent developments in both the Paris and London regions in terms of IT and telecommunications infrastructures illustrate an increasing variety of ways in which these infrastructures are being territorially deployed and 'fixed'. We have already analysed the development of some of these new networked spaces in the first section of this chapter, where we looked in particular at the strategies of La Défense, the Sipperec project and the London boroughs of Lewisham and Camden. These were held to exemplify some of the close links between telecommunications developments and the territorial market regulation of local governance institutions. We built on this in the last section by discussing the relations between operators and local authorities. In this section, we look at some other examples of how operators and territorial authorities are shaping and constructing new networked spaces in Paris and London, and thereby further configuring and articulating the territorial contexts and specificities of local places and telecommunications developments in parallel.

The Docklands: wiring in the east The emergence of the Docklands as a new networked urban space in London evidently went hand in hand with the regeneration of the Docklands as a whole, and in particular the development of the area around Canary Wharf as a key business area in the capital to rival the City of London. Even in the mid 1980s, the development of the Docklands included an aim 'to attract high technology and high data users' (Ward, 1987, p.291). The LDDC ICT strategy was that "The large office developments in the Docklands were designed to be filled by communications-dependent international corporations over-spilling from the City of London. This helped to make the case with the telecom companies to install the most sophisticated electronic infrastructure" (Cox, Spires and Wylson, 1994, p.89). Foster quotes the chief executive of the LDDC, who provides a technological perspective on the way in which the overall Docklands project prioritised virtually any kind of development, especially for the benefit of the private sector, as we saw earlier in chapter 5:

Docklands was the last remaining sector in London which was actually totally free of electromagnetic interference... It suddenly occurred to me, why couldn't we promote Docklands as the new telecommunications centre in the UK?... This happened to be at the time when British Telecom were actually looking to build satellite earth stations... I went to them and said 'You know the normal planning problem?' I said 'Come into Docklands and you can build as many antennas as you like and as big as you like.' And so in a very short period of time we persuaded British Telecom to build the satellite earth station in North Woolwich. Mercury was being promoted as a competitor so it wasn't too difficult to play Mercury off against British Telecom. We got the Mercury satellite built at Wood Wharf in the middle of the West India Docks... By the end of [19]83 we had actually persuaded British Telecom to build a complete fibre optic ring main around Docklands when in a sense you can say there wasn't a single end user in sight [as the development had hardly begun] (Reg Ward; quoted in Foster, 1999, pp.83-84).

London was, then, the first European city in the 1980s to have two teleports. Both are located in the Docklands and have been operational for many years. The British Telecom International teleport at Woolwich / Silvertown, the first operational teleport in the world, offers numerous important satellite connections, notably a diverse range of digital transatlantic links (Henry and Thépin, 1992). It handles hundreds of major broadcasting events each year. The Docklands East London agency (DEAL) also suggests that 'locating a call centre close to the teleport can lead to lower operating costs for call centres than might otherwise be the case' (DEAL, 2000, p.2). The Mercury teleport, or earth satellite station and telecommunications centre, at East Wood Wharf which was initiated after regulation changes, provided a private fibre optic network, and this developed on a national level linking many of the country's largest cities. In addition, the earth satellite station provides transatlantic voice and data communications, and the telecommunications centre houses the British end of Mercury's fibre optics cable to France and the equipment for their three earth stations linked with North America and Europe. Cox, Spires and Wylson suggested that the developments of BT and Mercury in the Docklands 'were replicating the marketing and technology upgrading efforts they have made in the City of London's financial centre' (Cox, Spires and Wylson, 1994, p.89).

The customer network design team of BT was able to develop a pro-active strategy to cope with the emergence of the Docklands as a key new networked space, as they knew roughly what the demand was going to be in terms of both quality and quantity (David Ellis, BT, interview with author, January 2001): 'Docklands was not like anything we'd seen before! Hong Kong and Shanghai Bank wanted 6350 lines just like that!' (Rob Tapping, BT, interview with author, January 2001). The area is served by the Poplar and Albert Dock BT exchanges, and Rob Tapping had to make a crucial decision on how many copper cables to deploy to serve the Docklands area: 'The 4800 pair is the biggest, but it costs £150,000 a time. We just put four straight in at the Docklands... we were confident that that was going to be taken up.' This was a calculated financial risk: 'we couldn't afford capacity to be dormant because it plays havoc with the overall budget!' (Rob Tapping, BT, interview with author, January 2001).

Fibre is provided directly to big customers in London via 'dedicated lines'. BT's customer network design team evaluates what the customer needs, and fits this with their products (David Ellis, BT, interview with author, January 2001). Concentrations such as the Docklands tend to use most of the fibre provision, but this is usually over short distances, commonly in 500-1000 metre shorthaul links, although the maximum is 2000 metres. Therefore, BT can roll out bulk fibre, from their main exchange at Bishopsgate for example, into a specific area, and place it where they think the demand is most likely to be focused. At the moment, companies are both 'creeping towards Shoreditch' and moving in to the wharf areas of London, and this can lead to a kind of 'snowballing effect' in telecommunications provision. The Docklands area also benefits from the local authorities imposing only low rents and low underground rates (David Ellis, BT, interview with author, January 2001). David Ellis stressed that it is very important for BT not to be seen as anti-competitive, and that there is a need for 'natural market forces' to shape developments.

The development of BT's networks in local sectors within London can have more widespread territorial consequences. As an unnamed interviewee in the Local Futures Group report observes with regard to the key sector of the City of London:

> National telecommunications tariffs have come down to the City's impact on competition. But we can not differentiate tariffs by region. Price competition is driven out of the City and other business tariffs come down as a result, across the whole country. London is the driver, but the benefits are national (quoted in Local Futures Group, 1999a, p.11).

This illustrates further the multiple scalar levels implicated in local telecommunications developments. The report goes on to state:

> Whilst BT can not vary its tariffs geographically, its competitors generally can do so (Oftel can impose similar restrictions as it does on BT if their market share in the local market is significant). Thus, it is safe to conclude that the City of London's buying power in telecommunications − ...35% of the national telecommunications market is driven by financial services − has benefitted the rest of London's economy, especially small businesses. These 'externalities' are the pecuniary benefits (cost savings in telecommunications) for other businesses that arise from and spread out of the City of London's highly competitive telecommunications market (Local Futures Group, 1999a, p.11).

Here then, explicating from the implications of the local development of BT networks and services within London, we have a brief example of how the national context of telecommunications regulation combines with a local context of intensive telecommunications requirements in the City of London and local operator strategy to produce regional and national implications in the tariffs of telecommunications markets, and broader economic development and territorial cohesion benefits. In this way, parallel multiscalar contexts and specificities are bound up in the deployment of telecommunications networks and services, and in turn influence territorial development at multiple scalar levels.

There are held to be not many real differences in the ways telecommunications networks and services are developing in the Docklands business area compared to the City, 'although the developer of Canary Warf did lay duct nests during the initial construction, and this is a practice that we will encourage' (Colin Jenkins, GLA, personal communication, June 2001). Nevertheless, the 'hub status' of the London Docklands as a networked urban space is suggested to be the result of at least four key elements (Pullen, 2001):

- The history, the language, and the location of the Docklands.
- The position of being a springboard for transatlantic capacity – London-New York at 26.7 Gbps; London-Paris at 24.3 Gbps.
- The presence of a concentration of media and financial business communities.
- The position of London as being a top Internet bandwidth city with up to 86.5 Gbps.

These elements constitute a variety of territorial specificities, in terms of scale, historical continuity, economic context, and technical configurations, which, it is argued, together give the Docklands area a competitive advantage over other areas in this domain. Indeed, for Richard Elliott of Band-X:

> If you listen to the streets in East India Dock you will hear them hum. It is likely that a higher value of commerce passes through Telehouse on an average day than was traded from the entire London Docks during their peak (quoted in Pullen, 2001).

The Docklands has also more recently become one of the main European centres for telehousing. As Phil Jackson of the operator Level 3 says: 'Right now, London is number one in Europe because there are a concentration of transatlantic cable companies here. Docklands is the communications equivalent of Heathrow' (quoted in Lamb, 2000). Market deregulation and the availability of cheaper land are given as the principal reasons for the location of telehouses in the Docklands. This is a major part of what appears to be quite substantial evidence of a spatial shift in the pre-eminent location of telecommunications activities within London. Of the 1.2 million square feet of carrier neutral hotel space in London currently, 58% is located in the Docklands against 42% in the 'traditional' location of West London. The overall vacancy rate of this space is 36%, but only 2% of carrier neutral hotel space in the Docklands is unused, whereas in West London the figure is 83%. Furthermore, the trend is towards an increasing dominance of the Docklands area in this domain in the future. By the end of 2002, for example, the total size of the carrier neutral hotel market is predicted to be around 3 million square feet, with 71% in the Docklands to 29% in West London (Pullen, 2001). This spatial shift illustrates a number of territorial specificities giving the Docklands a locational advantage in this domain. The DEAL board highlights four of these (DEAL, 2000, p.3):

- Buildings which meet their technical demands in terms of floor loadings and floor to ceiling heights.

- Existing telecom 'loops' whether fibre optic or copper.
- Availability of back-up power supply from separate parts of the national grid.
- Proximity to core markets – City / West End at reasonable rents.

They also point out the suggestion that the market in West London might be becoming 'overheated', thus attempting in one sense to play up their own potential intra-regional competitive advantages.

Thus, with teleports, telehouses, fibre optic networks and proposals for a networked village, it is hard to imagine a more intensive parallel reconfiguring of telecommunications and urban space than in the Docklands.

Telehousing / data centres, or the local territorial fixing of operator activities? Telehousing or data centres (or carrier hotels or switch centres) are fully equipped and serviced buildings aimed at Internet and multimedia businesses which require secure, high speed telecommunications connections in their own small spaces close to the centre of global cities.[82] Paris and London have therefore concentrated many of these developments.

They are usually located on the fibre networks of numerous operators, thereby giving their clients a choice of carrier. Heavy investment is required in order to equip these centres, but they have become successful and profitable activities, so, because of both these points, it is unsurprising that it is many of the main telecommunications operators in Europe which have branched out into this market, which therefore becomes an important part of the local territorially fixed element of their glocal infrastructure networks. The focusing of operator strategy on data centres seems to be symbolic of the increasing importance of value added services in an overall telecommunications sector (Jean Nunez, Cegetel, interview with author, September 2000), within which heavy competition and a glut of suppliers in many markets (technical and territorial) suggest a need to concentrate on niche market services in order to make a profit.

In London, this form of telecommunications development has already been constructed for a number of years. Telehouse Europe built its first facility in the Docklands in 1990. As Lamb states:

> Telehouses were first introduced into the UK five years ago as a way of reducing the cost and improving the performance of internet connections. Most internet links within the UK were routed via the US. To speed communications and to make it easier to create and break links between one another, a group of UK ISPs and telcos formed the London Internet Exchange and installed their equipment in the same building. The

---

[82] Longcore and Rees have outlined the intensive needs of these 'smart' buildings:
Heating and cooling systems (HVAC) must be automated to control environments required by electronic equipment, and space must be found for back-up water tanks for air conditioning. Security of uninterrupted power supply demands redundancy and duplication, substantially raising a building's power service needs and demanding heavy generating equipment on site. Space for reinforced slabs for upwards of four diesel generators must be provided, as well as room for fuel storage tanks (Longcore and Rees, 1996, p.356).

close proximity of the systems cut costs and made it easier for the partners to do business with each other (Lamb, 2000).

The Docklands area of London has become a focus for these activities,[83] leading to the board of the Docklands East London inward investment agency becoming increasingly informed about local development opportunities arising from them. A board paper summarised the phenomenon:

> A typical data centre is 300,000 sq ft with build costs between £160-£200 sq ft making for a £50-60 million investment. It is reported that a company would typically recoup its investment in 12-18 months. The ability to do this is driven by the ability to charge rents which may vary between £30-120 sq ft depending upon the level of services provided. Equipment owners effectively pay rent for using neutral space, which has strict requirements on temperature, power telecoms and network connections, security and engineering support. For example in the case of a 300,000 sq ft building a rent of £100 sq ft would result in an income of £30 million per annum (DEAL, 2000, p.1).

Telehousing in the Docklands is also making this area highly communications intensive at the eastern end of the cross-London 'corridor of fibre'. As Kevin Still of TeleCity explains:

> Global companies need telehousing because it's easy for telecoms companies to be multinational, but not so easy to be multi-local. They need to be close to their users, to develop a virtual point of presence: telehousing provides a neutral exchange point (quoted in Lamb, 2000).

This development of points of presence or proximity also seems to be the case for users:

> although data centres can be considered to be a back office function there is evidence that occupiers are keen to be near their servers, and indeed data centres may provide networking opportunities for dotcom companies to do business with each other (DEAL, 2000, p.2).

This suggests, then, that telehousing is characterised by a two-way desire for both provider-user and user-provider propinquity.

Another Docklands perspective on telehouse developments though suggests that this local territorial fixing of operator activity is solely aimed at increasing the customer base of their global infrastructures: 'The main driver isn't to make money out of providing buildings and space, it is to encourage people to use our 4000 km network' (Charles McGregor, Fibernet Group, quoted in Lamb, 2000). DTZ nonetheless note two main factors underpinning the local territorial fixing of co-location facilities:

---

[83] Telehouse occupies two buildings at East Side Dock, Global Switch is located in the former FT building, and TeleCity also operates in the Docklands.

operators like to locate facilities in or near the centre of major urban areas, driven by the need to access as many fibre networks as possible and to be close to clients; there is a strong preference for a local presence in separate markets throughout Europe, leading to multiple locations being required to ensure comprehensive coverage (DTZ, 2000b).

Facilities are increasing in size, as with the cases of Global Switch at East India Dock House in the Docklands (159,000 square feet), and the developer Markley Stearns, who have been building a 120,000 square metre (1,300,000 square feet) co-location facility in Paris to open in 2001 (DTZ, 2000b).[84] DTZ have also stressed the importance of local territorial aspects to the co-location market in London:

> Until recently all the significant facilities were concentrated in Docklands, close to the City and having not only the requisite concentration of fibre networks, but the most important internet exchange as well. It also had a readily available stock of buildings and sites left over from the recession of the early 1990s. As demand for office space increased, so the availability of suitable properties that could be used for co-location premises began to decline. As a result the focus has now switched to locations in West London, notably around Heathrow, where there is an adequate number of clustered fibre networks and available property, especially warehouses suitable for conversion (DTZ, 2000b).

There is a difference in the principal territorial locations of telehousing or data centre activity in London and Paris. While in London, we have seen its focus on the Docklands area, in Paris, there are numerous examples of locations in the peripheral edges of the city: 'a godsend for the border communes of the Paris region which are seeing the birth of these factories of the Internet era on their old industrial land' (Girard, 2000, p.50, author's translation). GTS has a 4,500 square metre data centre in Clichy, WorldCom have located their 18,000 square metre data centre in Saint-Denis, Completel are to be found in Aubervilliers, Level 3 has 3,500 square metres of space in Nanterre, while Colt bucks this trend with a 3,500 square metre centre in Paris itself (Girard, 2000). As we can see, the operators who have moved into the co-location services market tend to be ones who have

---

[84] The increasing size of facilities necessitates increased levels of security and resilience. These issues have taken on new resonance in the light of the terrorist attacks on New York in September 2001. The Global Switch facility in the Docklands, for example, advertises 'peace of mind' in three ways (Global Switch website):

- Power – Global Switch / London draws energy directly from the National Grid. In the event of a total power failure, on-site generators take over and can operate for 7 days without refuelling.
- Security – Manned security operates every hour, every day, every week of the year. In addition to the latest colour CCTV digital cameras, the latest security entry systems will be located at the entrances to ensure swift access to authorised personnel.
- Resilience – In addition to physical security measures and power loss alternatives, Global Switch / London has been designed to withstand localised damage without complete shutdown of all facilities.

deployed metropolitan fibre optic networks 'in order to transport the flows generated by the Internet and to sell capacity' (Girard, 2000, p.52, author's translation). The major absentees from these telehouse developments, however, are the incumbents, BT and France Télécom. For Sue Uglow of the Ovum consultancy, 'BT and France Télécom don't like these telehouse facilities. For them, it would mean using their exchanges, and they think that the co-location of other operators at their exchanges will make them lose customers' (Sue Uglow, Ovum, telephone conversation, October 2000).

Extending the networked core areas   As well as the ways in which telehousing developments are contributing to the intensity of networking in certain areas of Paris and London, there is also evidence of how the networked core areas in the two cities are being extended. In this regard, we can explore briefly how, for example, 'clusters' of start-ups and IT firms are developing in the *quartiers* adjoining the Paris Bourse, and in London's City Fringe.

Start-ups in Silicon Sentier   The principal cluster of start-up Internet companies in Paris has developed in the 2nd *arrondissement*, close to the original main business sector of the city around the Bourse. This is now known, in a similar fashion to many IT / telecom quarters around the world, as Silicon Sentier. The Sentier was formerly a district of Paris best known for its textile industry, but was steadily invaded around the end of 1999 and the start of 2000 by new Internet businesses, looking to profit from the proximity of the quality telecommunications infrastructure of the Bourse (Belot, 2001). It was a very 'spontaneous' kind of development, with 'no influence from the public sector at all' (Stéphane Lelux, Tactis, interview with author, October 2000). Françoise Courtois-Martignoni of the Mairie de Paris concurs, but elaborates on the territorial specificities of Silicon Sentier:

> The Sentier emerged due to the sector of the Bourse because the infrastructures were there; because there was a lot of property space left over from the textile industry leaving for Aubervilliers, so the places weren't expensive.[85] But the Sentier also has an address effect – the centre of Paris, the proximity of clients – and a community effect for ICT businesses – in the Sentier, you're in a community, you feel good – and that's also an important factor. [...] Unfortunately, companies don't always find available places which correspond to their growth in the Sentier and move into the rest of Paris and also in the near suburbs (Issy-les-Moulineaux, Montreuil etc). [...] The last criteria of the Sentier: the importance of infrastructures; effectively, in the last Parisian poles, we don't always have them, and the whole of Paris is not yet covered by broadband (Françoise Courtois-Martignoni, Mairie de Paris, quoted in IAURIF, 2001b, author's translation).

Competitive telecommunications networks are prevalent in the area, with the presence on the main street in the quarter, the rue des Jeûneurs, of Telehouse,

---

[85] Annual rents of FF1200 per square metre in the Sentier compare favourably, for example, with the FF1500 – 1800 in the area towards the Opéra (Courtois-Martignoni, 1999, p.10).

which concentrates and houses much of the operators' infrastructure. The presence of Telehouse has been important in the development of the sector:

> With regard to the Sentier, you recall that four years ago, there weren't any telecoms, Internet or IT activities there. Telehouse located itself on the rue des Jeûneurs, and very quickly (in two years), the 2000 square metres of premises filled up, and Telehouse created a telecoms directory which interested a large number of actors. Which is how the term Silicon Sentier has been able to be used... Telehouse doesn't select broadband. The concept is to gather together in the same place different actors or businesses in the telecoms and IT field. In encouraging this centralisation, the immediate interest of the different actors is to be close to each other and to be able to work together by creating a partnership.

> The objective is thus to offer access to electrical, space and climatisation resources and to optic fibres. Today, thirty or even forty optic fibres arrive in the buildings from about fifteen operators. These operators bring their resources into the buildings... The term 'directory' could be used [to describe the role of Telehouse] (Olivier Caron, Telehouse, ODEP, 2000, author's translation).

The relative success of the Sentier as a concentration of ICT companies has, though, inevitably led to an increase in average rents for properties. For Françoise Courtois-Martignoni of the Mairie de Paris, the fact that the Ville de Paris cannot control this is leading to a major problem (Françoise Courtois-Martignoni, Mairie de Paris, interview with author, October 2000). Nevertheless, as with elsewhere, following the plummeting technological stock prices of the last year or so, many of the companies have folded or have hit hard times (Belot, 2001). The territorial implications for the extended networked core area of the Sentier are still to fully unfold.

'Clusters', start-ups and the new economy of the City Fringe   The areas immediately surrounding the City of London, such as Clerkenwell, Shoreditch, Spitalfields and Hoxton, have likewise gradually developed into a concentration or concentrations of multimedia and IT firms (Local Futures Group, 1999a; Pete Large and Pauline Irwin, Corporation of London Economic Development Unit, interview with author, August 2000). The Corporation of London had a 'scoping' study about 'clusters' done by James Simmie, 'mapping the concentrations of different sorts of businesses in inner London' (Pete Large and Pauline Irwin, Corporation of London Economic Development Unit, interview with author, August 2000). In addition, in August 2000, Pauline Irwin of the Economic Development Unit of the Corporation of London was:

> actually in the middle of writing a committee report to get a significant amount of funding for us to do further work on clusters, to see whether the linkages and networks in the City, in the kind of City's critical mass, do kind of influence those growths of other service companies around, but we haven't done it yet (Pete Large and Pauline Irwin, Corporation of London Economic Development Unit, interview with author, August 2000).

Nevertheless, a Corporation of London brochure fully plays up the 'new economy' clusters forming in London:

> Like Silicon Alley in Manhattan, a growing cluster of young companies involved in digital media is forming on the northern edge of the City where large premises at affordable rents are to be found. This is now known as London's Digital Triangle. Here the internet industry is forging ahead faster than anywhere in Europe, galvanised by the City's need for the most advanced information technology to deal with a global market. Already, London ranks among the world's top three financial software industry centres. The software and services market already employs more Londoners than does telecoms. In the City today, there is more than one way to network (Corporation of London, 2000, p.29).

The e-business advisor to the GLA posits some of the implications of these clusters for London:

> The LDA's encouragement and support of clustering is important for SMEs. Companies that cluster tend to be highly innovative and the critical mass that clustering produces can lead to economies of scale in a number of areas, including broadband access to the internet. The LDA should consider using the new money being made available from the government's broadband fund to stimulate take up of broadband by existing clusters of SMEs (Jenkins, 2001b, p.13).

What the Corporation of London's Economic Development Unit has been trying to do is to identify properties for IT firms:

> I mean, I think for that and for a couple of other projects we've been trying to find out about properties in the City Fringe because we know that various companies are buying up huge properties just to fill with hardware, kind of switch centres and whatever else and laying cables, which isn't great for economic development because obviously the buildings aren't full of people. But again we don't know too much about it at the moment (Pete Large and Pauline Irwin, Corporation of London Economic Development Unit, interview with author, August 2000).

Here again, though, is an illustration of both the obvious need for fairly central locations in the ICT sector, and the permeable boundary of the territorial concerns of the Corporation of London. Given the limited and expensive nature of property space in the City, this latter aspect is perhaps inevitable:

> and the thing is that the rents in the City are very high, and in the City Fringe they're not so much, not as high, but we've found over the last kind of 12 or 18 months that there's been inward investors coming in from abroad as well as from the rest of the UK that actually want big buildings with huge amounts of space in them, bang in the middle of the City just to fill with machines... and they're prepared to pay for it. But, a lot of that, none of these companies will tell us very much because it's so commercially sensitive to give that information... I mean inward investment enquiries in the last 12 months have predominantly been financial IT companies from America. If we can set them up in the City Fringes, because they can't afford City rents and there's no reason why they should be in the City, but by being in the City Fringe they can be 15 minutes

walk from their clients and have a physical presence, but that also has a positive knock-on effect for the areas that are even further back than that (Pete Large and Pauline Irwin, Corporation of London Economic Development Unit, interview with author, August 2000).

The Corporation, for example, bought up a piece of land outside the City boundary:

> largely to use as a sort of Internet incubator... I think in this particular instance it was for an economic development reason, to actually set up a business incubator for IT firms and to make it the right size for the right sorts of equipment to, you know, facilitate growth in the City Fringe for the good of the City (Pete Large and Pauline Irwin, Corporation of London Economic Development Unit, interview with author, August 2000).

There is also the example of the old Truman's Brewery in E1, which has been turned into a space for Internet and multimedia businesses. Rent is around £15 per square foot, which is seen as very reasonable. DEAL notes how:

> A contributory factor to the success of Truman's Brewery is considered to be the proximity of bars, restaurants and other creative businesses which offer a conducive environment to young software and internet experts. The establishment and growth of dotcom companies in these areas helps to draw in other activities which we may need or want (DEAL, 2000, p.3).

The Corporation of London is part of the City Fringe Partnership, the main aim of which is to regenerate the area around the City of London. This can be seen as an illustration of the way in which '[s]ince the abolition of the GLC, the City has come to play a wider regional role...' (Newman and Thornley, 1997, p.978).

The Docklands, then, might be able to take advantage of the proximity of these developments in the City Fringe. Equally, 'there is the opportunity to create additional office supply within the Docklands and East London market to cater for dotcoms which have successfully passed through their start-up phase as well as locations for satellite offices for more established businesses' (DEAL, 2000, p.4).

The Development of Corridors of Fibre Across Paris and London

From what we have seen so far, in this chapter and the previous ones, it is clear that operator activities, and particularly the deployment of infrastructures, are largely concentrated in certain intensive sectors of Paris and London. The general picture being painted of telecommunications developments in both Paris and London is one of operators locating themselves in the main financial and technological activity areas and leaving the incumbent as the sole operator in less profitable, less dense areas. Within our territorial perspective on telecommunications developments, one of the most interesting aspects relates to the spatial distribution of networks within the two cities, and more specifically, the ways in which corridors of fibre can be seen to have developed, focused in particular between key new and old networked urban spaces – between La Défense and central Paris, and the Docklands and the City in London. The establishment of each of these axes is, in turn, in many ways crucial to telecommunications developments in both cities, with regard to the existence and parallel importance of new and old networked urban spaces in Paris and London. It is crucial in the sense of the axes being at once a presupposition, a medium and an outcome of telecommunications networks. In the first instance, telecommunications infrastructures would not have served the City – the Docklands or central Paris – La Défense without the overall urban projects which created the axes as a broader focus for economic activity and territorial organisation in the regions. Secondly, the axes can be viewed partly as the means by which the deployment of telecommunications infrastructures links the City – the Docklands or central Paris – La Défense. In the third instance, the telecommunications infrastructures now create, in turn, part of the pertinence or the *raison d'être* of the axes within urban economic activity and territorial organisation.

With regard to the key sectors as 'new' and 'old' networked urban spaces, these terms are used here in a very general and relative manner. There is nothing particularly new about La Défense, where development has been ongoing since the 1950s, as we saw in chapter 5. Indeed, one interviewee in Paris suggested that many people in the telecommunications and IT domain would now see La Défense as an old networked space, and central Paris with its *quartiers* of start-up companies as a new networked space (Stéphane Lelux, Tactis, interview with author, October 2000). However, the aim is to explore and compare telecommunications developments in different key spaces of Paris and London, so it can be argued to make sense to focus on the relatively older networked spaces of central Paris and the City of London, and also the relatively newer networked spaces of La Défense and the Docklands. New and old could, therefore, be seen to be interchangeable with core and peripheral, or traditional and contemporary, as was discussed in chapter 4.

New strategic networked spaces in Paris and London, such as La Défense and the Docklands, are thus beginning to dominate urban telecommunications markets. While they have increasingly become part of dense corridors of fibre connecting them to the more traditional networked spaces in the centres of the cities, at the same time their lack of relational links to surrounding urban spaces makes them resemble 'glocal enclaves', or 'premium network spaces' which have 'delinked' from their hinterlands in order to privilege connecting their own zones to primary global networks.

In both London and Paris, it is suggested, for example, that demand for space is firstly increasingly concentrated most on or in proximity to key local fibre optic networks, and secondly, decreases with distance from these networks (IAURIF, 2000a; DEAL, 2000). IAURIF has highlighted three main locations of IT and telecommunications activity in the Ile-de-France region (IAURIF, 2000a). Firstly, increasing numbers of centres housing servers can be found along the fibre optic networks of operators other than France Télécom. They are therefore benefitting from liberalisation by looking for competitive tariffs which are often much lower than those proposed by the former monopoly operator. Secondly, these centres seem to have an impact on the location of start-ups in the region. Beneficial telecommunications costs and the quality of maintenance are two advantages of the centres. The sectors of the Sentier (in the 2nd *arrondissement*, near the Bourse) and République in Paris are examples of this development. Thirdly, e-commerce companies require locations advantageous to their logistics, as for deliveries. The Roissy – Charles de Gaulle airport area has thus become an important centre for these companies with the presence of a Federal Express centre and the fibre optic networks of several operators (IAURIF, 2000a). Meanwhile, alternative operators can mainly be found in the west of Paris (the 8th, 17th, 7th and 15th *arrondissements*), La Défense and its neighbouring areas Suresnes, Rueil-Malmaison, Nanterre, Puteaux, Courbevoie and Neuilly Levallois, the Val de Seine (Issy-les-Moulineaux, Boulogne-Billancourt, Saint-Cloud), and the Cergy-Pontoise pole (Abbou, 2000, p.59). Three areas to the east have been identified, in contrast, as standing out in terms of a lack of alternative operators: Roissy and Tremblay-en-France, Aubervilliers and Marne-la-Vallée. Alternative networks are being planned for Saint-Quentin-en-Yvelines, the plateau de Saclay and between Marne-la-Vallée and Paris (Abbou, 2000, p.59).

Emilio Tempia of the DREIF certainly viewed the infrastructure of the Ile-de-France region as being dominated by a corridor from La Défense through central Paris and out towards Marne-la-Vallée. The main beneficiaries of competition in telecommunications have been the west of Paris, the Hauts de Seine and the *villes nouvelles* (Emilio Tempia, DREIF, interview with author, September 2000).

In turn, the Local Futures Group report (1999b, p.21) suggests that there are three main areas in London where the 'knowledge-driven information economy' is most significant: the City and the Docklands; a south-western arc beyond the M25; and Croydon. The Heathrow area has been highlighted as particularly important within the London region, and it is suggested that the development of Terminal 5 can only increase this importance. At one time, indeed, Heathrow perhaps accounted for as much as a third of all BT profits (Rob Tapping, BT, interview with author, January 2001).

A concentration of telecommunications developments in London in a corridor or 'massive spoke' (Matt Cochrane, Cable & Wireless, interview with author, January 2001), running from the West End through the City to the Docklands leads though to significant territorial variations in London in the telecommunications and IT domain. For Colin Jenkins of the GLA:

The real problem is the last mile. The main fibre networks (Metropolitan Area Networks or MANs) are in the West End, the City and Docklands and there is sufficient critical mass in terms of major businesses for a number of telecoms operators to build networks targeted at customer premises. The offerings on these networks are high quality and competitively priced. Outside these areas there is insufficient economic justification even though there is a strong small and medium enterprise (SME) presence and fairly dense housing. There are a number of fibre networks that touch the main centres of outer London – for example there are 5 network operators whose fibre networks have a point of presence (POP) in Croydon - but the last mile connection is invariably made by BT (Colin Jenkins, GLA, personal communication, June 2001).

In this way, the process of unbundling the local loop and allowing alternative operators to deploy their own last mile broadband connections becomes absolutely crucial. The GLA, as we have seen, is becoming increasingly concerned with the wider implications of such territorial variations in infrastructure deployment. Their Broadband London objectives are liable to be jeopardised unless an effective way to increase infrastructure deployment beyond the core areas is found:

> This lack of broadband penetration outside central London is worrying because:
> to remain competitive, SMEs also need to embrace the advantages of e-business. Many of them are in highly innovative industries and so need affordable access to technology; there is an increasing tendency to work from home or on the move, in which case fast access to the office server is becoming essential;
> there is a whole new entertainment industry which is being inhibited because of a lack of affordable broadband access (Colin Jenkins, GLA, personal communication, June 2001).

Nevertheless, within a national context, much of London is relatively well off in terms of telecommunications infrastructures, compared to other parts of the UK:

> Outside London the problem is quite often that there is insufficient critical mass to justify long distance connections. Places like Birmingham, Manchester, Glasgow, Edinburgh, Bristol… are connected on a long distance basis and have competing MAN operators, again this is because there is a critical mass of large businesses. However, smaller centres, say Ipswich, and places in between these centres do not even have the advantages of a Croydon (Colin Jenkins, GLA, personal communication, June 2001).

Beyond the central areas of London, and the corridor of fibre between the West End, the City and the Docklands, there is not much territorial access to fibre, except for the network of BT. Subsequently:

> Broadband access beyond the central rings will have to await DSL roll out, deployment of fixed Broadband Wireless Access (BWA), and widespread deployment of cable modems. The pace of roll out of these technologies in the UK is somewhat behind that in the rest of Europe (Jenkins, 2001a).

This suggests that spatial or territorial variations in connection to telecommunications infrastructures are, in some ways, wider in London than in Paris.

The means for connection Corridors of fibre across and between the networked spaces of Paris and London inevitably necessitate the use of (usually) subterranean pathways for laying the required cables and fibres. Gaining access to these key local markets in the most intensely urbanised and crowded parts of Paris and London is a tricky and expensive business. As Graham notes: 'Fully 80 per cent of the costs of a network are associated with this traditional, 'messy' business of getting it into the ground in highly congested and contested, urban areas' (Graham, 1999, p.937).

Nevertheless, Jean Nunez of Cegetel observed that 'it is easy to deploy loops in Paris. The métro and the sewers turn Paris into something like a Gruyère cheese. Putting cable in to the main activity zones is simple and not expensive, in comparison with other cities' (Jean Nunez, Cegetel, interview with author, September 2000). Operators most commonly use the drains and sewers, and metro tunnels, under dense urban areas to deploy their fibre networks. In Ile-de-France, MCI WorldCom has in addition laid 60 kilometres of fibre optic cable at the bottom of the river Seine. This is primarily for long distance traffic, but some metropolitan communications as well. This part of their network starts at La Défense and runs to the *ville nouvelle* of Melun to the south. It was apparently quick to sort out with WorldCom only needing to negotiate with one agency in order to get permission, and is much cheaper than having to lay new cables under roads (Marc Bourgeois, MCI WorldCom, interview with author, November 2000). In any case, in central Paris, there are limited possibilities for infrastructure deployment in the public domain:

> In effect, the Mairie de Paris makes it quite clear that passage by the sewers is the only means to deploy a network rapidly. The other possibilities [such as the catacombs] would, how can we put it, meet with a certain administrative frostiness. In this way, even if you have to make a large detour, it would still be in your interest to use the sewers, as it would happen much quicker! (Jean-Raphael Hardy, Colt France, quoted in Géomatique Expert, 2002, p.31, author's translation).

Operators in London benefitted from being 'able to utilise redundant pipelines formerly used for hydraulic power to provide routes for glass fibre telecommunications links without severe disruption of the urban fabric' (Kennedy, 1991, p.141). For example, Mercury bought the old London Hydraulic Power Company in the early 1980s in order to have access to its networks of cast-iron pipes:

> [t]he huge, single benefit of owning the old network of pipes is that Mercury has an automatic right of way in busy city streets that are already densely packed with services. The pipes neatly follow the routes that Mercury needs to reach its customers, most of whom are in the City and in the West End. For long stretches, Mercury could simply feed its fibre-optic cable along the old hydraulic pipes, many of which were in excellent condition (Aldhous et al, 1995, p.7).

The London Underground system runs much deeper underground than the Paris métro, which makes the latter both easier to manage as a whole, and also easier to use as a means for deploying cable and fibre throughout the metropolitan area (Rob Tapping, BT, interview with author, January 2001). Avoiding the disruption of digging up roads is very important, because one of the major problems for operators in large metropolitan areas is gaining access to certain parts of their cable or fibre networks, particularly in traffic intensive areas. For example, to do work on or under Oxford Street in London requires a 28 day notification to the local authority. This kind of delay can be crucial to operators: 'it plays havoc with delivery times' (Rob Tapping, BT, interview with author, January 2001).

There is a problem if BT suddenly discovers a fault in their network in such an area, and cannot repair it quickly, thereby potentially leaving an important customer disconnected for days, in spite of the short delivery times in which BT tries to get the service up and running again (Rob Tapping, BT, interview with author, January 2001). Still, BT has its own deep-level tunnel system across London, which was already there, so they do not really need to use the tunnels of the London Underground, for example. The duct routes are fairly congested, however, as copper and fibre are laid in the same ducts (Rob Tapping, BT, interview with author, January 2001; David Ellis, BT, interview with author, January 2001).[86]

Rights of passage tariffs and wayleaves are national level issues which have a major influence on telecommunications developments in the two cities (Filliatre, 2000). In Paris, the municipality has configured the rights of passage tariffs (the *redevance d'occupation*) to which it is entitled for telecommunications infrastructures in such a way as to promote increased local coverage of networks. This was in direct response to the new telecommunications legislation of 1998. The Ville de Paris has therefore fixed a favourable and evolving tariff schedule within which in 1998 an operator paid FF25 per linear metre of their network in the first years, up to around FF50 per linear metre from the ninth year. As Filliatre continues:

> It is on the basis of a double mechanism (linear arteries counterbalanced by the diameter of the cables or covers, or subsidiary equipment) that the tariff applicable for deployment in the sewers has been calculated, in such a way as to urge operators: a) to deploy under municipal control by the instigation of a tariff schedule, which is not only progressive with time, but which also slides in scale in space; b) to not just content themselves with connecting certain zones or certain buildings which are particularly profitable, but to multiply local coverage of buildings in Paris by the application of a particular tariff schedule for end connections (Filliatre, 2000, author's translation).

Up to September 2000, 25 operators had concluded deals with the Ville de Paris for the deployment of a total of 2150 kilometres of network cable in the city (Filliatre, 2000). These rights of passage tariffs for Paris (FF25-50 per linear metre) compared favourably, for operators, with those of London, where it is necessary to pay FF190 per linear metre in the Underground system. This London tariff is unsurprisingly probably the highest in the whole of Europe (Filliatre, 2000).

Intra-regional competition between key networked spaces Besides the question of the role of telecommunications and IT in the competitiveness of the London and Ile-de-France regions on a European scale, it is perhaps somewhat surprising to find that there is also significant competition within the regions between the different networked spaces which are connected by important corridors of fibre.

---

[86] This duct ownership is a major difference between incumbents and competitive operators. The fact that BT has its own duct network makes it 'more regulated', whereas the likes of Colt and MCI WorldCom share ducts sometimes in the deployment of their networks (David Ellis, BT, interview with author, January 2001).

Such intra-regional competition has important implications for the territorial cohesion of the city-region as a whole.

In the Ile-de-France region, this competition is most intense between the two richest *départements*, Paris and Hauts-de-Seine (Françoise Courtois-Martignoni, Mairie de Paris, interview with author, October 2000). In this case, it has been to some extent a question of creating and keeping jobs, as we will see later in the chapter. Paris itself and the Ile-de-France region as a whole can perhaps best be seen as a kind of competitive dialectic. Paris has very few of the problems that characterise the region, including widespread social and territorial incohesion, so therefore the Mairie de Paris would often rather go it alone with developments and strategies for its own *intra muros* area than be a part of regional schemes which frequently have main territorial rebalancing and redistributive objectives (Françoise Courtois-Martignoni, Mairie de Paris, interview with author, October 2000). There is a curiously similar view held by the commune of Issy-les-Moulineaux, regarding the proximity of Paris. While this is felt to be an advantage in terms of transport, Issy necessarily feels it has both a generally 'separate identity', partly because, with 53,000 inhabitants, it is quite a large town in itself, and almost complete economic separation (Florence Abily, Issy Média, interview with author, July 2000).

Intra-urban competition has also been an explicit territorial mechanism in the London region, as we saw with the example of the Docklands development and the responses of the City of London to this in chapter 5. The view from the Economic Development Unit of the Corporation of London on the competition between the City and the Docklands dilutes somewhat the idea of fierce competition between the two areas as key networked spaces:

> We tend to see the City and Docklands as offering different things, being suitable locations for different types of operation. Now there is a huge chunk of IT infrastructure down in Docklands, but that is not necessarily to our detriment. I mean if the infrastructure is available to City companies what it means is that the very expensive properties, the high rents in the City are being pushed even higher by these operations which can be located outside. There's a few large operations, City-type operations have gone down to Docklands in the last few years, but they tend to be back office operations or commoditising trading operations rather than the kind of face-to-face corporate finance sort of function, very high value added functions which will have to stay in the City. So the way we see it is that hopefully Docklands or spillout over into the City Fringe is acting more as a safety valve on the property market, in other words it's stopping the differentials in rents becoming unsustainably high, which is what happened of course in the late 80s (Pete Large and Pauline Irwin, Corporation of London Economic Development Unit, interview with author, August 2000).

This all boils down to the fact that the Corporation of London has an important role on the wider London stage, as well as being the local authority for the City:

> As far as inward investment goes, we're one of the funders of London First Centre, which is the inward investment agency for Greater London. We pay a certain amount of money, as do a number of other bodies, to get a certain amount of enquiries, and

Docklands also, well Docklands East London is actually part of London First Centre. And so, any companies which come in, as long as they go to London rather than Frankfurt, that's really the bottom line. I mean, we'd rather they were in Docklands than in Frankfurt. That's the ultimate thing (Pete Large and Pauline Irwin, Corporation of London Economic Development Unit, interview with author, August 2000).

The territorial concerns of the Corporation of London are, then, very much, multiscalar, as we might expect, given the global reach of the City in the world economy:

The Corporation of course is the local authority for the Square Mile, and it should be our constituents which we are primarily concerned with, but of course the critical mass argument, which is ultimately on what the City's health depends... Critical mass doesn't stop at our boundaries. It's still very strong as you go into the City Fringe, and still strong if you go into Docklands or into the West End, so we can boost London generally by boosting the City by hopefully boosting London generally (Pete Large and Pauline Irwin, Corporation of London Economic Development Unit, interview with author, August 2000).

In this way, then, in the cases of both regions, as the regional Chambers of Commerce of Ile-de-France summarise: 'The capital region is a complex space, both strong and fragile, characterised as much by an important wealth production as by internal imbalances and increasingly marked transformations' (Chambres de Commerce et d'Industrie de Paris – Ile-de-France, 1999, p.15, author's translation).

Connected and Disconnected Urban Spaces, and Territorial Cohesion in Paris and London

This chapter has focused mainly on how different urban spaces within Paris and London are being connected by telecommunications developments. It is important, however, not to separate the notions of connection and disconnection, which means that as well as exploring these connected urban spaces, we need to look at how other (peripheral) spaces in Paris and London are being left disconnected by telecommunications developments. This focus on the disconnected urban spaces of Paris and London is essential for us to be able to consider, as an important adjunct to our empirical study, an overall view of how telecommunications developments relate to the processes and practices of territorial cohesion in global cities as a whole.

Spatial differentiation in the type and level of connections between different spaces within the Paris and London regions appears to reflect their wider intrinsic 'internal imbalances' and augment the importance of overall territorial cohesion. There seems to be difficulty in balancing the dual and parallel aims of telecommunications developments of economic development and urban competitiveness on the one hand, and socio-territorial cohesion on the other hand.

The general territorial implications of restructuring in the telecommunications industry are particularly grave on a local geographical level. A recent report by the

Yankee Group Europe, for example, summarised the pessimistic implications of operator restructuring to focus on business sectors (The Yankee Group Europe, 2001). Their recommendations, however, clearly only took the positions of operators into account:

> The Yankee Group believes competitive carriers that have not yet responded to the new market realities, particularly those that are dependent on consumer voice services for the bulk of their revenue, are most likely to have difficulty in the months ahead. Actions like those of Viatel and Interoute, where consumer business has been sacrificed to improve the bottom line, or of XO and KPNQwest, where network investment has been deferred, are appropriate and necessary in the current environment (The Yankee Group Europe, 2001, p.11).

In light of this, we can ask what the future might hold for the plain old residential consumer if the 'market' and operators view reducing connection, or even disconnection, as 'appropriate and necessary'? This perhaps illustrates the need for further local authority intervention in telecommunications.

Given the lack of market interest in many peripheral territories, it may subsequently be a question of a somewhat paradoxical need for new 'natural' monopolies in these disconnected territories in order to obtain broadband coverage. As a participant at the recent OTV seminars questioned: 'How can unprofitable territories for one operator be profitable for many operators? France must get rid of this nonsense' (Roger Mezin, premier adjoint au Maire d'Amiens; quoted in OTV, 2001d, author's translation).

The example of operators in the mobile sector has shown a possible way to territorially regulate the market to encourage operators in the unbundled broadband sector to deploy more widely than the business areas of Paris (FEDIA, 1999). Mobile operators have crucial *aménagement du territoire* concerns to take into account in the configurations of their national networks, but it is probable that mobile and broadband are too different for similar forms of territorial regulation.

Social and territorial inequalities or disconnections in relation to telecommunications developments take different forms in Paris and London. One of the main sources of these inequalities relates to the fact that the ICT sector is dominated by highly qualified professionals. The 1999 ODEP report recognises that training and the creation of new functions can work in the favour of less qualified individuals, and therefore avoid a new category of exclusion, that of the information society (Courtois-Martignoni, 1999, p.9).

Authorities in both cities have thus a very broad notion of 'infrastructure' in the telecommunications and IT domain. Besides networks, applications, services and actual computers, there is a great deal of focus on the importance of skills and training as part of this important infrastructure. In London, this is perhaps part of the greater public sector focus on IT than market-driven telecommunications. Indeed, the Local Futures Group report for the Corporation of London argues that 'The City's technological leadership derives less from telecommunications and more from IT' (Local Futures Group, 1999a, p.12). The report goes on to suggest that the quality of urban telecommunications infrastructures *per se* is becoming standardised between cities, while the availability of ICT skills can provide a city such as London with an increasing competitive advantage, presumably to make up for the steadily disappearing competitive advantage in infrastructures (Local Futures Group, 1999a, p.5). However, it is also noted that there is a great shortage of ICT skills, not just in London but around the world (BH2, 2000; Karen Sieracki, BH2, interview with author, October 2000; Alex Bax, GLA, interview with author, February 2001; Colin Jenkins, GLA, personal communication, June 2001). The opportunities for increasing the level of ICT skills among the workforce are recognised, but again discussed via token statements without concrete suggestions following up:

> It is frustrating to note, for example, that young people in areas of high unemployment in London are quite adept at text-messaging on mobile phones, operating game machines and programming video recorders. If these hidden skills could be recognised and harnessed, employment prospects could be greatly enhanced (Jenkins, 2001b, p.10).[87]

The overall shape of the development of the ICT domain in Paris and London can be viewed as characterised broadly by a 'core' and 'periphery' spatial differentiation: 'The very favourable conditions remain limited at the current time to central sectors and developments of the offer for less dense or less favoured territories are giving rise to interrogations about the appearance of a new form of territorial inequality' (Cahiers de l'IAURIF, 1999, p.61, author's translation). Innovation in the IT sector in London, for example, seems to spread itself on a global rather than local scale, 'flowing between financial capitals rather than spilling over into the surrounding regional economy' (Local Futures Group, 1999a, p.29). In terms of local 'spill-over' benefits, there only really seems to be evidence of companies in the City offering outdated IT equipment to the voluntary sector. A more positive benefit would be if these companies were to employ local people, but they are reluctant to take on applicants who do not already possess the required skills and experience, because being in a fast-moving business environment restricts the available time for training.

Socio-spatial inequalities within the telecommunications and IT domain have also begun to emerge with regard to the distinction between companies and businesses in this domain. In relation to central London, for example, DTZ conclude: 'Take-up by internet companies reveals a polarisation between those

---

[87] For reasons of scope and space, this study is concerned primarily with economic development aspects of telecommunications networks, rather than the implications of the latter for social exclusion. Nevertheless, ICT strategies have been developed in both Paris and London to address the broad 'digital divide' policy area, within which objectives are to improve access, availability and take-up of ICTs particularly in poorer communities, through, for example, skills and training initiatives. With regard to Paris, the recent Guerquin report has brought 'digital divide' questions on to the regional agenda (Guerquin, 2001), while a series of neighbourhood-based 'social' ICT programmes and projects have been initiated across the London region tied in to the work of the Social Exclusion Unit of the government and its Policy Action Team (PAT 15) for information technology.

leasing new space in core locations and others taking poorer space, often in fringe locations' (DTZ, 2000a).

The western arc of Ile-de-France seems to be the 'hot spot' of telecommunications developments. Within Ile-de-France, the *départements* of Paris and Hauts-de-Seine dominate in the domain of telecommunications and IT. For example, two thirds of the employees of the region in this sector are based in those two *départements*.[88] It has also been noted that growth in the numbers of employees in Paris is restricted to telecommunications and IT services, whilst in Hauts-de- Seine it is the whole ICT sector which is growing (Courtois-Martignoni, 1999, pp.6-7). This is another reason for the intensity of the competition between the two *départements*, which we saw earlier in the chapter (Françoise Courtois-Martignoni, Mairie de Paris, interview with author, October 2000).

Social and territorial cohesion in Ile-de-France is one of the key objectives of ARTESI (Agence Régionale des Technologies de l'Information et de l'Internet) through the promotion of the development of ICT use. In particular, the Agence is now concerned with ensuring that Internet access in the region is available to all, and does not become characterised by divisions between those groups and areas which have access, and those which do not (Yannick Landais, ARTESI, interview with author, October 2000; ARTESI, 2000; ARTESI website). ARTESI brings together public and private actors concerned with information society objectives in Ile-de-France.[89] The main imbalance in the ICT domain was suggested to be between the west and east of the Ile-de-France region, where the latter was behind the former (Yannick Landais, ARTESI, interview with author, October 2000).

The creation of such partnerships can be seen as a way of ensuring a certain degree of overall territorial cohesion, and of territorially extending the benefits of the telecommunications domain. The Corporation of London also participates in a number of partnerships with authorities and organisations beyond the City:

> It is acutely aware that though the Square Mile houses the richest concentrations of economic business in the world, its neighbours include some of the country's poorest communities. For example, the City's neighbours include the London Boroughs of Tower Hamlets, Hackney, Islington, Camden, Westminster, Lambeth and Southwark – five of which feature in the Government's list of the eight most deprived communities in Britain (Corporation of London website).

In the telecommunications domain, an early attempt at ensuring a kind of regional territorial cohesion in Ile-de-France in terms of infrastructure was a project called La Francilienne des Télécoms, which was originally closely tied in with the Téléport initiative, which we have discussed at some length. The name of the project draws parallels with the road infrastructure of the region, because the

---

[88] The Hauts-de-Seine *département* (particularly Puteaux, Courbevoie and Nanterre), with 38% of regional ICT employees, actually outstrips Paris (29%) in this regard (Courtois-Martignoni, 1999, p.6).
[89] The Guerquin report for the CESR (Guerquin, 2001) has also discussed the widespread use of ICTs in the Ile-de-France region, and means to improve access to these in order to reduce the possibility of social and spatial 'digital divides' in the region.

main ring road around Paris is called La Francilienne, so the aim was clearly to have an equivalent information highway for the region. The investment from the Conseil Régional for this network came to an end as competition became inevitable. Subsequently, whilst noting the increasing numbers of infrastructures in the region following liberalisation, the Conseil Régional and IAURIF has posed the question of how to ensure all sectors (and therefore all companies) of the region benefit from this and not just the traditionally attractive areas:

> If we really observed major imbalances in the positioning of operators in Ile-de-France, we would perhaps have to think about creating rebalancing mechanisms. Normally that wouldn't mean networking the whole of the north east part of the Essonne, but instead proposing development points where we could eventually attract operators and institutions, so that there were conditions fairly close to what we find in the western sectors of Ile-de-France. But for the moment, we haven't reached this point... On the other hand, there is the Sipperec initiative, which is already a reaction (Daniel Thépin, IAURIF, interview with author, September 2000).

Nevertheless, territorial cohesion in the development of telecommunications infrastructures is a major part of a 'planning role' for the Conseil Régional d'Ile-de-France identified by Jean-Baptiste Hennequin:

> In Ile-de-France, there are a lot of places, the south of the Essonne, Yvelines, the Val d'Oise, where the population density is small and the operators just won't go. At the moment, it's not too serious because even if the operators are not present, the telephone service is, likewise the television, without problem. We're in a competitive system today and I don't know if the historic operator will continue to look after its network as well as in the past... There are many zones where operators won't go. So who's going to pay? Territorial authorities, in other words the communes, the departments, the Région, and this will happen in the short term, the medium term, a year, two years, I don't know... The government can say that universal service includes broadband Internet, but operators are not going to accept losing money. Therefore, in sparse zones, it's necessary to invest to create a market... (Jean-Baptiste Hennequin, Conseil Régional d'Ile-de-France, interview with author, October 2000).

The Institut is also thinking about how ICT development can create opportunities for social and economic integration. More large-scale projects are intended as:

> Ile-de-France has excellent potential to make use of, but also probably an image deficit. Global competition in the 'new economy' requires focusing on the constitution or the reinforcing of a number of quality 'poles' around which economic development can crystallise (IAURIF, 2000a, author's translation).

Issy-les-Moulineaux seems to be a particularly good example of a commune for which urban cohesion and competitiveness are joint objectives in telecommunications developments. As well as having a high quality telecommunications infrastructure to attract external companies, citizen connection has also been high on the agenda. All schools were connected by 1996, and by

1998 it was reckoned that the whole commune had access to new technology, either privately or via public access machines at the médiathèque, or the various projects which have specifically targetted certain groups (Florence Abily, Issy Média, interview with author, July 2000).

In London as well, economic development and social cohesion are allegedly parallel objectives of actions and interventions in the telecommunications and IT domain. Whether these are realistic parallel objectives, and whether they require different approaches and different agencies to be in charge of them is far from clear, particularly when discussed in a token and banal manner: 'A tension must exist between the two. The tension is around the allocation of resources, both money and manpower. It is a question of where that tension is managed. Even within social cohesion areas there are competing priorities' (Colin Jenkins, GLA, personal communication, June 2001). For Rob Tapping of BT access network planning in London, the key issues for the development and deployment of telecommunications infrastructures and services in London in the future seem to perhaps imply a combination of market and cohesion elements: 'there is a need for realistic forecasting of what is wanted by customers, and the anticipated take-up of our services, and constant re-evaluation, as well as generating a feeling for the pulse of the community' (Rob Tapping, BT, interview with author, January 2001).

There is evidence, however, that local and regional authorities are trying to direct economic development towards the east of the London region. As Ken Livingstone, Mayor of London, has said: 'London is initiating major changes which are going to reverse its historical development of the west towards the 'new east'' (quoted in Les Echos, 2000c). The Docklands East London agency is talking to boroughs further east of its territory (Gill Marshall, DEAL, interview with author, October 2000). There are important economic development initiatives in the east of London, such as the East London Business Alliance and the ExCel exhibition centre. Demand for property developments is allegedly outstripping supply in east London (Gill Marshall, DEAL, interview with author, October 2000). This reorientation towards the east of London has been on the agenda for a number of years. As Lord Young, former Trade Secretary, argued:

> London is bursting at the seams and we have to find somewhere to accommodate that growth. The only way to go is east. That is what the whole of Docklands, not just Canary Wharf is about... the whole balance of London will change... the effect will be to maintain London as the second financial centre of the world and to consolidate it as the financial centre of Europe (quoted in Crilley, 1993, p.132).

With skills and training becoming seen as an increasingly important part of telecommunications and IT infrastructure, local intervention in this sector could become a key aspect of the territorial regulation of the local telecommunications domain. The Docklands East London agency, for example, advocates taking advantage of the widespread provision of telecommunications infrastructure in East London to promote the area as a prime location for call centres, which can create important new employment for the local area (DEAL, 2000, p.2).

Concluding Remarks

In this chapter, we have seen how telecommunications developments in Paris and London are bound up in a number of territorial processes and practices at the local or intra-urban level, in addition to those of both the national level which we focused on in chapter 6 and the urban regional level which we focused on in chapter 7. We explored and analysed these local or intra-urban territorialities in three ways. Firstly, we looked at the territorial regulation role of local government in the shaping of certain telecommunications markets within Paris and London. This introduced aspects of the second section, in which we concentrated more on the largely market-led construction of new types of networked space and development of important corridors of fibre, which connect, in particular, these new spaces with the more traditional networked spaces of Paris and London. Thirdly, following on from all the discussions and analysis of our empirical chapters, we briefly explored the implications of telecommunications developments in Paris and London for territorial cohesion, and their incremental binding together of territorial contexts, specificities and practices from multiple scalar levels.

These analytical angles have drawn out crucial elements both for our comparison of telecommunications developments in Paris and London, and in relation to our theoretical discussions, in particular in chapter 4 regarding political economies of urban governance, rescaling and economic development. We need to consider all this in more detail now.

Firstly, the construction of particular local networked spaces varies greatly both within and between Paris and London, according to differing territorial contexts, specificities and practices. In particular, we tried to illustrate how processes and practices of local governance are bound up in the development of these networked spaces. Thus, the shaping of the 'new' networked space of La Défense was largely related to the way in which the local authority agency EPAD attempted to configure a changing telecommunications regulatory context for the deployment and provision of a quality and coherent infrastructure for the large companies located in the zone. The Sipperec initiative has been dependent on the construction of an intercommunality between communes in the inner *couronne* of Paris, and therefore on a willingness on the part of local mayors and authorities to participate in the initiative. In contrast, local governance has been a peripheral element in local telecommunications developments in the London region, although the London Boroughs Information Society Network (LBISN) did briefly invoke a coordinated inter-borough approach to the ICT sector, without the infrastructure focus of the Sipperec initiative. This approach has continued into the more recent and larger scale London Connects project. Throughout, it has been Lewisham, Camden and Newham, which have been the most active local boroughs in the ICT domain in London, and we were able, for the first two, to highlight some of the key aspects behind their enthusiasm and action. Leadership and financial support seem to have been crucial factors in the development of local initiatives, and both were more in evidence in Lewisham than in Camden, although, in turn, their joint leadership positions within London Connects promises to intensify interest and action in relation to ICTs from the actors of local governance in the London region.

The discussion of the complexities of the telecommunications market of La Défense appears to underline the substantial differences that remain between French and UK contexts for implementing telecommunications developments. In particular, difficulties arose because of the lack of experience of the local authority in configuring its territorial regulation in an emerging competitive context. Whilst wanting to improve the quality and extent of infrastructures and services available at La Défense, they emphasised retaining a 'coherence' to their development by designating a single concessionary company to manage and rent the dark fibre network. Although this took place in 1996, it proved to be completely at odds with the coming market liberalisation, particularly given that the prohibitive pricing introduced by Fibres Optiques Défense proved very costly to competitive operators, who needed to reach some of their biggest clients. Thus, the 'entrepreneurial' local governance strategy undertaken by EPAD became heavily criticised and eventually deemed anti-competitive, which, in a way, was the opposite to what it was trying to achieve in the first place. This does illustrate, however, the greater power of French local authorities in a more decentralised political context, compared to those in the UK, where territorial regulatory influence remains strongest with central government. Telecommunications developments in the Docklands, for example, are heavily shaped by the market and its key operators such as BT. Authorities such as DEAL have had little regulatory input so far, even if they are beginning to reflect on possible territorial strategies linked to telehousing and call centres. It is clear though that these operators must adapt their network deployment strategies to the territorial specificities of the Docklands, which we saw outlined by Pullen. It is these specificities, and especially the intensive redevelopment project and its growth as a major business area (which we saw in chapter 5), which have meant that for telecommunications operators, Docklands is not like anything they have seen before. In this way, then, differing regulatory and governance contexts mean that telecommunications developments in the two major 'new' networked spaces of Paris and London respectively have been shaped, configured and implemented differently, although the resolution of the legislative problems at La Défense suggest a shift towards a dominantly market-led form of territorial infrastructure deployment, as in the Docklands.

Secondly, although telecommunications developments in both cities can be argued to be dominated by vast corridors of fibre and copper connections linking key networked spaces, the nature, context and influence of these corridors differs in the two cities, according to varying territorial specificities and practices. We highlighted how the main networks in Paris and London seemed to extend from west to east across the central sectors of the two cities, with a number of spokes extending to other 'pockets' of economic activity. This does not, however, mean that telecommunications networks are being deployed in both cities in an overwhelmingly homogeneous manner. The dominance of the City of London as the key networked space in the London region, linked to factors of historic tradition and culture, remains important, and has not diminished despite the presence of intensive economic activity and telecommunications infrastructure in both the West End and the Docklands. The corridor of copper and fibre in the London region is overshadowed by the City and its concentration of global – local connections. The Paris region does not have such a dominant networked space. The nearest equivalent would be La Défense, but we saw how the construction of a networked space here was fraught with territorially-specific difficulties. In many ways, the original territorial focus of the Téléport outlined over ten years ago remains relevant today. Central Paris has a substantial concentration of economic activity and telecommunications infrastructure, while spokes of fibre extend in each direction towards the sites of the Téléport – west to La Défense, east to Marne-la-Vallée, north to Roissy, and south to the Plateau de Saclay. This 'spoke' structure varies completely to London and its central dominance of infrastructure.

Nevertheless, we did look more closely at interrelated practices of telecommunications development in both regions, which are reinforcing the importance of certain local sectors in the telecommunications domain – telehousing and start-up activities. In the former case, however, while the Docklands has become the prime location in London for these local territorially 'fixed' developments, in the Paris region, data centres are being located in many different areas, generally around the edges of central Paris. In the case of start-up companies, there are more similarities between the two cities, as the main concentrations have tended, rather spontaneously, to juxtapose the 'core' networked spaces of the City of London and the Paris Bourse sector respectively. These developments could be viewed as important extensions to the traditional core areas, particularly given the way in which, in both cities, territorial authorities have become increasingly concerned with their promotion (the Mairie de Paris and the Corporation of London).

Thirdly, all this can be seen to have led to another territorial juxtaposition, of wider regional importance, namely a reinforced double (scaled) contrast between these planned, coherent 'core' networked spaces and both the incoherent 'peripheral' disconnected spaces of both metropoles, and the subsequent overall socio-territorial incohesion and disjunction of both cities as a whole. However, the shape and nature of these territorial contrasts again varies between Paris and London. This variation is all the more intense because the ways in which telecommunications developments in the two cities are bound up with overall territorial cohesion are themselves bound up in all the differing, complex territorial contexts, specificities and practices from different scalar levels that we have focused on in our empirical chapters.

It has been stressed all the way through our empirical exploration of telecommunications developments in Paris and London that probably the principal territorial element of these developments is the contrast between the key (business) zones which have concentrated infrastructure deployment and the more 'peripheral' zones where the extent of infrastructure is restricted or even non-existent. While our focus in this study has been largely on the former, we were also concerned not to neglect the disconnected areas and groups present in both regions, in terms of what this means for these local areas and groups, and what these parallel (and in some cases, juxtaposed) connections and disconnections mean for the overall territorial cohesion of Paris and London as global cities. Socio-territorial cohesion in the ICT domain has only recently made it on to the agenda of local and regional authorities in London, for example as a key aspect of the London Connects project. In the Paris region, one senses that this has always been an important part of ICT developments, through the agendas and actions of bodies such as ARTESI and Sipperec, and local developments like those at Issy-les-Moulineaux.

In chapter 4, we discussed the common dual purposes or objectives of urban policies and developments in general, namely economic development and social and territorial cohesion. The difficulties in achieving both in parallel were highlighted, with the Docklands redevelopment project being, in some ways, emblematic of these difficulties. From the evidence and analysis of this chapter, and the previous ones, we can only conclude that it is at least equally difficult for urban telecommunications developments to achieve similar dual goals, perhaps especially in global cities. The specific telecommunications strategies which we have looked at all have either a clear and undiluted focus on economic development and competitiveness objectives, or have wider strategic objectives, of which economic development and competitiveness are a major part. In some ways, the friction or conflict in the landscape of telecommunications developments in Paris and London between competitiveness and cohesion objectives is virtually synonymous with the friction between the market and public sector regulation or intervention. Nevertheless, these observations merit some qualification. It was never the intention of this study to focus specifically on the role of telecommunications in *socio*-territorial cohesion in global cities. We have been very clearly most interested in examining closely the complex and multiple ways in which telecommunications developments are initiated and produced to implicitly and explicitly 'add value' to Paris and London as global cities. This focus might be viewed as being overwhelmingly economic, but, as we have seen, it has importantly necessitated exploring the wider implications of the undoubted privileging of economic development and urban competitiveness goals in telecommunications developments, namely the ways in which the construction of networked urban spaces is bound up with the bypassing of other, often juxtaposed, spaces and communities. The 'economic' objectives of telecommunications developments in global cities almost necessarily preclude social cohesion objectives, because to achieve both within the same development is either financially, institutionally, technically or territorially nigh on impossible. Yet, the increased intensity of territorial disjunctions within global cities, both on a local and a regional level, as a result of telecommunications developments, might arguably eventually re-impinge on the economic development and urban competitiveness aims of both telecommunications developments and global cities as a whole. This would, however, require further empirical study.

Our exploration in this chapter of the relations between telecommunications developments in Paris and London and particular local or intra-urban contexts and specificities has made up the third layer of our exploration of the parallel and incremental multiply scaled territorialities of these developments. However, as we underlined at the end of both chapters 6 and 7, these contexts and specificities are not bounded, independent and unilateral influences on these developments, as they cannot and should not be considered independently of the other layers of territorialities that we looked at in both these chapters.

We have now come to the end of our main empirical chapters presenting the elements of our research on the territorialities of telecommunications developments in Paris and London. Through the organisation of this empirical work into three specific themed chapters, we have been able to highlight, analyse, and compare the main features of these telecommunications developments in relation to the ways in which they are bound up in varying, but highly intertwined and interdependent, scalar territorialities, processes and practices. The analytical and comparative elements of the study have, thus, been structured so far according to the three main strands of the research as a whole. It is clearly necessary, therefore, to bring these elements together in a further detailed discussion of the study and the research strands as a whole. Only then will we be able to illustrate *both* the crucially intertwined and interdependent nature of the various scalar territorialities of telecommunications developments in Paris and London, and the similarities and differences in this between the two cities. This final analytical and comparative discussion is the basis of chapter 9.

PART III
ANALYSIS AND CONCLUSIONS

Chapter 9

Analytical Conclusions

Introduction

In the second part of this study, we embarked upon an empirical investigation of the territorialities of telecommunications developments in global cities through two thematically structured case studies of developments in Paris and London. The juxtaposition of the discussion of developments in both cities in the three thematic case study chapters permitted more direct comparative analysis than would have been the case if they had been separated into different chapters. In this chapter, we need to continue to relate this comparative analysis and the findings of the case studies back to the theoretical framework we outlined in chapters 2-4, thereby illustrating in some detail the empirical contribution of this study. We can then also look again at the theoretical framework in the light of the empirical analysis, to demonstrate that the relationship between the theoretical and the empirical should be a two-way process.

To complete this task, we shall continue to use the three thematic strands as a basic foundation, but it is now necessary to illustrate further the ways in which the strands, and the varying territorial specificities that they broadly focused on, interact and are mutually constitutive in how they influence and are influenced by the construction and packaging of telecommunications developments in Paris and London. From this, we can highlight the analytical conclusions and theoretical implications of the study as a whole.

The next section concentrates on the comparative and theoretical analysis related to the first thematic strand of our research, namely national policy influences on telecommunications developments in global cities. The following section then looks at our second thematic strand, the territorialities of telecommunications and global cities, before the third thematic strand on telecommunications, differing urban spaces and territorial cohesion is analysed. Following on from these individually stranded analyses, we draw the conclusions together as a whole, and consider in detail the theoretical implications of telecommunications developments in global cities from what we explored in this study. The final section brings the chapter to a close and looks ahead to the final concluding chapter.

National Influences on Telecommunications Developments in Global Cities

The first strand of our study was derived from the theoretical discussions of cross-national policy contexts in chapter 2, and has been concerned with exploring how telecommunications developments in Paris and London are influenced by both telecommunications policies and regulatory practices, and political cultures, intergovernmental systems, and urban and economic development policies in France and the UK.

The Continuing Resonance of National Policy Contexts

From our empirical explorations and comparative analyses, particularly in chapter 6, we have been able to see how, in spite of the development of increasingly high quality and ultramodern telecommunications infrastructures and services, and their promise of widespread glocal connections, as part of an apparently overarching global logic, their development and deployment in global cities such as Paris and London remain filtered and refracted through nation states and their urban and telecommunications policies and practices. All actors involved in the construction and packaging of telecommunications developments in Paris and London, whether public or private, situate themselves and their actions firmly in relation to national telecommunications and urban policy contexts. This suggests that we can only possibly hope to understand fully the dynamics of the development of telecommunications infrastructures and services in global cities if we take fully into account the relevant policies, actions and contexts formulated and implemented at the national level.

It is also important to understand the historical trajectories of these national policy contexts, which we explored in chapter 5. Contemporary French and UK telecommunications and urban policy, governance systems, and cultures of territoriality have steadily evolved, rather than emerged overnight. We saw, for example, how the current focus in Paris and France in general on relating the deployment of telecommunications infrastructures such as mobile and broadband to *aménagement du territoire* policy has its roots in the way in which the French state of the late eighteenth century tried to configure the development of the visual telegraph with the organisation of the national territory. More recently, in London and the UK, the continuing dominance and maturity of the free market in the deployment of infrastructures and the development of services in central London must be linked back to the neo-liberal and *laissez-faire* policies of the UK government in the early 1980s when telecommunications liberalisation began and the former PTT was privatised. Institutional differences between nation states such as France and the UK, then, lead to policy differences, and therefore, as we have seen, there are inevitably 'unique national paths of development' (Thatcher, 1999, p.2).

This all demonstrates the continuing importance of the nation state in territorial developments, and not the emergence of a 'phantom state', or a 'hollowed out' state, overtaken by the hegemony of economic dynamics of globalisation (Thrift and Leyshon, 1994; Jessop, 1994). The 'communicative power of electronic networks' still resides in and is regulated primarily by the nation state, just like urban policies, economic development practices and governance regimes as a whole.

The development of telecommunications infrastructures have to be placed within the 'national' context of the two cities as major capitals and historical centres. A clear general point which needs to be emphasised here is the profusion of infrastructures and services in Paris and London, compared with their respective national counterparts. This is directly related to their status as capitals, historical centres, and economic hubs, but equally the quality of their telecommunications infrastructures in turn reinforces this status. Nevertheless, we have been able to identify that this quality is sometimes overemphasised or overplayed, leading to the assumption that Paris and London have all the telecommunications networks and services that they may require. For example, while the national report on broadband development suggested that London has 100% broadband coverage, the regional report of the e-business adviser to the GLA implicitly refuted this and suggested instead that socio-territorial access to broadband should be one of the principal objectives of the future. We can also highlight the concern in the London Connects strategy for the boroughs of the capital to work together in the partnership to deliver e-government services by the 2005 target, particularly as a 'demonstration of good practice' to be set by London local authorities for the rest of the country to try to follow. The aim of London Connects to create a communication portal and regional network also takes into consideration the capital and global city status of London as a whole.

Despite an apparent shift away from nation state-dominated development of infrastructure networks towards the construction of concentrated, market-driven, elite networked spaces in key cities (Graham and Marvin, 2001), remnants of national, integrated practices of infrastructure deployment are still present. The examples of the nationally-based logics underpinning the development of broadband networks in the Paris and London regions, as well as mobile and wireless networks in France as a whole, demonstrate this quite clearly. We are not yet in a situation in which 'the traditional 'modern' notion that nationhood is partly defined by the ability to roll-out universally accessible infrastructural grids to 'bind' the national space is completely destroyed by the new infrastructural logic' (Graham, 1999, p.947). In France especially, we have seen how the development and deployment of any kind of telecommunications infrastructure is immediately set within a context of and requirement to obey the nationally 'binding' logic of *aménagement du territoire*. The wish for universal broadband access in both countries illustrates that building infrastructure networks with complete territorial coverage remains part of the role of the French and British states. Even in the market-dominated telecommunications environment of London, national policy focusing on the social and territorial development of broadband and ICT infrastructures as a whole has filtered down and powerfully shaped strategies such as London Connects and Broadband London. Indeed, it is particularly hard to imagine how any global 'infrastructural logic', however overwhelming, could bypass or overtake the historically embedded ideal and practice of French *aménagement du territoire*, which is itself gleaned from the egality ideal at the heart of the French nation. Graham and Marvin (2001) are quite right then to remind us of the continuing central influence of national regulation with regard to the reconfiguring of urban space and infrastructure networks in parallel. In sum, telecommunications developments in both Paris and London continue to be consistently shaped, and even dominated, by the distinctive, traditional territorial network cultures and contexts of France and the UK respectively. Thus, we have continually seen throughout the study the tenacity and path-dependent nature of French and UK approaches to territory – network regulation. Developments in Paris are bound up with the traditional French approach to territoriality, dating back, as Mattelart showed, at least as far as the Revolution, and in recent decades formulated under the policy of *aménagement du territoire* and its state-led programme of promoting the balanced spatial development of *l'hexagone*. In contrast, developments in London are bound up with the more deterritorialised and neo-liberal approach of the state to the UK space economy, within which a notion of informational regulation dominated by the market emerged quickly, and whereby intrinsic conceptualisations of territoriality were stunted or shunned in favour of a pragmatic, 'spatially selective' strategy focus.

Paris: The construction of decentralised and partnership-based strategies in a relatively early and unhabituated context of liberalisation From our analysis of national telecommunications policy contexts in chapter 2, we were able to see how France and the UK had formulated completely different national responses in the last thirty years to industrial and market changes in the telecommunications sector. There were some similarities, but on the whole, the French government's protectionist, *dirigiste* approach up to the mid 1990s was at the opposite end of the spectrum from the Thatcher government's neo-liberal market *laissez-faire* approach. Some of the territorial implications of these approaches have been evident from investigating telecommunications developments in Paris and London.

Given the lengthier period of state control and continuing monopoly in telecommunications policy in France, we would perhaps expect telecommunications developments in Paris to show a greater state influence than those in London, and this was borne out to a certain extent. The strategy of France Télécom in Ile-de-France is still necessarily influenced by the fact that the operator is still more than half-owned by the state, and we can suggest that this situation continues to partly overshadow the landscape of telecommunications developments in the region (and undoubtedly in France as a whole), not least through the extremely slow process of local loop unbundling. Some, notably AFOPT, the association of alternative operators in the French market, have seen in this snail-like process, the continuing level of 'protection' afforded to France Télécom by the French government and regulator. The incumbent was also allowed to get its ADSL service up and running in the Paris region a long time before unbundling even got anywhere near started in earnest (see e-territoires, 2001a).

We have also seen how the development of broadband infrastructures in Paris is being shaped, to some extent, by 'information society' policies at the national level (the *Loi sur la société de l'information* of the Jospin government), although less so than in London, because the development of broadband networks in Ile-de-

France in the last couple of years for the research communities of the region has been shaped and funded primarily by the Conseil Régional. Nevertheless, the importance of the national framework for ensuring the present and future roll-out and socio-territorial regulation of broadband in the Paris region is growing, and will dictate, at least partly, future intervention and action on the part of the Conseil Régional d'Ile-de-France in the broadband domain.

Chantal de Gournay's (1988) observation of the limited roles given to local authorities by the state in the telecommunications domain has evolved to a certain extent. Within telecommunications developments in Paris and London, there did not appear to be as explicit a state strategy to limit the extent of local authority regulation in this domain as her illustration of the deliberate lack of territorial fit between regional telecommunications offices and local administrative boundaries in France. However, following the liberalisation of French markets in 1998, as we saw in chapter 6, the state drew up regulations to limit the role of local authorities in local telecommunications markets and network management. By making local authorities overcome a series of regulatory obstacles to gain the right to construct networks (the need to undertake a 'publicity procedure' in order to demonstrate a 'deficiency' in the existing territorial offer; and the need to amortise investments within the short time span of eight years), the state clearly wanted to ensure it retained control of competitive telecommunications markets, and, some might say, protect the position of France Télécom. The state was obliged, however, eventually to give in to complaints from local authorities that these regulations were unrealistic, and relax them somewhat, although local authorities are still prevented from undertaking any role that even resembles that of an operator. Elements of this situation were able to be seen in chapter 8 in what happened at La Défense, where the local authority there bypassed these regulations and gave a monopoly concession to a private operator for infrastructure provision for the whole business area. The fact that the administrative tribunal of Paris ruled this to be anti-competitive shows that the French state is now keen for the benefits of competition to become widespread. Many elements of telecommunications developments in Paris seem to reflect, then, the still fully emerging context of liberalisation, and the difficulties of a shift from a protected state monopoly to open competition.

In parallel, and closely bound up, with this backdrop, we have also been able to draw links between telecommunications developments in Paris and the French system of governance, particularly the decentralisation measures introduced over the last twenty years, and its relatively strong set of intergovernmental relations and negotiations between communes, *départements*, *régions* and *l'État*. Before the shift from monopoly to liberalisation, for example, the construction of the Téléport Paris – Ile-de-France seemed to represent a typical decentralised economic development strategy, conceived and planned as it was by the Conseil Régional and IAURIF in the light of devolved administrative powers from the state downwards in the mid to late 1980s. The strategic role allotted to the regional government level in France in economic and territorial development has continued, and can now be seen in the lead the Conseil Régional has taken in financing and supporting the development of broadband infrastructure networks for regional research communities.

At a lower level, the extensive action undertaken in the commune of Issy-les-Moulineaux in the telecommunications domain has also been influenced by national governance arrangements. This action has had much to do with the leadership of the mayor André Santini, a former central government minister for communication, thus illustrating well the power attributed to political leaders in the French administrative system. In addition, the increasing emphasis in French territorial governance on intercommunality is perfectly exemplified by the Sipperec initiative, in which many of the communes of the inner *couronne* of Paris have joined together to attempt to regulate the territorial market of telecommunications infrastructure in a more favourable and beneficial manner than could be achieved individually.

London: The dominance of a laissez-faire market, mature competition, and territorial strategies founded on centralised policy The contrast between the UK government's neo-liberal market *laissez-faire* approach to the telecommunications sector starting in the early 1980s, and the French *dirigiste* and protectionist stance to the sector could be seen in some elements of our exploration of telecommunications developments in London.

Given the earlier introduction of market dynamics in telecommunications policy in the UK, we would perhaps expect telecommunications developments in London to show a greater level of market-led influence than those in Paris, and this was borne out to a certain extent. However, there are relative similarities between France and the UK in the domain of local loop unbundling, which has been almost as problematic in London and the UK as in Paris and France. This is in spite of differences between France Télécom and BT in terms of operator – state – regulator relations. The argument that BT could be being 'protected', in the same way as France Télécom is suggested to be, holds little sway because the former is completely privatised, whereas the latter is still majority controlled by the state.

We have also seen how the discursive and prospective material development of broadband infrastructures in London is being very largely shaped by 'information society' policies at the national level. While the GLA and their e-business adviser have been discussing and formulating proposals in some detail recently, in a similar way to the Conseil Régional in Paris, these proposals and further interventions are being placed firmly in relation to the original policy document of the government's e-envoy. Without this national framework, it is doubtful whether the GLA would have placed such emphasis on ensuring the roll-out and socio-territorial regulation of broadband in the London region.

In parallel, and closely bound up, with this backdrop, we have also been able to draw links between telecommunications developments in London and the UK system of governance, particularly in the dominance of centralisation measures, and a relatively weak set of intergovernmental relations, which have favoured top-down approaches to economic development practices and strategies in the last two decades. This has been most evident in the relative lack of public sector strategies in the telecommunications domain in the London region up until recent years. The overwhelming centralisation of the UK governance system, with no regional level of government between 1986 and 2000 and limited intervention possibilities for local government, meant that telecommunications developments in London were influenced by the dominant neo-liberal state approach of creating the right conditions for the market and the private sector to shape developments. Beyond the influence of the government decision on the liberalisation of UK telecommunications markets, the development of infrastructures and services in the key business areas of the City of London and the Docklands was overwhelmingly shaped by the market, with little or no role for local authorities. While these primary networked spaces are still dominated by competitive offers from multiple operators, in other respects, the restructuring of territorial governance in the UK appears to be intrinsically bound up in the emergence of telecommunications strategies in London shaped more by territorial authorities than the market *per se*. The reintroduction of a regional government authority, the GLA and the Mayor, and the slightly less centralised, top-down nature of intergovernmental relations, have been at the heart of current strategies for configuring territorial broadband network deployment across the London region, in which the GLA and its e-business advisor have been taking the lead, and for London Connects, a veritable intergovernmental initiative promoting social cohesion and economic development, in which certain boroughs such as Lewisham, Camden and Newham have been able to participate very actively. This restructuring of national governance arrangements appears, then, to have reterritorialised telecommunications developments in the London region, and opened the possibility that while the market can continue to take care of key networked spaces, elsewhere, local and regional authorities can territorially regulate the markets of peripheral areas in order to try to ensure widespread broadband access, or even access to simple e-government services.

The Parallel Territorialities of Global Cities and Telecommunications

The second strand of our study was derived from the discussions of the theoretical conceptualisations of the relations between global cities and telecommunications in chapter 3, and has been concerned with exploring how telecommunications developments in Paris and London are discursively and materially constructed and packaged to meet and respond to the requirements of Paris and London as global cities.

The Construction and Establishment of Territorially Specific Global Localisation Strategies by Telecommunications Operators

From our empirical explorations and comparative analyses, particularly in chapters 7 and 8, we have been able to see how on the surface, there appears to be substantial convergence in the type and nature of telecommunications infrastructures and services being deployed in major European cities such as Paris and London. However, a more detailed investigation of this reveals that there are still significant differences between cities because telecommunications operators are constructing differing regional and local strategies according to the economic, political, social and territorial specificities and assets of these cities as a whole, and the urban spaces within them. This suggests that we can only possibly hope to understand fully the dynamics of the development of telecommunications infrastructures and services in Paris and London if we take fully into account the differing territorial specificities and contexts common to each city. There are two points to be made here in relation to the strategies of telecommunications operators.

Firstly, this means that *different* operators are constructing 'glocal scalar fixes' of telecommunications developments in *different* ways in Paris and London in response to the different contexts and specificities of these two cities. This would probably be expected though because of varying corporate strategies, cultures and focuses. Thus, we saw in chapter 7 how the territorial strategies of the competitive operators Colt and MCI WorldCom differed in certain elements both for the Paris and the London regions. Equally, the strategy of France Télécom for the Paris region differs significantly to that of BT for the London region, not merely as a direct result of them being two different companies, but also more with regard to the territorial contexts and specificities of the two regions.

Secondly, and far more revealingly, we have seen important elements of the ways in which the *same* operators are constructing 'glocal scalar fixes' of telecommunications developments in *different* ways in Paris and London in response to the different territorial contexts and specificities of these two cities. This is a key element in the whole landscape of global city telecommunications, because it demonstrates the crucial importance of territoriality in the development and deployment of infrastructure networks and services, which is otherwise held out to be globally convergent and increasingly territorially homogeneous. Thus, we also saw in chapter 7 how the territorial strategy of Colt in the Paris region differs from the territorial strategy of Colt in the London region. While, on the one hand, Colt proclaims itself to be a pan-European operator with numerous metropolitan fibre networks located all over the continent and a fairly homogeneous set of voice and data network services available anywhere, on the other hand, in order to be able to offer these networks and services, it is clear that they must still formulate and implement particular strategies for each metropolitan area such as Paris and London based inherently on the territorial contexts and specificities of these cities. These important differences in territorial strategy also appeared to be present in the activities of MCI WorldCom, but our evidence here was less extensive. Nevertheless, we can hypothesise that MCI WorldCom must take into account many of the same contexts and specificities as Colt in both regions (for example, tariffs for rights of passage; overall regulatory context; 'hub and spoke' geographies), and therefore adapt their strategies according to differences in these between Paris and London. They proclaim a 'local – global – local' strategy to their activities, which may include elements of homogeneous network and service provision, but which must, as they highlight 'local', equally include a distinctively territorial basis to the deployment of their infrastructures in both regions.

The strategies of telecommunications operators are therefore increasingly shaped by the parallel influence of global logics and territorial specificities. By having to construct, deploy and run integrated pan-European, regional and local infrastructure networks respectively in parallel in and between Paris and London, it is clear that, for many telecommunications operators in Europe, the influence of multiply scaled and mutually constitutive territorialities on their activities and strategies is absolutely crucial.

The Historical, Institutional and Territorial Shaping of Regional and Local Authority Action in Telecommunications Developments

In addition to the territorial strategies of telecommunications operators, from our empirical explorations and comparative analyses, particularly in chapters 7 and 8, we have also been able to see how there exist varying responses by regional and local authorities to telecommunications developments in Paris and London according to varying historical, institutional and territorial contexts. These contexts become crucial, then, in influencing the differing ways in which these authorities 'shape' the 'space of flows' to meet their particular economic development and / or socio-territorial cohesion concerns.

Equally, the different groups and actors involved in telecommunications developments in Paris and London inevitably have differing views on and approaches to the wider territorialities of these cities. We discussed the views and approaches of operators in the previous section, and saw how they differed both between operators and for the same operators. Regional and local authorities develop differing views and approaches between themselves as well. The concerns of these authorities in the telecommunications domain differ primarily according to their differing territorial concerns.

It is clear, then, from our study that telecommunications developments are being discursively and materially constructed and packaged in a wide variety of ways to meet and respond to the requirements of Paris and London as global cities. This situation is unquestionably partly the result of the wide variety of actors and agencies involved to differing degrees in telecommunications and IT in both cities. There are telecommunications developments in both cities which involve predominantly the public sector (local borough IT strategies in London; reconfigured broadband research infrastructures in Ile-de-France), predominantly the private sector (the dominance of the free market in infrastructure deployment in the City of London, and more recently at La Défense), and partnerships between the two (London Connects and the Sipperec initiative in Paris).

The comparative advantages of cities such as Paris and London in terms of telecommunications infrastructures are continually said to be weakening, as similar investments are made in other urban regions, as operators try to configure competitive global networks with numerous homogeneous nodes and seamless end-to-end connections. The construction of many parallel pan-European networks in recent years seems to bear this out. In the light of this study, what appears to be becoming even more important, then, is not so much the nature and extent of urban telecommunications infrastructures in Paris and London *per se*, although, as we have seen, even these remain highly differentiated on their own terms, but the binding together of these infrastructures with a variety of territorial mechanisms, processes and practices from the national, urban and local levels in an interacting exercise of parallel and mutually constitutive shaping and reconfiguring. It is, therefore, the territorialities of telecommunications developments in Paris and London which will continue to mould, transform and determine the advantages, in relation to telecommunications infrastructures, that the cities retain, lose and regain within the global competitive context.

Paris: The configuration of developments by an abundance of actors to shape the 'space of flows' within a crucial competitiveness agenda   Telecommunications developments in Paris currently involve or concern an increasingly wide variety of groups, agencies and organisations. Strategies in the Ile-de-France region are most commonly being framed within a recognised need to increase the regional competitiveness of Paris in relation to other European cities. We can suggest that the proposed development of a multisite Téléport in the Paris region at the end of the 1980s can be connected with the observation of Lipietz (1995) on how some French planners and agencies were beginning at that time to try to configure a Parisian megapolis to prevent the capital region becoming marginalised within Europe. Certainly the regional planning agency, IAURIF, was centrally involved in the strategy, and made increasing regional competitiveness a main objective of the Téléport. This competitiveness agenda has, nevertheless, not been particularly well problematised or detailed. There appears to be an element of 'inevitability', in terms of both the need for Paris to improve its European standing, and the role of telecommunications networks within this. This was perhaps reflected in the way in which the original potential costly proposals from the regional Chambres de Commerce for intervention in the telecommunications domain in the context of the Contrat de Plan between the state and the region were narrowed down to a much smaller focus merely on research infrastructure networks. While local and regional authorities in the Paris region view telecommunications infrastructures as broadly necessary and important for competitiveness objectives, they are perhaps not quite sure exactly why this is the case, or they cannot see the direct results of investment in these infrastructures as clearly as they can for other types of investment. Nevertheless, given that the Ile-de-France region perceives itself to be an important centre for research and development sectors and industries, and wants to maintain this advantage, we saw how the Conseil Régional has rarely hesitated to support and provide financial assistance for the deployment of broadband infrastructures for this community in several locations around the region. They have understood that a lack of broadband access would handicap these sectors and industries, which would handicap the economic development of the Ile-de-France region as a whole.

London: The predominance of global finance, a scarcity of strategic actors, and a prospect of new regional government action   In contrast to the situation in Paris, telecommunications developments in London involve or concern a limited number of groups, agencies and organisations. While the Téléport development in the French capital highlights the key role taken there by the regional arm of the state in telecommunications strategies as part of urban and economic development policy, we can compare this to the minimal role of the Corporation of London, and their Economic Development Unit, in telecommunications strategies for the City of

London. Here, throughout the 1990s, the market has dominated telecommunications developments, mirroring the importance of the market as opposed to the state in UK economic governance as a whole during this time. The reintroduction of a regional level of government for London, and the key recent role of the e-business adviser to the GLA in helping shape regional public intervention in the telecommunications domain in London (for example, in broadband access) is perhaps an indication of a slight shift in the nature of economic governance in relation to telecommunications.

London's position as a centre of global finance, and as a global city, is undoubtedly one of the key elements in shaping the landscape of telecommunications in the city. The phenomenal demand of financial and business institutions located in the City and in the Docklands for high quality infrastructures and services has not only been bound up in the overall evolution of UK telecommunications policy in the last twenty years or so, but has equally structured market provision for the London region as a whole.

The development of a strategic regional stance and approach to the telecommunications domain is only just now beginning to take shape, following the creation of the GLA. This approach has, then, been predominantly discursive and exploratory up to now. Recent strategies such as broadband network deployment and London Connects are attempting to configure a relatively inchoative territorial or socio-territorial focus to a telecommunications domain in the London region traditionally oriented around market-led network deployment. The GLA and local boroughs in London have therefore inevitably had to take time to develop their skill and knowledge to implement such an approach. The implications of their actions in these strategies will only become clear in time, but the emphasis placed on public-private and intergovernmental partnerships within these strategies seems to have provided a solid base to work from, ensured a wide variety of viewpoints are included, and helped resolve problems and tensions quickly and effectively.

Telecommunications, the Differing Urban Spaces of Global Cities, and Overall Territorial Cohesion

The third strand of our study was derived from the theoretical discussions of economic governance, rescaling and urban restructuring in chapter 4, and has been concerned with exploring how telecommunications developments in Paris and London both interact with processes of local governance and economic development, and relate to wider concerns with territorial cohesion in global cities.

The Inherent Intertwining of Territorial Fragmentation and Telecommunications Developments

From our empirical explorations and comparative analyses, particularly in chapter 8, we have been able to see how the construction and packaging of telecommunications developments in Paris and London is bound up in attempts at increasing territorial cohesion as a whole, but also in a general intensification of metropolitan fragmentation. The development of telecommunications infrastructures and services illustrates the size and nature of the difficulty of ensuring territorial cohesion in global cities such as Paris and London. Whilst they are often viewed as a 'fix' to this problem, their deployment seems instead to create a kind of zero-sum game in which the divide between connected, core strategic glocal 'premium network spaces' and disconnected, peripheral 'local' spaces in these city regions is reinforced, despite their paradoxical juxtaposition. This suggests that we can only possibly hope to understand fully the dynamics of the development of telecommunications infrastructures and services in Paris and London if we take fully into account the ways in which the creation of networked urban spaces implicitly or explicitly creates less connected, or even disconnected, 'interstitial' urban spaces, and subsequently problematises the territorial cohesion of the city as a whole.

Telecommunications developments in global cities connect and disconnect, and they do so at and between differing scales, and with differing intensities. It is not the case either that economic development and territorial cohesion objectives for telecommunications developments are completely separate and unrelated. The privileging of economic development goals through, for example, the creation of 'premium network spaces' to increase the competitiveness of global cities as a whole, might potentially rebound on the economic development and competitiveness of global cities if overall territorial disjunction or incohesion becomes overly problematic or neglected. Strategic networked spaces within a city can benefit the city as a whole, but by the same token, disconnected spaces within a city can harm the city as a whole. Global cities, then, cannot be seen as a homogeneous 'whole', given the great socio-spatial differences within them.

The territorialities of telecommunications developments in Paris and London have been most complex in relation to our third strand. Although this thematic strand was conceived to focus on the intra-urban or local levels within global cities, it is clear that the complexity derives essentially from the fact that the territorialities of telecommunications developments of the previous strands ('national' and 'urban regional' specificities and influences) are also present at the intra-urban or local level, and combine with particular local specificities and influences here to create a situation which exemplifies most intensely the multiply scaled nature of, and the mutual constitution of scalar levels within, telecommunications developments in global cities such as Paris and London.

Paris: Increasing territorial responses at the level of local governance, and relatively limited spatial variations in connection Telecommunications developments in Paris are being increasingly constructed and shaped at the local level, by authorities, communes or groups of communes. We can suggest that there are more territorially networked spaces in the Paris region than in London. In addition to the main business areas of central Paris and La Défense, we have explored the packaging of infrastructures and services at Issy-les-Moulineaux, throughout the inner *couronne* in the Sipperec initiative, in the technopoles and science parks to the south of Paris (broadband research infrastructures), and at

Marne-la-Vallée, which has the 'remnants' of the Téléport strategy. Despite the identification of territories in the region which lie beyond the networks of the main operators, looking at the Ile-de-France region strategically, suggests slightly less intensive spatial variations in telecommunications provision and network access than in London.

The Sipperec strategy, for example, is a more inclusive type of 'local bypass', resulting from the construction of a new parallel infrastructure network. It can even be seen as a 'strategy of resistance' to the overpricing of local connection for communes to the networks of France Télécom. The communes of the inner *couronne* of Paris, through the Sipperec group, are thus very much 'asserting the continued publicness' of urban telecommunications developments (Graham and Marvin, 2001, p.392-393). Even if we cannot really characterise these communes as particularly poor and disconnected local spaces, the counterhegemonic foundation of the Sipperec strategy offers nonetheless an example of the ways in which 'resistance practices can help shape the creation of places in the interstices between the logics of premium network spaces' (Graham and Marvin, 2001, p.394). In this way, it also illustrates that such resistance practices in contemporary cities can be subtle, innovative and collective attempts based on the territorial regulation and control of the market, and not merely explicit and demonstrational organisation of social unrest to wrestle control from the hands of hegemonic powers.

London: Limited local intervention, market-led deployment, and distinct variations between networked and disconnected spaces We have already observed the dominance of market-led developments in telecommunications in London. The local authorities in and around the City and the Docklands area, the two most intensively networked spaces, have very limited roles in shaping the deployment of infrastructures and services there, beyond ensuring that the conditions for competition between operators are optimal. We can draw parallels between the corridor of fibre that extends between the Docklands and the City and on to the West End, and that which extends between La Défense and central Paris and goes on towards Marne-la-Vallée in the Paris region. However, although we can suggest that the London corridor is probably more intensively networked, beyond these highly networked spaces and corridors in London, the situation is much different. While local communes and authorities are becoming heavily involved in telecommunications provision in their territories in Paris, there is very limited local authority intervention in London, even in the case of boroughs which are only located on BT networks. The spatial variations in telecommunications provision and network access, between connections and disconnections, in the London region is, therefore, much greater than in Paris. The City of London and the Docklands area concentrate much of the infrastructure provision of the whole region. For example, while the development of data centres has taken place across the central Paris and inner couronne areas, in London it has become focused largely on the Docklands zone.

Theoretical Implications of Telecommunications Developments in Global Cities

Telecommunications developments in Paris and London are illustrative of a proliferating demand for and use of communications technologies throughout western society. In France and the UK, this demand and use is heavily concentrated in their capital cities.[90] As we have seen, the full territorial provision of telecommunications networks and services is much less of an issue in these two cities than in provincial areas of the two countries, despite the prevalence of territorial disconnections in the less favoured sectors of Paris and London.

We can link increasingly widespread telecommunications developments in Paris and London, and the proliferating demand and use of communications technologies that they demonstrate, to the nodal status and role both cities have in a globalising economy. The quality and quantity of telecommunications networks and services of the City of London, the Docklands, central Paris and La Défense are emblematic of, as well as being a major factor in, both cities, although London in particular, being key urban financial centres. In this respect, we can reaffirm the view of Sassen (1991; 2000) on how information and communications technologies are one of the key supports for global cities to act as major centres of command and control in the world economy.

Continuing the discussion about the relationships between Paris and London as global cities, their telecommunications developments, and their roles in a globalising economy leads to more tenuous ground, however. This study has sought to show how telecommunications developments in global cities are bound up in a variety of continuing and even reinforced territorial specificities and contexts, which are most noticeably, although not exclusively, present on a national level. The differences between the territorialities of telecommunications developments in Paris and the territorialities of telecommunications developments in London are sometimes quite stark, and reflect, above all, divergent French and UK policies, traditions, cultures and contexts. This conclusion goes adamantly against the grain of many theoretical arguments and prognoses, which suggest, implicitly or explicitly, that processes of globalisation have rendered obsolete or 'hollowed out' the importance of the nation state, and that global cities are becoming or have already become territorially homogeneous entities or nodes on territorially apathetic, space-annihilating global networks of exchange and control. Indeed, given that it is the progress in 'global', 'space-transcending', 'place neglecting' information and communications technologies which is frequently held up to be a major factor in the supposed decline of the nation state and of the importance of territory, there is something highly paradoxical and ironic in the ways in which telecommunications developments in Paris and London are revealed to be both utterly bound up in differing national level processes and practices, and, relatedly, characterised by great territorial variations at multiple, but telescoped, scales.

---

[90] The Paris and London regions have been the traditional centres of the French and UK computing and communications industries (see Swyngedouw, Lemattre and Wells, 1992).

In this way then, both the implementation and implications of telecommunications developments in Paris and London need to be thought of as a 'scale question' as much as a (traditional) 'urban question' (Brenner, 2000). Their impinging scalar complexity is composed of both the forms and processes which necessarily influence the formulation and implementation of these strategies from above the urban scale (global, national, regional) and below it (local), and the subsequent post-implementation interactions of these strategies with practices above and below the urban scale, which in turn re-influence the forms and processes that influenced the strategies to start with. It is therefore a two-way incremental interaction with logics and policies from a combination of scales both influencing and being influenced by the telecommunications strategies.

The Need for a Territorial Basis to Conceptualisations of Global / World Cities

There is a need for recognition of the multiply scaled and mutually constitutive processes and practices bound up in relations between telecommunications and global cities. It is clear that we cannot privilege perspectives purely based on the nature of telecommunications as necessarily space-transcending and distance-annihilating technologies, in which their need for 'fixity' or a territorial basis is forgotten. Neither can we overly concentrate on global city conceptualisations, in which telecommunications are both solely seen to be bound up in the reinforcement of these cities as nodes of the global economy, and viewed as territorially homogeneous infrastructures, synonymous with singular, automatic connection to global networks of exchange and information. There seems to be an overwhelming tendency in research on global cities to focus largely, if not exclusively, on the 'global', and to ignore or to forget about the 'city', and its specificities. In other words, there is a concentration on the 'metageography' of relations *between* cities, and a certain ignorance of the original geography of complex relations *within* cities (Beaverstock, Smith and Taylor, 2000; see also Taylor, 2000). Inter-city economic flows and links are held up to illustrate the primacy of New York, London and Tokyo among other cities, while the territorial specificities, contexts and cultures of each of these cities, which are the absolute basis of their 'global' primacy, seem to be background material. The office location strategies of producer service firms are a particular favourite in global city research, but the analysis of these strategies stops at lists of cities where a firm is located, and does not attempt a more detailed understanding of the territorial specificities of the cities which influences the location strategy of a firm. The fact that New York is a global city seems to be the inherent reason for a firm to have an office there. In this study, we have attempted to go beyond the implication that the quality and quantity of telecommunications infrastructures and services in Paris and London is simply related to their status as important global cities, and develop an understanding of how territoriality both underpins telecommunications deployment in the first place, and is then, in turn, shaped by this deployment of infrastructures and services. Telecommunications developments in Paris and London illustrate thus the crucial importance of 'local conditions' as well as, perhaps even more than, the 'world economy' to the form of the two cities

(Longcore and Rees, 1996, p.355). Command and control are therefore underpinned primarily by local territoriality before the flows and networks of the global economy. The global city cannot be seen merely as 'a function of a network' as Sassen would have it (Sassen, 2000, xiii), but more as a function of parallel multiply scaled territorial processes, practices, mechanisms, specificities and contexts.

Telecommunications Developments in Paris and London: Shaping the 'Space of Flows'

Looking at our empirical discussions and analyses of telecommunications developments in Paris and London suggests interesting links back to the work of Castells on the relations between global cities and telecommunications networks, particularly his theories expounding the emergence of a dominant 'space of flows', which we discussed in chapter 3. Our case study chapters have revealed the existence in Paris and London of the three layers of physical supports making up Castells' space of flows: a circuit of electronic impulses such as telecommunications; a network of place-based nodes and hubs; and the spatial manifestations of dominant groups and interests. The first and second layers, the telecommunications infrastructures of the two cities and their place-based nodes and hubs, are relatively evident compared to the third layer, the territorial configurations and articulations of local and regional authorities and telecommunications operators. While the latter are very much the essence of the subject of this study – the comparative territorialities of telecommunications developments in Paris and London – it is clear that it is in the differing ways in which these configurations and articulations are bound up with, or shaped in parallel to, the other layers, particularly the first layer, that key territorial differences within and between global cities are produced. In this way, our study has illustrated how there is a fundamental set of hard physical foundations to the space of flows, constituted by various types of complex, fixed infrastructures, which must be woven into the urban fabric and wired from a to b. This foundational infrastructure deployment intrinsically manifests a parallel set of often long-established, path dependent and proper regulatory regimes and national and local development procedures and practices, which remain firmly and obstinately rooted in spite of pressure from apparently increasing globalising logics. Thus, the territorial strategy of an operator such as Colt (and to a lesser extent, MCI WorldCom) differs between the Paris and London regions, according to the differing set of contexts and specificities bound up in the deployment of telecommunications infrastructures in the two regions.

While we have seen how the technological infrastructures of the first layer of the space of flows and the location of Paris and London within global webs of place-based nodes and hubs of the second layer suggests a degree of convergence between the two cities *per se*, we have focused most emphatically on the differing territorial contexts, specificities and cultures which influence and are influenced by the spatial configurations, articulations and requirements of dominant groups in terms of telecommunications developments in the two cities. It is therefore the third layer of physical supports in the space of flows that shapes and is subsequently shaped by the first layer, and these articulations then determine the place of the nodes and hubs of a city in the global network architecture of the second layer. The layers of the space of flows must therefore be seen as completely mutually constitutive.

In the light of our research, where we can suggest the conceptualisation of the space of flows needs to be deepened is in the highly differentiated nature of the third layer. Here, rather than discussing a rather homogeneous layer of the spatial configurations, articulations and requirements of dominant groups, we need to open this out to reveal the multifarious and sometimes highly conflicting and discordant nature of these configurations, articulations and requirements. As we saw in chapter 8, the shaping of the space of flows in the business quarter of La Défense was characterised by a set of conflicting positions and viewpoints, between the local public planning agency and its decision to allocate a territorial market monopoly to Fibres Optiques Défense, other telecommunications operators and their legal objections to the monopoly concession, and the administrative tribunal of Paris which eventually upheld the objections.

At the same time, as Castells makes clear, we need to be aware of the continuing importance of the space of places in parallel with the emergence of a dynamic space of flows. As emphasised in chapter 3, this whole study has been very much situated at the crossroads between these two logics, attempting to explore the interplay both between the two cities as part of the space of flows and the two cities as territorially specific places, in terms of their telecommunications developments, and between telecommunications developments in the two cities as an infrastructural support to the space of flows and telecommunications developments as place-based infrastructures. Here, our theoretical foundation of cross-national policy contexts has helped to demonstrate the limits to the emergence of a space of flows in Paris and London, as in both cases, as we saw in chapter 6, the influence of national urban and telecommunications policy processes and practices on urban telecommunications developments remains highly significant. Subsequently, we can question whether the shaping of the space of flows in global cities such as Paris and London depends on the mutually constitutive physical supports of telecommunications infrastructures (first layer) and the spatial configurations, articulations and requirements of dominant groups (third layer), but also on a more pronounced 'second layer' of territorially specific national and urban processes and practices, rather than a network of place-based nodes and hubs, as Castells suggests. It may seem slightly paradoxical to posit that a set of relatively *fixed* territorial processes and practices forms more of a physical support to the emergence of a space of *flows* in a global city than a network of nodes and hubs in a global economy, but this study has illustrated the absolutely critical role national policy contexts and cultures continue to play in the emergence of a space of flows in global cities, as represented by urban telecommunications developments. In other words, French and UK urban and telecommunications policies and cultures tend to have more influence on and are more influenced by telecommunications developments in Paris and London than the position of (key hubs in) the two cities in relation to networks of a global economy. In this way, even telecommunications developments, with their global-local connection implications, help to illustrate how Paris and London as global cities remain very much places instead of processes.

Telecommunications Developments in Paris and London: The Construction of Nationally Specific 'Glocal Scalar Fixes'

The telecommunications developments of Paris and London have illustrated quite well the resonance and the complexity of the scale question, as discussed in the theoretical chapters. What we have analysed throughout the study is a selection of some of the types of telecommunications developments which exist side by side in the Paris and London regions. One of the key ways in which they can be differentiated, however, is in the scale of the connections they produce or construct. While the infrastructures being deployed in the business areas of the City, the Docklands and La Défense are primarily about connecting these concentrated local zones to global networks of exchange and transactions, which are also connected in turn to other local business areas around the world, other strategies imply, at least in the first instance, smaller-scale connections. The Sipperec project and London Connects initiative are concerned with increasing the level of connection of local communes or boroughs to regional networks, either of infrastructure in the former, or of e-government practices in the latter. Together, they produce a picture of the way in which telecommunications developments in global cities both exist at, rely on, and reinforce differing geographical scales in parallel, as well as involving interaction in a number of ways across these scales (through, for example, partnerships between different actors). In this way, we can at least initially suggest that telecommunications developments in Paris and London begin to illustrate Brenner's observations from chapter 4 about how '[t]he territorial organisation of contemporary urban spaces and state institutions must be viewed at once as a presupposition, a medium and an outcome of th[e] highly conflictual dynamic of global spatial restructuring' (Brenner, 1999, p.432). The ways in which telecommunications infrastructures and services are being (differentially) 'fixed' locally, whilst at the same time connecting to networks at the global scale suggests that they can be viewed as a fine illustration of Brenner's (1998b) notion of the 'glocal scalar fix'. The varying ways in which this process is taking place in both cities, through the constructing and packaging of telecommunications developments, means that we are necessarily exploring the different types of glocal scalar fixes in Paris and London in the telecommunications domain.

   The changing nature of geographical scale, or the politics of rescaling, has been a core component of this study of telecommunications developments in global cities. As we saw in chapters 3 and 4, this was derived from both the intrinsic characteristics or connotations of, and the relational characteristics or connotations between, telecommunications and global cities as multiply scaled social constructions. Telecommunications and global cities depend on parallel mechanisms of territorial fixity and 'global' network connection. It is therefore clear that neither can be 'contained' at a particular scalar level, but that scale is an important 'component' in their development, and in the wider production of space that is engendered by their being shaped in parallel. All this has been clearly seen and foregrounded throughout the study through a structure based on three intertwined thematic strands, focusing on the national, urban and local levels. The emphasis has not been on suggesting that certain elements in the shaping of telecommunications developments in Paris and London are strictly 'contained' in the national level (chapter 6), urban level (chapter 7), or local level (chapter 8), but that these telecommunications developments are cumulatively shaped by, and in turn shape, a series of interrelated and parallel territorial specificities, processes and practices from these differing, but intertwined, scalar levels. In this way, our study of telecommunications developments in Paris and London has illustrated well how 'spatial scales constitute a hierarchical scaffolding of territorial organisation upon, within, and through which the capital circulation process is successively territorialised, deterritorialised, and reterritorialised' (Brenner, 1998b, p.464). The 'hierarchical scaffolding' of national – urban – local specificities, processes and practices becomes at once the foundation for the territorialisation of telecommunications developments in the two cities (which are an increasingly crucial part of the overall capital circulation process), and the means by which this territorialisation occurs, as well as being reconfigured itself as a result of this territorialisation. Brenner has elsewhere described this tripartite function as 'presupposition, medium and outcome'. However, we have also elaborated on his linked notion of the 'glocal scalar fix', as a way of analysing how the interrelations between this 'hierarchical scaffolding' of multiscalar practices and the development of telecommunications networks lead to the 'fixing' in place of particular strategies and infrastructures. Thus, we have been able to see how telecommunications developments in Paris and London are marked by the construction of nationally specific 'glocal scalar fixes' through the juxtaposition of particular cultures and contexts of regulation, governance and territoriality, and particular global localisation strategies of telecommunications operators. If the circulation of capital is dominated by a contradiction between fixity and motion, as it tries to overcome spatial constraints from territorially fixed structures, then firstly, the 'glocal scalar fix' appears to be a crucial concept in our attempts at understanding economic restructuring processes, and secondly, within this, telecommunications developments appear to be one of the foremost illustrations of this concept, based as they are too on territorially fixed infrastructures over which planetary communications circulate. Given the parallels between them, it is therefore perhaps unsurprising that the circulation of capital and telecommunications infrastructures are so increasingly interrelated and dependent on one another.

The recent emergence of local, regional and national government concern in both Paris and London for assuring and extending the territorial coverage of broadband infrastructures can be seen as a further configuration to the notion of telecommunications developments as forms of 'glocal scalar fix'. This intergovernmental concern is as much borne out of socio-territorial objectives as it is out of economic development ones, given that the dominant feeling is that the market will not be able or will not want to deploy these infrastructures in the more peripheral, less profitable zones of the two cities. Such a socio-territorial component to the 'glocal scalar fix' seems thus far to have been bypassed by overwhelmingly dominant economic components, through which particular places have attempted to construct a competitive and attractive 'node' for investment and location purposes. In contrast, then, the potential 'glocal scalar fixes' of broadband infrastructures in some parts of Paris and London are a way of promoting territorial cohesion, albeit that part of the interest in regional and national territorial cohesion can sometimes be seen as ensuring that less favoured, disconnected areas do not 'disfigure' the competitiveness and attractiveness of the region or nation as a whole. Nevertheless, this socio-territorial component to the 'glocal scalar fix' raises in turn the question of the homogenisation of the concept. Does the construction of a 'glocal scalar fix' not imply a notion of spatial or strategic selectivity? In other words, if all the varying telecommunications developments in Paris and London that we have looked at in this study can be viewed as 'glocal scalar fixes' – from private operator strategies through regional infrastructure deployments to the smaller scale initiatives of communes and boroughs – then, surely the advantage gained through their construction is diminishing, and Brenner's interpretation of 'globalisation' as 'a multidimensional process of rescaling in which the scalar organisation of both cities and states is being reterritorialised in the conflictual search for 'glocal' scalar fixes' (Brenner, 1998b, p.462) will mean that cities, states and telecommunications operators are going to have to *continually* reterritorialise or reconfigure their scalar organisation in order to find competitive advantage. The search for niche markets is though already a strategic given in the telecommunications industry, and will perhaps increasingly be the case for cities and states.

This study, then, through its focus on telecommunications developments, has demonstrated quite clearly some of the aspects of Brenner's multiscalar theories. In particular, we have seen how constructing urban telecommunications strategies in global cities necessitates 'a 'glocal' rescaling of state territorial organisation' (Brenner, 1998a, p.16), and a reconfiguration of the regulatory level of urban governance in relation to processes and practices beyond the scale of the municipality (Brenner, 1999). The intrinsic global-local logics and interplays of telecommunications infrastructures and services illustrate these multiscalar *enjeux* or contentions more saliently than virtually any other territorial process.

Limited 'Hollowing Out' of the State in the Domain of Telecommunications Governance and Regulation

It can be argued that the study has affirmed the emergence of mechanisms of reregulation, rather than deregulation, of the telecommunications sector, as it is recognised that market-led approaches are not enough to promote the complete territorial coverage of infrastructures in global cities, which can meet both economic development and territorial cohesion objectives in parallel.

This has become even more the case in the last couple of years, as states have recognised the inherent social, economic, political and cultural importance of 'the information society'. Consequently, we saw the emergence in France and the UK of 'the reconfiguration of the role of the state', in particular with a focus on broadband infrastructures and services. We explored particular strategies in Paris and London which are being shaped at least in part by this national context, such as the development of broadband networks for the research communities of Ile-de-France, the more discursive (so far) negotiations revolving around territorial broadband coverage in the London region, and the broader London Connects initiative, in which the boroughs of London are responding to the e-government directives of central government, but which has brought together a series of public and private partners, as well as a type of local – regional – national intergovernmental relationship, which has been far from common in the British political context in the last two decades or so.

The key element here then is the continuing central role of the state. The governance or regulation of telecommunications developments in Paris and London is influenced by policies and practices from the supranational and subnational levels, but it still appears to be central government primarily setting or shaping the agendas and contexts for local and regional developments. In the telecommunications domain, the 'hollowed out' state of Jessop is far from becoming dominant.

Placing Telecommunications Developments in Paris and London: The Resonance of Historical Contexts, Cultures and Traditions

The construction of a historical positionality for this study was deemed an important prerequisite to the investigation of contemporary telecommunications developments in Paris and London. It is difficult to comprehend how any study of these developments can bypass placing them in their historical context. For all the current talk of 'the rise of the network society' (Castells, 1996), we saw in earlier chapters how the French writer Armand Mattelart, in particular, has demonstrated the historical origins of the networking of the world, and how 'networks have never ceased to be at the centre of struggles for control of the world' (Mattelart, 2000, viii). Hence, much of chapter 5 was concerned with providing a historical overview of the development of communications technologies, as well as planning and governance practices, and national cultures of territoriality in France and the UK. The aim was not only to 'set the scene' for the principal empirical chapters on the territorial construction of recent telecommunications developments in Paris and London, but as importantly to demonstrate the importance of these contexts, cultures and traditions in the shaping of recent developments. The territorial management implications of the development of the visual telegraph in France, for example, are surely the roots for the strong contemporary links between telecommunications networks and *aménagement du territoire* in France. We saw this in the regional territorial development objectives of the Téléport Paris – Ile-de-France, and also in the way the Sipperec project has embarked on an alternative territorial coverage of the communes of the inner *couronne* of Paris. Contemporary telecommunications developments in Paris and London tend, then, to relate quite closely to historical developments in, and differences between, France and the UK in the domains of telecommunications, planning and governance, and national traditions of territoriality. From a consideration of the work of Mattelart, this importance can be suggested to take two related forms. First, in the drawing of parallels between the historical development and deployment of early communications networks and the development and deployment of recent networks. Second, in the continuing resonance of cultures and traditions formulated in earlier times.

In the first case, we saw how Mattelart argued that the realpolitik of infrastructure development remains fairly similar today to that of two centuries ago. If the undersea cable laid between Dover and Calais in the mid nineteenth century, primarily to connect the London Stock Exchange and the Paris Bourse, was a demonstration of Victorian hegemony at that time, then we can suggest a similar symbolic element to the global deployment of fibre networks in recent years. The operators which are the main players in global telecommunications today can only be seen to be main players if they are deploying vast amounts of infrastructure across the world. This ties in with Kaika and Swyngedouw's discussions about the inherent prestige and authority of infrastructure networks, and indeed in some cases how these came to be seen as an 'embodiment of progress' (Kaika and Swyngedouw, 2000). Hegemony, prestige and progress continue to be key words in this domain, particularly in the ways we saw how the French and UK governments are each attempting to construe a national advantage over their western counterparts in the development of broadband networks. These national contexts clearly diffuse to policymaking at the regional level in Paris and London, as we also saw, for example, in the concern of the Conseil Régional d'Ile-de-France in ensuring the roll out of these broadband networks for its research communities, which are themselves crucial for regional economic development and regional prestige.

Mattelart's historical perspective also identified differences between French and British communications regulation throughout the nineteenth century, which seem to have persisted up to recent years. British undersea cable, for example, was managed by private companies at that time, while French cable was under the control of the state, which clearly contributed to the differing national cultures of respective private and public dominated forms of telecommunications governance, which, as we have seen, have continued in the last twenty years, and have, in particular, differently shaped telecommunications developments in Paris and London in recent years.

In the second case, we investigated how Mattelart has shown the historical foundations of specific French national cultures and traditions which we have looked at through their influence on telecommunications developments in Paris. As suggested earlier in the chapter, the dominant one has perhaps been the notion and practice of *aménagement du territoire*. Mattelart was particularly interested in the role of early forms of communications technology such as the visual telegraph system in helping to shape and organise a coherent and unified territoriality for France, which the French state has since gone on continuously to reconfigure under policies of *aménagement du territoire*. Traditional French thinking on networks has also remained influential, up to the current concerns with telecommunications infrastructures. Mattelart linked the notion of *réseau* to Enlightenment concerns with exchange and reason, as well as to the development of French territoriality.

In some ways, the starkest differences between telecommunications developments in Paris and London appear within the context of national traditions of territoriality. On the one hand, there is a set of developments in Paris strongly influenced by the tradition of *aménagement du territoire* in France. On the other hand, developments in London are seemingly notable for the relative absence of any real kind of interaction with the notion of 'territoriality'. This can perhaps be closely related to the relative importance of the role of the state in economic governance in both countries – a high level of state-led development in France, with the state firmly concerned by territorial approaches through its DATAR arm, and a much greater market dominance in the UK, with the state most concerned with ensuring the conditions for market-led development are present. Again, comparing the early 1990s strategies in both cities, the Téléport in the Paris region, with its five purposely chosen and strategically configured and balanced sites around the region, with the *laissez-faire* development of a competitive telecommunications market in the City of London, illustrates the differing approaches to 'territoriality' taken in France and the UK.

Historical perspectives have been, then, an important part of our comparisons of telecommunications developments in Paris and London, both in offering an early comparative component to the empirical part of the study in chapter 5, and in distinguishing traditional differences in national territorial network cultures and their influence on these developments. As Chant suggested, this is all about 'the intractability of two [territorial] fabrics that [are] quite distinct historical palimpsests' (Chant, 1999, p.221). The work of Mattelart has been very useful in this regard, and illustrates forcefully the relevance and the resonance of a historical positionality, even to a study focusing on something apparently as 'new' and 'innovative' as contemporary telecommunications infrastructure networks.

The Need for Relational and 'Multiplex' Readings of Urban Development

Linked to the need for territorially-based conceptions of global or world cities, it becomes necessary to view the city as a 'multiplex' of different spaces, with varying levels of connection and disconnection to each other, which tends to diminish the possibility of an overall territorial cohesion to metropolitan regions. We discussed some of the relational urban theories that have been developed in this regard towards the end of chapter 4, and we were able to apply some of their ideas to the empirical context of our study of the territorialities of telecommunications developments in Paris and London particularly in chapter 8, where we focused on local or intra-urban interactions between telecommunications infrastructures and their territorial contexts, specificities and processes.

In chapter 4, it was observed that some of the literatures concerned with spatial differentiation or fragmentation within cities have begun to reconfigure the traditional core – periphery model of urban organisation and form, as new socio-territorial inequalities take shape in urban regions, producing a 'splintered' metropolis of juxtaposed dominant and dependent spaces. It can be argued,

however, that the cores and peripheries of the contemporary city are more
functional in nature than the spatial ones of the traditional model in which the core
was necessarily the centre and the periphery the suburbs or outskirts. Thus, Picon
wrote about the central urban areas which are more peripheral than the transport or
business nodes on the outskirts of cities, that are well served by infrastructure
networks, and the case study by Longcore and Rees of the financial district of
Manhattan showed how firms are beginning to relocate outside the traditional
downtown core, suggesting that the Wall Street area may be becoming more of a
'subdistrict'. Similarly, then, our study of Paris and London has also highlighted
the ways in which telecommunications infrastructures are differently deployed
within the two urban regions, not always in keeping with the traditional dominance
of their centres compared to their peripheral zones. Thus, the business quarter of
La Défense to the west of central Paris is now the most 'networked' space of the
French capital, and concentrates the most interest and investment in deployment of
networks and spaces of telecommunications operators. This was especially well
illustrated by the regulatory conflict which focused on the alleged 'anti-
competitive' territorial regulation of the market at La Défense by its local planning
authority, which led to several operators taking their case for an ability to deploy
their own infrastructures in the zone to the administrative tribunal of Paris. Such a
conflict over territorial deployment suggests the primary status of La Défense
within the territorial strategies of telecommunications operators for the Paris
region. This is a major difference compared to the situation in London, where
although the Docklands business zone to the east has become a crucial focus for
the telecommunications market, primary status in the London region for operators
remains with the 'core' area of the City. A further territorial difference emerges in
a comparison of the traditionally more 'peripheral' communes of the Paris region
and the boroughs of the London region. In the former case, we saw how the
Sipperec initiative is attempting to develop a fibre network for the communes of
the inner *couronne*, where competitive private offers in telecommunications
networks are deficient, whereas in the latter case, boroughs such as Lewisham are
still dependent on the network of the incumbent BT and have had little chance so
far of attracting competitive networks within their territories. From studying the
territorialities of telecommunications developments in the Paris region, then, we
can suggest that the 'core' networked area is not strictly defined or bounded, and
that there is evidence of this area being extended towards the inner *couronne*, for
example. This contrasts with the telecommunications market of the London region,
which seems very well set in a defined 'core' networked corridor running from the
West End through the City to the Docklands. To use Christine Boyer's notions,
beyond this 'figured' part of the city, the spatial peripheries of the London region
are also technologically peripheral or 'disfigured', because the density of their
infrastructures and access to services is relatively low. These territorial inequalities
represent the 'power-geometry' of telecommunications developments – within
London, the authorities, agencies and businesses of the City and the Docklands
have greater power 'in relation to the flows and interconnections' (Massey, 1993)
of telecommunications networks than the authorities, businesses and communities
of more peripheral areas, who have limited power to shape their

telecommunications markets and attract competitive offers to that of the incumbent. Thus, our study of telecommunications developments in global cities seems to uphold in some ways the theoretical foundations provided by the likes of Massey, Allen and Swyngedouw, for whom cities are now shaped by their differing spaces and the differing concentrations of social and economic power held within these spaces in relation to hegemonic structures and processes. Swyngedouw distinguished here between the ability and the inability to command place.

It is clear, then, that given the socio-territorial differences within Paris and London in relation to telecommunications developments, we need to draw on relational and multiplex conceptions of the city in order to be able to identify these differences and analyse the factors behind them. We can suggest that work in geography, planning and urban studies has been very pertinent in recognising the presence of varying spaces and power arrangements within cities, but has perhaps yet to really get to grips with the variety of territorial contexts and specificities which shape such intra-urban 'power-geometry'. Reconfigured core – periphery models are unlikely to provide the answers in this regard, because of the profusion of differing territorialities bound up in the development of individual places, so more comparative empirical work seems to be required for us to be able to understand the different facets behind notions such as 'power-geometries', 'figured' and 'disfigured' cities, and 'spatial selectivity'.

Table 9.1   A synthesis of the differing parallel territorial contexts, specificities and outcomes of telecommunications developments in Paris and London

PARALLEL SCALAR CONTEXTS	NATIONAL	URBAN-REGIONAL	LOCAL/INTRA-URBAN
THEORETICAL BACKDROPS	CROSS-NATIONAL URBAN & TELECOMS POLICY CONTEXTS	TERRITORIALITY OF GLOBAL CITIES & TELECOMS	POLITICAL ECONOMIES OF URBAN GOVERNANCE, RESCALING & ECONOMIC DEVELOPMENT
EMPIRICAL STRANDS	NATIONAL INFLUENCES ON TELECOMS DEVELOPMENT IN GLOBAL CITIES	TELECOMS NETWORKS & SERVICES IN THE GLOBAL CITY 'PACKAGE'	THE CONSTRUCTION OF NEW NETWORKED SPACES IN GLOBAL CITIES, & THE TERRITORIAL COHESION IMPERATIVE
MAIN CASE STUDY ELEMENTS	Paris: The construction of decentralised and partnership-based strategies in a relatively early and unhabituated context of liberalisation. London: The dominance of a laissez-faire market, mature competition, and territorial strategies founded on centralised policy.	Paris: The configuration of developments by an abundance of actors to shape the 'space of flows' within a crucial competitiveness agenda. London: The predominance of global finance, a scarcity of strategic actors, and a prospect of new regional government action.	Paris: Increasing territorial responses at the level of local governance, and relatively limited spatial variations in connection. London: Limited local intervention, market-led deployment, and distinct variations between networked and disconnected spaces.

Table 9.1 (continued)

| KEY ANALYTICAL CONCLUSIONS | The continuing influence of historically resonant national telecoms policy contexts and regulatory practices in the face of an apparently overarching global logic.

The continuing influence of historically resonant national policy contexts of planning, economic development and governance practices in the face of an apparently overarching global logic. | Continuing territorial variations in telecoms developments between global cities, due to the reinforcement of the social, economic, political, and territorial specificities and assets of these cities.

The same operators are constructing 'glocal scalar fixes' in different ways in response to these different contexts and specificities. | The impossibility of territorial cohesion in global cities – the deployment of telecoms infrastructures and services in connected, core strategic 'glocal' spaces widens the divide with disconnected, peripheral, less favoured 'local' areas. |
| | The presence of nationally specific 'glocal scalar fixes' as particular cultures of regulation, governance and territoriality combine with the global localisation strategies of operators. | | |

As synthesised in table 9.1, through the consideration and analysis of the three thematic research strands of this study, both individually and incrementally, our comparison of the territorialities of telecommunications developments in Paris and London has illustrated, above all, how these developments are marked by the construction of nationally specific 'glocal scalar fixes', through the juxtaposition of particular cultures and contexts of regulation, governance and territoriality, and particular global localisation strategies of telecommunications operators. The landscapes of the discursive and material construction and packaging of telecommunications developments in both Paris and London may be highly complex and technical, and marked by important relations and confrontations between different actors, but they are, and this explains in large part the differences between developments in Paris and London, primarily and eminently founded on territory and territoriality.

Dupuy's (1991) archetypal thinking, bringing together networks and cities, the reticulate and the urbanistic, summarised in the quote at the start of chapter 1, appears thus to have become even more crucial in the last ten years. The territorialities of telecommunications developments in Paris and London are increasingly complex, reticulate and multiscalar. The developments, the cities in which they are initiated, and the territorialities they engender, all necessitate thinking, analysis and action under the umbrella of 'network urbanism'.

Graham and Marvin recently highlighted one of the major research agendas for urban studies disciplines:

> to undertake detailed and comparative empirical investigations into the ways in which physical and sociotechnical shifts towards splintering urbanism, and unbundled networked infrastructures, are being politically and socially constructed in profoundly different political, cultural, economic and historical contexts. Such research needs to encompass developed nations, newly industrialising nations, developing cities, and post-communist metropolitan areas embedded within different state, political, cultural and urban traditions (Graham and Marvin, 2001, p.417).

In light of this, it is to be hoped that the comparative empirical investigation of this study, into the discursive and material construction and implementation of telecommunications developments in Paris and London, and its focus on differing political, cultural, economic and historical contexts, has offered a modest contribution to this emerging, but highly important, research agenda. Indeed, although this study has concentrated on the comparison of two large global cities in two developed western European nations, the extent of the variations in state, political, cultural and urban traditions and contexts as they have influenced (and are continuing to influence) telecommunications developments, even between apparently relatively 'similar' nations, would suggest that the above research agenda is even wider (and more crucial) than already portrayed.

Conclusion

In this chapter, we have completed our analysis of the empirical findings from the two case studies of telecommunications developments in Paris and London, by relating them back to the theoretical framework and comparing the cases according to our three thematic strands, both individually and as a multiply scaled whole. In this way, we have been able to draw out in some detail both the similarities and, more importantly, the differences between the two case studies, and the analytical – theoretical implications of the study of telecommunications developments in global cities. We have seen how the comparative, cross-national nature to the study, which we presented in the introduction as a major gap in urban telecommunications research thus far and therefore as a major 'trigger' for this study, has produced some interesting findings, and has especially allowed us to illustrate the continuing nationally specific contexts and specificities to the deployment and development of telecommunications infrastructures and services in global cities. The fact that the intertwined developments and articulations of telecommunications and global cities should, at the end of the day, still reside primarily in territoriality is a potent argument opposing those slightly blinkered views which only see in telecommunications and / or global cities a demonstration of the overriding placeless logics and processes of globalisation.

In the next, and final, chapter, we consider some of the limitations or weaknesses of the study as a whole, and identify and assess its implications both for future research and for policy making in the domain of urban telecommunications developments.

Chapter 10

Conclusion: Research and Policy Implications

Introduction

In the previous chapter, we attempted to pull out the main analytical conclusions from the case studies of telecommunications developments in Paris and London, particularly from the standpoint of the theoretical strands and overall framework of this study. In this concluding chapter, the remaining tasks of the study are to highlight and to consider the limitations of the study; to identify potential areas of research that could be fruitfully explored as a follow-up to the implications of the study; and to suggest some of the possible policy implications of this piece of research.

The Limitations of the Study

Few empirical studies get close to perfection in preparation, execution and explanation, and this study has been no exception, so whilst it has provided some success in offering a complex and comparative, cross-national analysis of telecommunications developments in Paris and London, it remains important to highlight its weaknesses and to bear in mind the ways in which it could have been further improved.

In particular, the benefits and the limitations of the thematic structure of the study are now evident. Each of the thematic strands was sufficiently well defined to have enabled the identification of the relevant theoretical bases, which came together to form the theoretical framework for the study. Each of these strands also enabled a tightly-focused targeting in both case studies of the relevant and most revealing empirical material, whether secondary, documentary, or for and from the meetings and interviews. As a result, we were able in the last chapter to draw out some of the most important analytical conclusions from the study, again using the three thematic strands as a way of refracting the empirical findings through the theoretical framework, and vice versa.

Nevertheless, the importance of analytical conclusions which seemed to cut across all of the thematic strands is perhaps also an illustration of the limitations of this thematic structure. It is unsurprising that some of these conclusions did not fit 'neatly' within one or other of the strands, given the slightly rigid scalar focus of the latter (national, urban, local or intra-urban). Nevertheless, we have emphasised all along how the territorialities of telecommunications developments in global cities require an inherent multiply scaled perspective, and, relatedly, how our case study chapters were necessarily mutually constitutive and incremental, rather than bounded. It is likely that a less complex, more traditional framework might not have led, directly or indirectly, to the same insights. For example, we can pose the question of whether the use of the simpler, more traditional structure for organising and presenting comparative, empirical case study findings – first case study chapter, second case study chapter, comparative analytical chapter – would have produced the same analytical and comparative findings and conclusions. The use of the more cross-cutting thematic structure for organising and presenting the case studies of telecommunications developments in Paris and London has been justified, despite its limitations or weaknesses.

Unquestionably the main restriction in carrying out this study was problems of access with potentially important interviewees. This was especially the case with telecommunications operators, and seemingly in particular with those based in London. The relative ease with which meetings were arranged with representatives of operators in Paris compared to (sometimes the same operators) in London was curious to say the least, and could reflect the way in which some researchers have suggested that it is far easier to be able to get to speak to people as a foreigner in a foreign country because one is perceived to be less 'threatening'. Nevertheless, the ability to obtain meetings or interviews with telecommunications operators in whichever country is always likely to be tested severely by the fact that they are working in an extremely competitive marketplace, where everything depends on innovative decision-making and projects, so that providing any detailed ideas of strategy and practice to outsiders is probably best avoided. We were able, to some extent, to make up for this loss of source of information by collecting, collating and analysing information from the websites of the operators and other secondary documents.

Following on from, and in some ways related to, the above, we should also recognise that the study would have been further improved by a more even treatment of the two case studies. It eventually became somewhat inevitable and unavoidable that the case study of telecommunications developments in Paris was going to be slightly richer and provide more detailed information and findings than the case study of telecommunications developments in London. Whilst again it can be emphasised that it has not overly affected the eventual success of the comparative nature of this study, we would have to admit that it would have been preferable to have two rich, detailed case studies to draw upon analytically, in order to create an even richer and fully detailed cross-national comparative study.

We should also note the slightly uneven exploration of the three research strands in our empirical analysis. Notably, the third strand is possibly not as rich and detailed as the first two. However, the choice of these strands (with national, urban regional, and local focuses) was dictated by the will to investigate and illustrate the interaction between telecommunications developments and a set of multiply scaled territorial specificities and contexts, which have to be seen as mutually constitutive. As such, it is important to view the theoretical and empirical elements of all the strands in relation to one another, so therefore the more detailed exploration of one strand over another is not as crucial as being able to see how each of the strands influences and is influenced by the others to create an interacting and incremental set of territorialities bound up in parallel in telecommunications developments in global cities such as Paris and London.

By focusing on telecommunications developments in global cities, the study has relatively neglected processes and practices at the micro scale. This means that we have not been able to investigate in detail the level of individual actors within specific telecommunications projects, for example. Nevertheless, to achieve this within case studies of telecommunications developments in global cities would have meant neglecting either other elements at other levels or dropping the comparative aspect of the study, because of space and resource restrictions. It was felt that because of the lack of previous empirical research on the relations between telecommunications and specific global cities, a comparative and cross-national study of these relations would offer a novel and revealing perspective, and that it was more crucial for this perspective not to lose the strategic elements and oversight that help to make up these relations, than for it to delve into the minute detail of the origins, shaping and implementation of individual projects using a separate theoretical tool such as actor network theory. This research avenue is definitely one that could be explored fruitfully in the future, however, with regard to telecommunications strategies in global cities. Graham (1996) has already offered a fine illustration of the exploration of comparative telecommunications projects in medium-sized cities using actor network theory.

Finally, we must note the lack of a particular demand-side perspective to this research. By focusing on telecommunications operators and local and regional authorities and organisations in Paris and London, we have ignored the users, business and residential, of the telecommunications developments we have explored. This again, though, would have extended the scope of the study far beyond its means and intentions, and would have necessitated the use of quantitative methods such as surveys to extrapolate detailed and coherent views and data. Again, given the lack of empirical investigation of telecommunications developments in global cities in previous research, it was seen as much more important to explore the implementation and 'supply' of telecommunications infrastructures and their implications. A demand-side perspective on telecommunications in global cities could be an interesting subject for future research, nonetheless.

Areas of Exploration for Future Research

This has been a highly complex study which has overlapped with numerous subject areas, so therefore suggesting topics for future research following on from this study is somewhat difficult. It can be argued, however, that perhaps the most important aspect for future research into telecommunications developments and cities to consider should be the continuing need for and benefits of a cross-disciplinary approach. This is somewhat inevitable anyway given the inherent differences between telecommunications and cities, and the fact that they have never really been studied in juxtaposition in any one discipline (planning is still only at the beginnings of such a process). Any study of telecommunications and cities should necessarily encompass the theories and practices of planning, geography, urban studies, communications studies, history, and political economy, and by doing so, becomes relevant and interesting to a far greater academic and practitioner audience than it would do by being solely focused within one discipline.

As well as privileging cross-disciplinary approaches to studying relations between telecommunications and cities, it is equally important perhaps to be able to situate telecommunications developments within the overall landscape of technical networks and systems in cities (Guy, Graham and Marvin, 1997; Coutard, 1999). Telecommunications are, after all, merely one element in urban infrastructure provision, in parallel with water and waste, electricity, gas, and transport networks. As Graham suggests: 'Only very rarely do single infrastructure networks develop in isolation from changes in others' (Graham, 2000a, p.114). An integrative and relational approach to urban infrastructure networks offers another crucially important perspective on urban development and restructuring, which explorations of individual infrastructures, as in this study, necessarily neglect, however detailed they may be (see Dupuy, 1991; Offner and Pumain, 1996; Graham and Marvin, 2001).

It is hoped, nevertheless, that this study has made a small contribution too to the still neglected domain of cross-national and comparative research into telecommunications developments and cities. Whilst on the one hand, there has been some effort in recent years to study cross-national similarities and differences in *national* telecommunications policies and regulation, and on the other hand, some examples of cross-national studies of cities and *urban* change, there have been very few studies which have put the two together and focused on cross-national comparative studies of telecommunications developments in specific cities.

The increasing importance of scalar dynamics and politics will thus undoubtedly be a major concern of future research both in studies of telecommunications and cities, and in planning and urban studies as a whole. It is not enough for us to succumb to such alleged future utopian promises of 'ascalar' and 'aterritorial' cybercities, in which virtuality completely dominates over materiality, as Mitchell's 'city of bits':

This will be a city unrooted to any definite spot on the surface of the earth, shaped by connectivity and bandwidth constraints rather than by accessibility and land values, largely asynchronous in its operation, and inhabited by disembodied and fragmented subjects who exist as collections of aliases and agents. Its places will be constructed virtually by software instead of physically from stones and timbers, and they will be connected by logical linkages rather than by doors, passageways, and streets (Mitchell, 1995, p.24).

Instead, we have discussed and analysed how the articulations and interactions between telecommunications developments and cities reflect very closely the political and economic restructuring and rescaling at and between every level from the local to the global. The ways in which urban telecommunications developments are shaped as 'glocal scalar fixes' for capital investment and economic development needs further investigation, and especially further empirical case study analysis. It would be interesting too to see studies of more 'peripheral' cities and regions around the world, to counterpoint the focus on the major western global cities (as in this study). Graham and Marvin (2001) point the way in this regard, with their truly global appraisal of the relationships between technologies and infrastructure networks, and both western urbanism and cities in developing parts of the world (see also Silva, 2000; Sussman, 1997, on the global dimensions of the information society). Indeed, counterpointing this is the relative lack of communications and other infrastructure networks in many parts of the world. A focus on western 'global' cities is necessarily a focus on cities with a tiny proportion of the world's population. Why is it, for example, that Manhattan has more telephone lines than the 49 countries of sub-Saharan Africa put together (Sussman, 1997, p.231)? In order to answer this question, we do not just need to know about the social, economic and political processes that are bound up in the connectivity of Manhattan. We also need to know about the social, economic and political processes that are bound up in the complete non-connectivity of those African countries.

This all becomes even more important in the light of the pessimistic views being tendered by some commentators. Mattelart, for example, speaks of how:

> The dynamic of economic models of globalisation risks leading towards a 'ghettoised' world, organised around a small number of megacities and regions, concentrated mainly in the 'North', with outliers in the 'South', serving as the nerve centres of worldwide markets and flows (Mattelart, 1999, p.189).

One of the key things for future research will be to continue to tread carefully between and avoid as much as possible the overhyped rhetoric of 'techno-utopian' discourses about communications technologies and their dreams of 'unmediated planetarism' through 'overcoming 'centrality', 'territoriality' and 'materiality''(Mattelart, 1999, p.190). This study has hopefully shown the continuing importance of these notions, both conceptually and materially.

Policy Implications of the Study

The focus of this study on telecommunications developments in the global cities of Paris and London rather precluded any assertive objectives for this study in the field of illustrating 'best practice', or the advantages and disadvantages of certain approaches to these developments, for policy makers. Firstly, the use of only two case studies makes generalisation of the findings undesirable. Secondly, it would arguably be hard, indeed virtually impossible, for policy makers from other European cities and regions to learn much from the detailed experiences of Paris and London that they could then apply to their own urban and regional contexts, because we are, after all, talking about the two major cities in Europe, and we saw, just in chapter 5, the important differences in historical and territorial contexts. As Parkinson suggests, 'The problems of global cities such as London or Paris are not those of medium-sized cities' (Parkinson, 2001, p.78). Indeed, Yin argues that case studies are not meant to be generalisable to 'populations', but only to one's 'theoretical propositions' (Yin, 1994, p.10).

Nevertheless, policy implications may be forthcoming from the more general aspects of the approaches to telecommunications developments in the two cities, and may have most relevance to cities not too far down the European urban hierarchy. It can be argued that the policy implications of a detailed qualitative study of two cities like this are far greater than those of a general quantitative study of many cities.

Perhaps the first element to note is that the case studies of telecommunications developments in Paris and London have demonstrated that policy makers at all levels can still have an important role to play in shaping the nature of public sector intervention in the telecommunications domain. There may exist an overriding impression that the market and the private sector are dominating this domain, and that either they are squeezing out the public sector, or there is simply no need for the public sector to intervene because everything is being covered and dealt with by the market. The recent discussions, negotiations and oppositions at all levels in both cities relating to the unbundling of the local loop and the roll-out of broadband DSL illustrate that this is not the case. Public sector intervention is required for ensuring increased territorial coverage of infrastructures, and so that these broadband Internet connections, which both the French and UK states have argued is crucial to socio-spatial equality in the information society, reach every level of the consumer market. As we saw in the previous chapter, it is abundantly clear from the viewpoint of Paris and London, that a *laissez-faire* approach with regard to the telecommunications market is simply not enough to promote the territorial development of infrastructures and services.

We can also highlight the apparent importance of partnership creation as a policy implication of our study. Needless to say, it appears that the larger the numbers and the more diverse the members within such partnerships in the telecommunications domain, the more chance it has of achieving its goals. The London Connects project has incorporated both regional and local levels of government, wider regional and local organisations, businesses and operators, and the apparent determination of these actors, both individually and as a whole, to see something come out of the project seems to have created a real forward dynamic, which could ensure it meets its ambitious objectives. Individual boroughs such as Lewisham and Camden seem to have a greater opportunity of meeting their telecommunications needs through this regional project than on their own. In Paris too, the Sipperec strategy has brought together local communes and an alternative operator, and ensured the deployment of a competing telecommunications infrastructure to that of France Télécom, and therefore more choice of network provision and a more competitive rate for services, which are important elements for communes for their territorial attractiveness to businesses and investors.

The question has, however, been raised of whether increasing public sector intervention in the telecommunications sector is characterised more by a real recognition of its potential for local development, or by an identification of a temporary 'fashionable' policy for improving a local image (Vinchon, 1998a). The strategies discussed and analysed in this study are unquestionably characterised by the former, but elements do seem to be present even here, particularly in France, in which telecommunications infrastructures are seen as a trendy, *en vogue* and 'spectacular' (Debord, 1995) territorial development practice. The Téléport Paris – Ile-de-France, for example, originated partly from the way that IAURIF had observed the rise of the teleport phenomenon in other cities around the world, and decided that the Ile-de-France region could develop something similar. Likewise, recent strategies in the commune of Issy-les-Moulineaux, such as the 'cyber-nursery' for small firms and the redevelopment of Fort d'Issy, can be considered as partly aimed at improving the image of the commune through an innovative and fashionable ICT policy.

A significant component of the policy implications of this study can be seen as the possible limitations and challenges to public sector intervention in urban telecommunications developments. We attempt to briefly highlight some of these in the next part.

Challenges for Policy-makers from the Realpolitik of Urban Telecommunications Developments

As was highlighted in chapter 7, in order to carve a role for themselves in the telecommunications domain (and therefore to avoid leaving developments totally in the realm of the market), public sector actors at both the local and the regional level require, amongst other traits, the ability (know-how and a lack of internal or external constraints), the willingness (desire) and the resources (financial or otherwise) to recognise the importance, potential benefits, and restrictions of telecommunications infrastructures and services for their territories. This is an important set of characteristics to be developed. We can suggest, therefore, at least six contemporary challenges and opportunities for policy-makers with regard to the *realpolitik* or actual development of urban telecommunications strategies.

1. Making the issues to do with telecommunications salient
Policy-makers face the challenge of understanding the real *raison d'être* of telecommunications networks and services. They must then try to highlight their positive implications for their territory and give reasons for the undertaking of a particular telecommunications strategy. This is far from evident, especially given that these implications and reasons may be harder to pin down than those for the traditional focuses of territorial policy such as building roads or business and commercial zones. There is also the challenge of using the most effective terminology to highlight the salience of telecommunications. As we saw, there are differences in the terminology and rhetoric used by the public sector in France and the UK for the telecommunications and IT domain, and this might explain some of

their uncertainty, if they struggle to grasp what the relations between cities and telecommunications concern or imply.

2. Dealing with arcane and fast-moving technology and terminology
Policy-makers face the challenge of developing both the high level of technical and technological expertise required, and especially the ability to follow the rapid pace of change in telecommunications regulation. For local and regional authorities, traditionally used to intervention in transport and land use planning, and their role in economic development, keeping up with highly complex technological and regulatory changes related to the telecommunications domain, which are happening virtually on a day-to-day basis, is extremely difficult and straining on often tight resources. This would probably be an area in which there are significant variations between cities depending on their place in the urban hierarchy. Most of the public sector bodies included in this study in both Paris and London did have either a specific delegate or a whole team working in this domain. This would probably not be the case in smaller cities and regions.

3. Getting telecommunications operators on board
Policy-makers face the challenge of making the most of competitive telecommunications markets, and forming partnerships with operators within their strategies. The potential difficulties here lie in the necessity for both sets of partners to be pulling in the same direction within an overall territorial strategy. There may be potential conflicts between an ICT strategy based on cohesion objectives and operator strategies based on profit-making and market development. A win – win partnership would need to be formulated.

4. Overcoming paradigm divisions
Policy-makers face the challenge of reconfiguring their own notions and practices of territorial planning and development, as well as perhaps of those to whom they are accountable. Telecommunications strategies imply and necessitate new forms of intervention, which may be substantially different to those traditionally employed by policy-makers. They are also difficult to place within set and bounded policy domains, as their inherently 'relational' nature makes them relevant to multiple disciplines and fields of territorial management.

5. Overcoming invisibility
Policy-makers face the challenge of actually 'seeing' or making visible elements of telecommunications developments. Communications infrastructure networks tend to be a hidden part of the urban landscape, usually tucked away underground. Territorial authorities are far more used to dealing with roads, buildings and activity zones – the most visible elements of towns and cities – that suddenly having to develop a policy agenda for something which cannot always be seen and which does not appear to have a significant 'impact' on the urban landscape is problematic. Furthermore, having 'seen' the general importance of telecommunications infrastructures, policy-makers then have the difficulty of making visible or obtaining knowledge of the telecommunications actors present

on their territories. For example, which operators run which type of networks and where?

6. Bringing different institutional cultures together

Policy-makers face the challenge of formulating and implementing strategies based on telecommunications on their territories in which are present both several different groups, and therefore several different sets of institutional viewpoints and traditions, and several national policy context elements. A telecommunications strategy needs to respond to and take into account the needs and views of the territorial authority itself, other territorial authorities from differing scalar levels, groups or organisations within the telecommunications domain or within the geographical domain of the strategy, telecommunications operators, large and small businesses and companies, and other users. In addition, as we saw particularly in chapters 2 and 6, it may be influenced by and bring together aspects of national telecommunications policy and regulatory practice, national urban or economic development policy, national political culture and system of intergovernmental relations. This is a quite substantial number of contexts, traditions and viewpoints to be taken into account by a policy-maker working in a relatively new domain.

Implications for Operators

As well as policy implications for territorial authorities, we can also briefly suggest some implications of this study for the activities and strategies of telecommunications operators. Again, however, as for cities and their authorities, it is important to underline that these implications relate to each operator in different ways depending on their nature and focus.

It is evident that the territorial strategies of telecommunications operators in global cities will be shaped to a large extent by the continuing fluctuations of the market. As the Yankee Group Europe observed:

> Only a year or two ago many carriers, incumbent and start-up alike, were being encouraged by investors and financial analysts to spend, spend, spend to build out, bulk up, and grow at all costs. Profit be damned (or at least postponed). [...] What a difference a year makes. Today, some carriers see their very survival threatened by a newly hostile equity market that demands profitability, high revenue growth, and service specialisation – now (Yankee Group Europe, 2001, p.1).

The various sectoral and territorial markets of global cities are increasingly likely to be divided both between different operators and within the strategies of the same operator, as these become concerned with the separation of their 'low-margin consumer voice telephony business from the faster-growing, higher-margin corporate data business' (Yankee Group Europe, 2001, p.2). Indeed, for the Yankee Group Europe, the successful telecommunications operators will now 'probably be carriers that will no longer try to be all things to all customers' (Yankee Group Europe, 2001, p.3). While incumbents like France Télécom and BT retain a presence in multiple sectoral and territorial markets, at least while national universal service policies remain in place, competitive new operators, focused completely on the most profitable markets, are likely to keep reconfiguring their strategies according to this focus, which may lead to the emergence of new niche sectoral and territorial markets, particularly given that the numbers of operators present in some markets make significant competitive advantage even harder to obtain.

Consequently, many operators may increasingly focus on building up 'one-stop shops' for their main clients to create integrated packages of telecommunications services with global coverage. These can be seen as kinds of intensive multiple mini 'glocal scalar fixes', offering seamless comprehensive voice, data, and mobile connections of the local spaces around the world where a large company may be located over the comprehensive glocal infrastructure networks of the operator to global economic and finance networks. In this way, these operators are offering multinational businesses a 'glocal infrastructural ideal' with maximum efficiency and minimum fuss.

While this market-led 'glocal infrastructural ideal' offers virtually limitless and distance-less connections for large companies located within the 'premium network spaces' of global cities, other juxtaposed urban spaces and communities, as we have seen, have nothing like the same access to telecommunications infrastructures and services. The future of national universal service policy appears to be critical in determining levels of socio-territorial connection in global cities. The prospect of universal broadband networks offering inexpensive potential access for all to fast Internet connections appears fraught with difficulties, regarding cost and market regulation, yet, as we saw in both Paris and London, some form of public sector intervention is needed to supplement and extend market developments. The implications of limited territorial regulation of broadband roll-out are otherwise likely to be increased socio-territorial inequalities within global cities, which may eventually impinge on the future overall economic development of these cities.

Final Remarks

The global economy will expand in the twenty-first century, using substantial increases in the power of telecommunications and information processing. It will penetrate all countries, all territories, all cultures, all communication flows, and all financial networks, relentlessly scanning the planet for new opportunities of profit-making. But it will do so selectively, linking valuable segments and discarding used up, or irrelevant, locales and people. The territorial unevenness of production will result in an extraordinary geography of differential value-making that will sharply contrast countries, regions, and metropolitan areas. Valuable locales and people will be found everywhere... But switched-off territories and people will also be found everywhere, albeit in different proportions. The planet is being segmented into clearly distinct spaces, defined by different time regimes (Castells, 1998, p.374).

It is hoped that this study has offered a small empirical illustration of the socio-economic and spatial logic that Castells identifies here. Whilst, by focusing specifically on telecommunications developments in Paris and London, we have not attempted to explore the allegedly fully global nature of this logic, we have managed to demonstrate some aspects of its selectiveness and territorial variations, and the subsequent contrasts that are developing both between and within the metropolitan areas of Paris and London. As a result, the territorialities of telecommunications developments in the two cities compare closely in their juxtapositions of 'valuable' and 'switched-off' territories and groups, 'defined by different time regimes', but differ inherently in the national specificities (and national responses to multiscalar specificities) that influence these juxtaposed territorialities:

> The state does not disappear, though. It is simply downsized in the Information Age. It proliferates under the form of local and regional governments, which dot the world with their projects, build up constituencies, and negotiate with national governments, multinational corporations, and international agencies. The era of globalisation of the economy is also the era of localisation of polity. What local and regional governments lack in power and resources, they make up in flexibility and networking. They are the only match, if any, to the dynamism of global networks of wealth and information (Castells, 1998, p.378).

Castells offers us a preliminary, broad, but logical and reflective 'prediction' of the global dynamics of the network society. What we now need to investigate much more closely are the territorialities and local implications of the network society. Furthermore, given the shaping of the network society, and its concentrations of nodes and hubs as its 'command centres', this investigation should surely emphasise the value of comparative and cross-national empirical research, as the principal means for reaching to the heart of this global logic and its differing local implications and articulations. It is to be hoped that this study has made a small contribution to this emerging and complex research agenda.

Bibliography

Abbou, M. (2000) *Le Rééquilibrage Est / Ouest de la Région Ile-de-France*, Paris: CCI de Paris, 16 November.

AFOPT (Association Française des Opérateurs Privés en Télécommunications) (2000) *Du pluralisme et de l'abolition des barrières en matière de télécommunications*, Paris: AFOPT.

Aldhous, P., Anderson, A., Coghlan, A., Mullins, J., O'Neill, B. and Spinney, L. (1995) 'Beneath your feet', *New Scientist (supplement)*, 1 April: 5-11.

Allen, J. (1999) 'Cities of power and influence: settled formations'. In Allen, J., Massey, D. and Pryke, M. (eds) *Unsettling Cities: Movement / Settlement*, London: Routledge, 181-227.

Amin, A. (ed) (1994) *Post-Fordism: A Reader*, Oxford: Blackwell.

Amin, A. and Graham, S. (1997) 'The ordinary city', *Transactions of the Institute of British Geographers NS* 22: 411-429.

Amin, A. and Graham, S. (1999) 'Cities of connection and disconnection'. In Allen, J., Massey, D. and Pryke, M. (eds) *Unsettling Cities: Movement / Settlement*, London: Routledge, 7-47.

Amin, A. and Thrift, N. (1992) 'Neo-Marshallian nodes in global networks', *International Journal of Urban and Regional Research* 16: 571-587.

Amin, A. and Thrift, N. (1994) 'Living in the global'. In Amin, A. and Thrift, N. (eds) *Globalisation, Institutions And Regional Development In Europe*, Oxford: Oxford University Press, 1-22.

Anderson, J. (1996) 'The shifting stage of politics: new medieval and postmodern territorialities?', *Environment and Planning D: Society and Space* 14: 133-153.

Andersson, G. (2000) 'Glassed over', *tele.com*, September 4: 74-80.

Aristote (1996) *Compte-Rendu du Voyage d'Études sur les Réseaux Métropolitains à Haut Débit au Royaume-Uni*, Paris: Aristote, December.

ART (Autorité de Régulation des Télécommunications) website, Available from: http://www.art-telecom.fr [Accessed 12 Apr 2000].

ARTESI (Agence Régionale des Technologies de l'Information et de l'Internet) (2000) *Promotional brochure*, Paris: ARTESI.

ARTESI (Agence Régionale des Technologies de l'Information et de l'Internet) website, Available from: http://www.artesi-ile-de-france.fr [Accessed 3 Feb 2000].

Ascher, F. (1995) *Métapolis ou L'Avenir des Villes*, Paris: Éditions Odile Jacob.

Ashford, D. (1989) 'British Dogmatism And French Pragmatism Revisited'. In Crouch, C. and Marquand, D. (eds) *The New Centralism: Britain Out of Step in Europe?*, Oxford: Blackwell, 77-93.

Atkinson, R. (1997) 'The digital technology revolution and the future of U.S. cities', *Journal of Urban Technology* 4(1): 81-98.

Augé, M. (1999) *Anthropology For Contemporaneous Worlds*, Stanford: Stanford University Press.

Bakis, H., Abler, R. and Roche, E. (eds) (1993) *Corporate Networks, International Telecommunications and Interdependence: Perspectives from Geography and Information Systems*, London: Belhaven Press.

Barkway, M. (2001) *The GLA and the Information Age*, London: GLA, 8 January.

Batty, M. (1990) 'Intelligent cities: using information networks to gain competitive advantage', *Environment and Planning B: Planning and Design* 17(2): 247-256.

Bax, A. (1999) *London Local Government in the Information Society* [online], London: London Research Centre. Available from: http://www.london-research.gov.uk [Accessed 12 Feb 2001].

Beauregard, R. (1995) 'Theorising The Global-Local Connection'. In Knox, P. and Taylor, P. (eds) *World Cities In A World-System*, Cambridge: Cambridge University Press, 232-248.

Beaverstock, J., Smith, R. and Taylor, P. (2000) 'World-City Network: A New Metageography?', *Annals of the Association of American Geographers* 90(1): 123-134.

Belot, L. (2001) 'Le 'Silicon Sentier' à l'heure du premier bilan', *Le Monde Interactif* [online], 15 February. Available from: http://www.lemonde.fr [Accessed 19 Feb 2001].

Bessières, H. (1989) 'La fièvre des téléports', *Telecoms Magazine* 24, May: 42-60.

Beunardeau, A. and Phan, D. (1992) 'Dix ans de libéralisation des services de télécommunications au Royaume-Uni: un tour d'horizon des changements institutionnels', *Flux* 8, Avril – Juin.

BH2 (2000) *EC does IT: Property Requirements of the IT/Telecoms Industry in the City and City Fringe*, London: Corporation of London, November.

Bijker, W. (1993) 'Do not despair: there is life after constructivism', *Science, Technology, & Human Values* 18(1): 113-138.

Boden, D. and Molotch, H. (1993) 'The compulsion of proximity'. In Friedland, R. and Boden, D. (eds) *Now/Here: Time, Space and Modernity*, Berkeley: University of California Press, 257-286.

Booth, P. and Green, H. (1993) 'Urban policy in England and Wales and in France: a comparative assessment of recent policy initiatives', *Environment And Planning C: Government And Policy* 11: 381-393.

Borja, J. and Castells, M. (1997) *Local and Global: The Management of Cities in the Information Age*, London: Earthscan.

Bourdier, J-C. (2000) *Réseaux à hauts débits: nouveaux contenus, nouveaux usages, nouveaux services – Rapport présenté à Monsieur Christian Pierret, Secrétaire d'État à l'Industrie* [online], Paris: Ministère de l'Economie, des Finances et de l'Industrie. Available from: http://www.telecom.gouv.fr/francais/activ/telecom/bourdier/rap-bourdier00.htm [Accessed 9 Jan 2001].

Boyer, C. (1995) 'The great frame-up: fantastic appearances in contemporary spatial politics'. In Liggett, H. and Perry, D. (eds) *Spatial Practices: Critical Explorations in Social / Spatial Theory*, Thousand Oaks, California: Sage, 81-109.

Boyer, C. (1996) *Cybercities: Visual Perception in the Age of Electronic Communication*, New York: Princeton Architectural Press.

Brechet, R. (1994) 'Plans et schémas directeurs en Ile-de-France', *Cahiers de l'IAURIF* 108: 49-61.

Brenner, N. (1998a) 'Global cities, glocal states: global city formation and state territorial restructuring in contemporary Europe', *Review of International Political Economy* 5: 1-37.

Brenner, N. (1998b) 'Between fixity and motion: accumulation, territorial organisation and the historical geography of spatial scales', *Environment and Planning D: Society And Space* 16: 459-481.

Brenner, N. (1999) 'Globalisation as reterritorialisation: the re-scaling of urban governance in the European Union', *Urban Studies* 36: 431-451.

Brenner, N. (2000) 'The urban question as a scale question: reflections on Henri Lefebvre, urban theory and the politics of scale', *International Journal of Urban And Regional Research* 24: 361-378.

Brindley, T., Rydin, Y. and Stoker, G. (1996) *Remaking Planning: The Politics of Urban Change: Second Edition*, London: Routledge.

Briole, A. and Lauraire, R. (1991) 'Les télécommunications dans l'aménagement du territoire: la nouvelle donne', *Netcom* 5 (1): 23-60.

Briole, A., de la Torre, L., Lauraire, R. and Négrier, E. (1993) *Les Politiques Publiques de Télécommunication en Europe du Sud: Service Public et Dynamiques Territoriales des Intérêts*, Montpellier: CEPEL.

Brotchie, J., Batty, M., Hall, P. and Newton, P. (eds) (1991) *Cities of the 21st Century: New Technologies and Spatial Systems*, Harlow: Longman Cheshire.

Brownill, S. (1990) *Developing London's Docklands: Another Great Planning Disaster?*, London: Paul Chapman.

Brunet, M.F. (2001) *L'accès des PME aux services haut-débit en Ile-de-France: quels besoins, quels moyens?*, Paris: CCI de Paris.

Budd, L. (1995) 'Globalisation, territory and strategic alliances in different financial centres', *Urban Studies* 32(2): 345-360.

Budd, L. (1997) 'Regional government and performance in France', *Regional Studies* 31(2): 187-192.

Budd, L. (1998) 'Territorial competition and globalisation: Scylla and Charybdis of European cities', *Urban Studies* 35(4): 663-685.

Buijs, S. (1994) 'Cities as networks in networks of cities: an interview with S.C. Buijs', *Flux* 15: 51-57.

Burgel, G. (1997) 'Paris: city of opposites'. In Jensen-Butler, C., Shachar, A. and van Weesep, J. (eds) *European Cities in Competition*, Aldershot: Avebury, 103-131.

Butor, P. and Parfait, Y. (1994) 'La Francilienne des Télécom', *Cahiers de l'IAURIF* 107: 58-61.

Cahiers de l'IAURIF (1999) *Enjeux économiques pour l'Ile-de-France: Du régional au local*, Paris: IAURIF, 124, 3rd trimester.

Calvino, I. (1979) *Invisible Cities*, London: Picador.

Carrez, J-F. (1991) *Le Développement des Fonctions Tertiaires Supérieures Internationales*, Rapport au Premier Ministre, Paris: La Documentation Française.

Cassé, M-C. (1995) 'Réseaux de communication et production du territoire', *Sciences de la Société* 35: 61-81, Mai.

Castells, M. (1989) *The Informational City: Information Technology, Economic Restructuring, and the Urban-Regional Process*, Oxford: Blackwell.

Castells, M. (1996) *The Information Age: Economy, Society and Culture Volume I – The Rise of the Network Society*, Oxford: Blackwell.

Castells, M. (1997) *The Information Age: Economy, Society and Culture Volume II – The Power of Identity*, Oxford: Blackwell.

Castells, M. (1998) *The Information Age: Economy, Society and Culture Volume III – End of Millennium*, Oxford: Blackwell.

Castells, M. and Hall, P. (1994) *Technopoles of the World: The Making of 21st Century Industrial Complexes*, London: Routledge.

Centre for Economics and Business Research Ltd (CEBR) and Observatoire de l'Économie et des Institutions Locales (OEIL) (1997) *Two Great Cities: A Comparison of the Economies of London and Paris*, London: Corporation of London, August.

Chambres de Commerce et d'Industrie de Paris – Ile-de-France (1999) *Les Entreprises de Paris – Ile-de-France dans le Contrat de Plan État – Région 2000-2006: Contribution des Chambres de Commerce et d'Industrie de Paris – Ile-de-France à la préparation du prochain Contrat de Plan*, Paris: CCI de Paris – Ile-de-France, January.

Chant, C. (1999) 'London and Paris'. In Goodman, D. and Chant, C. (eds) *European Cities & Technology: Industrial to Post-Industrial City*, London: Routledge, 178-224.

Chevrant-Breton, M. (1997) 'Selling the world city: a comparison of promotional strategies in Paris and London', *European Planning Studies* 5(2): 137-161.

Chinaud, R. (1999) 'Concurrence et aménagement du territoire dans le secteur des télécommunications: le rôle des collectivités territoriales', *Radiocommunications Magazine*, January.

Clarke, S. and Gaile, G. (1998) *The Work of Cities*, Minneapolis: University of Minnesota Press.

Cohen, E. (1992) *Le Colbertisme 'High-Tech': Économie des Télécom et du Grand Projet*, Paris: Hachette.

Cohen, M., Ruble, B., Tulchin, J. and Garland, A. (eds) (1996) *Preparing for the Urban Future: Global Pressures and Local Forces*, Washington: The Woodrow Wilson Center Press.

Collinge, C. (1999) 'Self-organisation of society by scale: a spatial reworking of regulation theory', *Environment And Planning D: Society And Space* 17: 557-574.

Colt Telecommunications (2000) *Colt en France, des gammes de services complètes à destination de 3 marchés cibles* [online], Press dossier, 3rd trimester, Paris: Colt. Available from: http://www.colt-telecom.fr [Accessed 11 Dec 2000].

Colt Telecommunications website, Available from: http://www.colt-telecom.com [Accessed 14 Sep 2000].

Comité Interministériel d'Aménagement et de Développement du Territoire (CIADT) (2001) *Réseaux haut débit: le CIADT confirme de nouvelles aides financières pour appuyer les projets des collectivités territoriales* [online], Press release, Paris: CIADT. Available from: http://www.premier-ministre.gouv.fr/ressources/fichiers/ciadt2001.pdf [Accessed 25 Sep 2001].

Conseil Régional d'Ile-de-France (2000a) *Contrat de Plan État – Région Ile-de-France 2000-2006* [online], Paris: Conseil Régional d'Ile-de-France. Available from: http://www.cr-ile-de-france.fr [Accessed 29 Oct 2000].

Conseil Régional d'Ile-de-France (2000b) *Création de l'Agence régionale de développement: la Région met en place un outil pour le développement économique de l'Ile-de-France* [online], Press release, 16 November, Paris: Conseil Régional d'Ile-de-France. Available from: http://www.cr-ile-de-france.fr/conseil/actu/compresse/ard.asp [Accessed 11 Dec 2000].

Conseil Régional d'Ile-de-France and IAURIF (1998) *L'Ile-de-France – Réalités Présentes, Questions d'Avenir*, Paris: Conseil Régional d'Ile-de-France and IAURIF, October.

Cooke, P. (1992) 'Global localisation in computing and communications: conclusions'. In Cooke, P., Moulaert, F., Swyngedouw, E., Weinstein, O. and Wells, P. (eds) *Towards Global Localisation*, London: UCL Press, 200-214.

Corporation of London (2000) *The Global Powerhouse: The City of London*, London: Corporation of London, January.

Corporation of London website, Available from: http://www.cityoflondon.gov.uk [Accessed 13 Mar 2000].

Costa, N. (2000) 'Cable & Wireless se renforce sur le marché des services IP en Europe', *Stratégies Télécoms & Multimédia* 192: 29, 27 July-8 September.

Coupland, A. (1992) 'Docklands: dream or disaster?'. In Thornley, A. (ed) *The Crisis of London*, London: Routledge, 149-162.

Courtois-Martignoni, F. (1999) *Les Nouvelles Technologies de l'Information et de la Communication à Paris et en Ile-de-France*, Rapport d'étape, Paris: Observatoire du Développement Économique Parisien, January.

Coutard, O. (1999) (ed) *The Governance of Large Technical Systems*, London: Routledge.

Coutard, O. (2002) 'Premium network spaces: a comment', *International Journal of Urban and Regional Research* 26(1): 166-174.

Cowan, R. (1997) *The Connected City: A New Approach to Making Cities Work*, London: Urban Initiatives.

Cox, M., Spires, R. and Wylson, D. (1994) *Electronic Highways: Advances in Business & Community Telecoms and the Implications for the South East*, Harlow: SEEDS.

Crilley, D. (1993) 'Megastructures and urban change: aesthetics, ideology and design'. In Knox, P. (ed) *The Restless Urban Landscape*, Eaglewood Cliffs: Prentice Hall, 127-164.

Curwen, P. (1997) *Restructuring Telecommunications: A Study of Europe in a Global Context*, London: Macmillan.

Curwen, P. (1999) 'Survival of the fittest: formation and development of international alliances in telecommunications', *Info* 1(2): 141-160.

Daniels, P. and Bobe, J. (1993) 'Extending the boundary of the City of London? The development of Canary Wharf', *Environment and Planning A* 25: 539-552.

DATAR (Délégation à l'Aménagement du Territoire et à l'Action Régionale) (1992) *Dossier Prospective et Territoires*, Paris: DATAR.

DATAR (Délégation à l'Aménagement du Territoire et à l'Action Régionale) (1993) *Débat National Pour L'Aménagement du Territoire*, Paris: La Documentation Francaise.

DATAR (Délégation à l'Aménagement du Territoire et à l'Action Régionale) (1999) *Pré-rapport au gouvernement sur l'état des disparités territoriales face au développement de la société de l'information*, Paris: DATAR.

DATAR (Délégation à l'Aménagement du Territoire et à l'Action Régionale) (2000) *Aménager la France de 2020: Mettre les Territoires en Mouvement*, Paris: La Documentation Française.

DATAR (Délégation à l'Aménagement du Territoire et à l'Action Régionale) and Préfecture d'Ile-de-France (1999) *Pour une métropolisation raisonnée: diagnostic socio-économique de l'Ile-de-France et du Bassin parisien*, Paris: DATAR and Préfecture d'Ile-de-France.

Dear, M. and Flusty, S. (1998) 'Postmodern urbanism', *Annals of the Association of American Geographers* 88(1): 50-72.

Debord, G. (1995) *The Society of the Spectacle*, New York: Zone Books.

de Gournay, C. (1988) 'Telephone networks in France and Great Britain'. In Tarr, J. and Dupuy, G. (eds) *Technology and the Rise of the Networked City in Europe and America*, Philadelphia: Temple University Press, 322-338.

Department of Trade and Industry (DTI) (1998) *Our Competitive Future: Building the Knowledge Driven Economy*, London: DTI.

Department of Trade and Industry (DTI) (1999) *Communications Liberalisation In The UK: Key Elements, History & Benefits*, London: DTI, January.

Docklands East London (DEAL) (2000) *The Dotcom and Telecom Opportunity*, Paper to the DEAL Board, 21 June.

Docklands East London (DEAL) promotional brochure, London: DEAL.

Doel, M. and Hubbard, P. (2002) 'Taking world cities literally: marketing the city in a global space of flows', *City* 6(3): 351-368.

Doulton, A., Harvey, M. and Wilson, R. (1997) *Telematics for London Workshop* [online], Proceedings report, February, London: London Research Centre. Available from: http://www.london-research.gov.uk [Accessed 18 Feb 2001].

DTZ (2000a) *The impact of the internet on the Central London office occupation market* [online], London: DTZ. Available from: http://www.dtz.com [Accessed 13 Dec 2000].

DTZ (2000b) *The Co-location Market In Europe* [online], London: DTZ. Available from: http://www.dtz.com [Accessed 13 Dec 2000].

Dufay, J-P. (1994) 'Telecommunications: invisible planning', *Cahiers de l'IAURIF* 107: 8.

Dupuy, G. (1991) *L'Urbanisme des Réseaux: Théories et Méthodes*, Paris: Armand Colin.

Dutton, W., Blumler, J. and Kraemer, K. (eds) (1987) *Wired Cities: Shaping the Future of Communications*, London: Cassell.

Eade, J. (ed) (1997) *Living the Global City: Globalisation as a Local Process*, London: Routledge.

east14.com (2000) *The Data Centre Project*, promotional brochure, London: RLD Ltd and Pyle Owen & Partners.

Eberlein, B. (1996) 'French centre-periphery relations and science park development: local policy initiatives and intergovernmental policymaking', *Governance: An International Journal of Policy and Administration* 9(4): 351-374.

Les Echos (2000a) 'Colt et France Télécom obtiennent le droit de s'implanter à la Défense', 24 January: 19.

Les Echos (2000b) 'France Télécom espère obtenir un allégement du contrôle des tarifs', 1 August.

Les Echos (2000c) *Dossier Immobilier: Londres*, 9 November: 59-61.

L'Eco d'Issy (2000) 'Issy, la cyber-cité', 10 (Summer): 8-11.

Edwards, M. (1996?) 'London's metropolitan economy: myth and reality', *City* 3-4: 150-157.

Eliassen, K. and Sjøvaag, M. (eds) (1999) *European Telecommunications Liberalisation*, London: Routledge.

Eliassen, K., Mason, T. and Sjøvaag, M. (1999) 'European telecommunications policies – deregulation, re-regulation or real liberalisation?'. In Eliassen, K. and Sjøvaag, M. (eds) *European Telecommunications Liberalisation*, London: Routledge, 23-37.

Elixmann, D. and Hermann, H. (1996) 'Strategic alliances in the telecommunications service sector: challenges for corporate strategy', *Communications & Stratégies* 24: 57-88.

EPAD (Établissement Public d'Aménagement de la Défense) promotional material, Paris: EPAD.

EPAD (Établissement Public d'Aménagement de la Défense) website, Available from: http://www.epaladefense.fr [Accessed 2 May 2000].

ESIS website, Available from: http://www.ll-a.fr [Accessed 2 Aug 2000].

e-territoires (2000) *e-territoires* [online] 8, 30 October 2000. Available from: http://www.afopt.asso.fr [Accessed 3 Nov 2000].

e-territoires (2001a) *e-territoires* [online] 18, 9 April 2001. Available from: http://www.afopt.asso.fr [Accessed 14 Apr 2001].

e-territoires (2001b) *e-territoires* [online] 22, 25 June 2001. Available from: http://www.afopt.asso.fr [Accessed 2 Jul 2001].

e-territoires (2001c) *e-territoires* [online] 23, 13 July 2001. Available from: http://www.afopt.asso.fr [Accessed 15 Jul 2001].

Ezechieli, C. (1998) 'Shifting boundaries: territories, networks and cities'. Paper presented at the *Telecommunications and the City* conference, University of Georgia, Athens, 21-23 March.

Fainstein, S. (1994) *The City Builders: Property, Politics, and Planning in London and New York*, Oxford: Blackwell.

Fainstein, S., Gordon, I. and Harloe, M. (eds) (1992) *Divided Cities: New York & London in the Contemporary World*, Oxford: Blackwell.

Fargette, Y. (1994) 'Introduction', *Cahiers de l'IAURIF* 107: 6.

FEDIA (Forum d'Études et de Développement des Infrastructures Alternatives) (1999) *Réponse à la consultation publique sur le développement de la concurrence sur le marché local*, Paris: FEDIA, 1 June.

Fibres Optiques Défense (FOD) promotional brochure, Paris: FOD.

Le Figaro (2000) 'Concurrence: affrontement sur les communications locales', 14 April: 3.

Filliatre, F. (2000) Presentation to the Table Ronde NTIC of the Observatoire du Développement Economique Parisien, 20 September, Paris: ODEP.

Le Fil MC des Télécoms (1997) 'Boucle locale: France Télécom, Cegetel, WorldCom et Colt Telecommunications s'unissent à Issy', 22 September: 15.

Financial Times (2000) *Europe's bandwidth bonanza: Intense competition among telecoms operators has led to a glut of fibre-optic capacity*, 10 September.

Finnie, G. (1998) 'Wired cities', *Communications Week International* 205, 18 May: 19-22.

Flanigan, B. (2000) 'Pan-European carrier market: supply and demand dynamics', *Telecoms World*, second quarter: 20-26.

Foster, J. (1999) *Docklands: Cultures in Conflict, Worlds in Collision*, London: UCL Press.

France Télécom website, Available from: http://www.francetelecom.fr.

France Télécom (2000a) *Au Service du Développement de l'Ile-de-France*, promotional brochure, Délégation pour l'Ile-de-France.

France Télécom (2000b) *L'Accès aux Nouvelles Technologies de l'Information et de la Communication en Ile-de-France: L'Offre Actuelle de France Télécom*, Presentation to the Commission NTIC du C.E.S.R., 31 May.

France Télécom (2000c) *Projet de réseaux Hauts Débits en Ile-de-France*, Presentation to the Commission NTIC du C.E.S.R., 31 May.

France Télécom (2001a) *France Télécom au Service des Collectivités Locales*, promotional brochure, Direction des Collectivités Locales.

France Télécom (2001b) *Expertel Services & FM: Histoires de Clients*, promotional brochure, Expertel Services & FM.

Friedmann, J. (1995) 'Where we stand: a decade of world city research'. In Knox, P. and Taylor, P. (eds) *World Cities in a World System*, Cambridge: Cambridge University Press, 21-47.

Friedmann, J. and Wolff, G. (1982) 'World city formation: an agenda for research and action', *International Journal of Urban and Regional Research* 6(3): 309-344.

Froment, E. and Karlin, M. (1999) 'La place financière de Paris en Europe'. In Gollain, V. and Sallez, A. (eds) *Emploi et territoires en Ile-de-France: prospective*, Paris: Éditions de l'Aube, 239-242.

Gadault, T. and Saget, E. (2001) 'Mon téléphone sans France Télécom', *L'Expansion* 646 (23 May-6 June): 56-58.

Garnham, N. (1990) 'Telecommunications in the UK', *Fabian Society Discussion Paper No 1*, London: Fabian Society.

La Gazette des Communes (2000) 'Les collectivités locales face à la fracture numérique', 26 June.

Géomatique Expert (2002) 'Tirer une fibre optique: une tâche pas si simple', 14 (March): 30-33.

Gille, L. (1995) 'La politique publique des télécommunications: 'service public' versus 'service universel''. In Musso, P. and Rallet, A. (eds) *Stratégies de Communication et Territoires*, Paris: L'Harmattan, 155-166.

Gillespie, A. (1991) 'Advanced communications networks, territorial integration and local development'. In Camagni, R. (ed) *Innovation Networks: Spatial Perspectives*, London: Belhaven Press, 214-229.

Gillespie, A. (1992) 'Communications technologies and the future of the city'. In Breheny, M. (ed) *Sustainable Development and Urban Form*, London: Pion, 67-78.

Gillespie, A. and Richardson, R. (2000) 'Teleworking and the city: myths of workplace transcendence and travel reduction'. In Wheeler, J., Aoyama, Y. and Warf, B. (eds) *Cities in the Telecommunications Age: The Fracturing of Geographies*, London: Routledge, 228-245.

Gillespie, A. and Robins, K. (1991) 'Non-universal service? Political economy and communications geography'. In Brotchie, J., Batty, M., Hall, P. and Newton, P. (eds) *Cities of the 21ˢᵗ Century: New Technologies and Spatial Systems*, Harlow: Longman Cheshire, 159-170.

Gillespie, A. and Williams, H. (1988) 'Telecommunications and the reconstruction of regional comparative advantage', *Environment and Planning A* 20: 1311-1321.

Girard, L. (2000) 'Les opérateurs investissent le marché de l'hébergement Internet', *L'Usine Nouvelle* 2753: 50-53, 26 October.

Global Crossing website, Available from: http://www.globalcrossing.com [Accessed 26 Mar 2000].

Global Switch website, Available from: http://www.londonswitch.net [Accessed 3 Feb 2000].

Gold, J. and Ward, S. (eds) (1994) *Place Promotion: The Use of Publicity and Marketing to Sell Towns and Regions*, Chichester: John Wiley & Sons.

Goodman, D. (1999) 'Two capitals: London and Paris'. In Goodman, D. and Chant, C. (eds) *European Cities & Technology: Industrial to Post-Industrial City*, London: Routledge, 73-120.

Goodwin, M. (1996) 'Governing the spaces of difference: regulation and globalisation in London', *Urban Studies* 33(8): 1395-1406.

Gorman, S. and Malecki, E. (2000) 'The networks of the Internet: an analysis of provider networks in the USA', *Telecommunications Policy* 24: 113-134.

Graham, S. (1994) 'Networking cities: telematics in urban policy – a critical review', *International Journal of Urban and Regional Research* 18(3): 416-432.

Graham, S. (1996) *Networking the City: A Comparison of Urban Telecommunications Initiatives in France and Britain*, Unpublished PhD thesis, Faculty of Science and Engineering, University of Manchester.

Graham, S. (1998) 'The end of geography or the explosion of place? Conceptualising space, place and information technology', *Progress In Human Geography* 22: 165-185.

Graham, S. (1999) 'Global grids of glass: on global cities, telecommunications and planetary urban networks', *Urban Studies* 36: 929-949.

Graham, S. (2000a) 'Introduction: cities and infrastructure networks', *International Journal of Urban and Regional Research* 24(1): 114-119.

Graham, S. (2000b) 'Constructing premium network spaces: reflections on infrastructure networks and contemporary urban development', *International Journal of Urban and Regional Research* 24(1): 183-200.

Graham, S. (2000c) *Bridging Urban Digital Divides? Urban Polarisation and Information and Communications Technologies (ICTs): Current Trends and Policy Prospects*, background paper for the United Nations Centre for Human Settlements (UNHCS).

Graham, S. (2002) 'On technology, infrastructure, and the contemporary urban condition: a response to Coutard', *International Journal of Urban and Regional Research* 26(1): 175-182.

Graham, S. and Guy, S. (2002) 'Digital space meets urban place: sociotechnologies of urban restructuring in downtown San Francisco', *City* 6(3): 369-382.

Graham, S. and Healey, P. (1999) 'Relational concepts of space and place: issues for planning theory and practice', *European Planning Studies* 7(5): 623-646.

Graham, S. and Marvin, S. (1996) *Telecommunications and the City: Electronic Spaces, Urban Places*, London: Routledge.

Graham, S. and Marvin, S. (1998) *The Richness of Cities: Working Paper 3 – Net Effects: Urban Planning and the Technological Future of Cities*, London: Comedia and Demos.

Graham, S. and Marvin, S. (2001) *Splintering Urbanism: Networked Infrastructures, Technological Mobilities and the Urban Condition*, London: Routledge.

Greater London Council (GLC) (1984) *The Future Of Telecommunications In London*, London: GLC.

Greater London Council (GLC) (1985) *London Industrial Strategy*, London: GLC.

Guerquin, E. (2001) *L'Accès aux Nouvelles Technologies de l'Information et de la Communication en Ile-de-France*, Paris: CESR, 8 February.

Guillaume, M. (2001) 'Entretien', *Dialogues (bimestriel de France Télécom)* 13: 5.

Guy, S., Graham, S. and Marvin, S. (1997) 'Splintering networks: cities and technical networks in 1990s Britain', *Urban Studies* 34(2): 191-216.

Hall, P. (1977) *The World Cities 2ⁿᵈ Edition*, London: Weidenfeld and Nicolson.

Hall, P. (1998) *Cities in Civilization: Culture, Innovation, and Urban Order*, London: Phoenix Giant.

Hantrais, L. and Mangen, S. (1996) 'Method and management of cross-national social research'. In Hantrais, L. and Mangen, S. (eds) *Cross-National Research Methods in the Social Sciences*, London: Pinter, 1-12.

Harding, A. and Le Galès, P. (1997) 'Globalisation, urban change and urban policies in Britain and France'. In Scott, A. (ed) *The Limits of Globalisation: Cases and Arguments*, London: Routledge, 181-201.

Harvey, D. (1985) 'The Geopolitics of Capitalism'. In Gregory, D. and Urry, J. (eds) *Social Relations and Spatial Structures*, London.

Harvey, D. (1989a) 'From managerialism to entrepreneurialism: the transformation in urban governance in late capitalism', *Geografiska Annaler B* 71(1): 3-17.

Harvey, D. (1989b) *The Condition of Postmodernity: An Enquiry into the Origins of Cultural Change*, Oxford: Blackwell.

Haywood, I. (1998) 'City Management Profile: London', *Cities* 15(5): 381-392.

Healey, P. (1997) *Collaborative Planning: Shaping Places in Fragmented Societies*, London: Macmillan.

Healey and Baker (2000) *European Cities Monitor: Europe's Top Cities – Executive Summary*, London: Healey and Baker.

Hebbert, M. (1998) *London: More by Fortune than Design*, Chichester: John Wiley & Sons.

Heinz, W. (ed) (1994) *Partenariats Public-Privé dans L'Aménagement Urbain*, Paris: L'Harmattan.

Hennequin, J-B. (2000) *La Région Ile-de-France et les nouvelles technologies: Définition, contexte et enjeux, programme régional d'action*, Paris: Direction du Développement Économique et de la Formation Professionnelle, Conseil Régional d'Ile-de-France, internal report, 18 October.

Henry, M. and Thépin, D. (1992) 'Téléports à l'européenne', *Cahiers de l'IAURIF* 100: 73-82.

Henry, M. and Thépin, D. (1994) 'Le téléport Paris – Ile-de-France: une force régionale', *Cahiers de l'IAURIF* 107: 13-23.

Hepworth, M. (1990) 'Planning for the information city: the challenge and response', *Urban Studies* 27: 537-558.

Hepworth, M. (1992) 'Telecommunications and the future of London', *Policy Studies* 13(2): 31-45.

Hillis, K. (1998) 'On the margins: the invisibility of communications in geography', *Progress In Human Geography* 22(4): 543-566.

Hirst, P. and Thompson, G. (1996) *Globalisation in Question: The International Economy and the Possibilities of Governance*, Cambridge: Polity Press.

Ho, K.C. (1999) 'Telecommunications and the competition for hub functions: Hong Kong, Singapore, and Sydney compared', Paper presented at the *Cities in the Global Information Society: An International Perspective* conference, Newcastle-upon-Tyne, 22-24 November.

Howkins, J. (1987) 'Putting wires in their social place'. In Dutton, W., Blumler, J. and Kraemer, K. (eds) *Wired Cities: Shaping the Future of Communications*, London: Cassell, 421-430.

Hubbard, P. and Hall, T. (1998) 'The entrepreneurial city and the "new urban politics"'. In Hall, T. and Hubbard, P. (eds) *The Entrepreneurial City: Geographies of Politics, Regime and Representation*, Chichester: John Wiley and Sons, 1-23.

Hugill, P. (1999) *Global Communications Since 1844: Geopolitics and Technology*, Baltimore: The Johns Hopkins University Press.

Hulsink, W. (1998) *Privatisation and Liberalisation in European Telecommunications: Comparing Britain, the Netherlands and France*, London: Routledge.

Humphreys, P. (1990) 'The political economy of telecommunications in France: a case study of "telematics"'. In Dyson, K. and Humphreys, P. (eds) *The Political Economy of Communications: International & European Dimensions*, London: Routledge, 198-228.

IAURIF (Institut d'Aménagement et d'Urbanisme de la Région Ile-de-France) (1998a) *Les téléports dans le monde: rappel des concepts fondateurs et évolution*, Paris: IAURIF, November.

IAURIF (Institut d'Aménagement et d'Urbanisme de la Région Ile-de-France) (1998b) *Les réseaux cablés en Ile-de-France: à l'heure des autoroutes électroniques et de la Société de l'information*, Paris: IAURIF, December.

IAURIF (Institut d'Aménagement et d'Urbanisme de la Région Ile-de-France) (2000a) 'Quelle place pour l'Ile-de-France dans la course à la 'nouvelle économie'?', *Note Rapide sur l'Economie* 181, June, Paris: IAURIF.

IAURIF (Institut d'Aménagement et d'Urbanisme de la Région Ile-de-France) (2000b) *Quelles place pour l'Ile-de-France dans la course à la 'nouvelle économie'?: Actes de la table ronde du 14 juin 2000*, Paris: IAURIF, October.

IAURIF (Institut d'Aménagement et d'Urbanisme de la Région Ile-de-France) (2001a) *40 ans en Ile-de-France: Rétrospective 1960-2000*, Paris: IAURIF.

IAURIF (Institut d'Aménagement et d'Urbanisme de la Région Ile-de-France) (2001b) *Les territoires du multimédia et de l'internet en Ile-de-France:Actes de la Table Ronde du 12 Décembre 2000*, Paris: IAURIF, February.

IDATE (1999) *Observatoire des Télécommunications et des Téléservices sur le Territoire: Rapport de travail no 1*, Paris: DATAR, February.

Ireland, J. (1994) *City Research Project – The Importance of Telecommunications to London as an International Financial Centre*, London: Corporation of London, September.

Jeannot, G., Valeyre, A. and Zarifian, P. (1999) 'Une transformation organisationnelle à France Télécom', *Flux* 36/37: 38-45.

Jenkins, C. (2001a) *Report on London's Competitiveness through the Impact of E-Business*, London: Greater London Authority, internal report.

Jenkins, C. (2001b) *E-London: An Outline of London's Opportunities and Challenges*, London: Greater London Authority.

Jensen-Butler, C. (1997) 'Competition between cities, urban performance and the role of urban policy: a theoretical framework'. In Jensen-Butler, C., Shachar, A. and van Weesep, J. (eds) *European Cities in Competition*, Aldershot: Avebury, 3-42.

Jessop, B. (1994) 'Post-Fordism and the state'. In Amin, A. (ed) *Post-Fordism: A Reader*, Oxford: Blackwell, 251-279.

Jessop, B. (1995) 'The regulation approach, governance and post-Fordism: alternative perspectives on economic and political change?', *Economy And Society* 24(3): 307-333.

Jessop, B. (1997a) 'The entrepreneurial city: re-imaging localities, redesigning economic governance, or restructuring capital?'. In Jewson, N. and MacGregor, S. (eds)

Transforming Cities: Contested Governance and New Spatial Divisions, London: Routledge, 28-41.

Jessop, B. (1997b) 'A neo-Gramscian approach to the regulation of urban regimes: accumulation strategies, hegemonic projects, and governance'. In Lauria, M. (ed) *Reconstructing Urban Regimes Theory: Regulating Urban Politics in a Global Economy*, London: Sage, 51-73.

Jessop, B. (1998) 'The narrative of enterprise and the enterprise of narrative: place marketing and the entrepreneurial city'. In Hall, T. and Hubbard, P. (eds) *The Entrepreneurial City: Geographies of Politics, Regime and Representation*, Chichester: John Wiley and Sons, 77-99.

John, P. and Cole, A. (1999) 'Political leadership in the new urban governance: Britain and France compared', *Local Government Studies* 25(4): 98-115.

Jones, M. (1997) 'Spatial selectivity of the state? The regulationist enigma and local struggles over economic governance', *Environment and Planning A* 29: 831-864.

Jospin, L. (1999) *Société de l'information: discours du Premier ministre à l'Université d'été de la communication* [online], Paris. Available from: http://www.premier-ministre.gouv.fr/PM/D260899.htm [Accessed 28 Aug 2000].

Jouve, B. and Lefèvre, C. (1999) 'Pouvoirs urbains: entreprises politiques, territoires et institutions en Europe'. In Jouve, B. and Lefèvre, C. (eds) *Villes, Métropoles: Les Nouveaux Territoires du Politique*, Paris: Anthropos, 9-44.

Kaika, M. and Swyngedouw, E. (2000) 'Fetishizing the modern city: the phantasmagoria of urban technological networks', *International Journal of Urban and Regional Research* 24(1): 120-138.

Kearns, G. and Philo, C. (eds) (1993) *Selling Places: The City as Cultural Capital, Past and Present*, Oxford: Pergamon Press.

Keating, M. (1991a) *Comparative Urban Politics: Power and the City in the United States, Canada, Britain and France*, Aldershot: Edward Elgar.

Keating, M. (1991b) 'Local economic development politics in France', *Journal of Urban Affairs* 13(4): 443-459.

Keating, M. and Midwinter, A. (1994) 'The politics of central-local grants in Britain and France', *Environment And Planning C: Government And Policy* 12: 177-194.

Keil, R. (1998) 'Globalization makes states: perspectives of local governance in the age of the world city', *Review Of International Political Economy* 5(4): 616-646.

Kennedy, R. (1991) *London: World City Moving into the 21st Century: A Research Project*, London: HMSO.

Kerbes, D. (2001) *The City of London ICT Infrastructure Review*, London: Corporation of London and PA Consulting Group, August.

Kern, S. (1983) *The Culture of Time and Space 1880-1918*, Cambridge, Massachusetts: Harvard University Press.

King, A. (1990) *Global Cities: Post-Imperialism and the Internationalisation of London*, London: Routledge.

King, A. (1993) 'Identity and difference: the internationalisation of capital and the globalisation of culture'. In Knox, P. (ed) *The Restless Urban Landscape*, Eaglewood Cliffs: Prentice Hall, 83-110.

Kitchin, R. (1998) *Cyberspace: The World in the Wires*, Chichester: John Wiley & Sons.

Knox, P. (1995) 'World cities in a world-system'. In Knox, P. and Taylor, P. (eds) *World Cities in a World System*, Cambridge: Cambridge University Press, 3-20.

Lacaze, J-P. (1994) *Paris, Urbanisme d'État et Destin d'une Ville*, Paris: Flammarion.

Lamb, J. (2000) 'Telehouses: London is leading centre among European cities', *Financial Times* [online]. Available from: www.ft.com/ftsurveys/sp9f0e.htm [Accessed 3 Jan 2001].

Land, N. (1995) 'Machines and technological complexity', *Theory, Culture and Society* 12: 131-140.

Latour, B. (1993) *We Have Never Been Modern*, London: Harvester Wheatsheaf.

Latour, B. and Hermant, E. (1998) *Paris Ville Invisible*, Paris: La Découverte.

Lecomte, D. and Gollain, V. (1992) 'La position concurrentielle de l'Ile-de-France en Europe, *Cahiers de l'IAURIF* 100: 9-23.

Lee, R. and Schmidt-Marwede, U. (1993) 'Interurban competition? Financial centres and the geography of financial production', *International Journal of Urban and Regional Research* 17(4): 492-515.

Lefebvre, A. (1998) 'Le territoire revisité par les technologies d'information et de communication'. In Lefebvre, A. and Tremblay, G. (eds) *Autoroutes de l'Information et Dynamiques Territoriales*, Toulouse and Québec: Presses Universitaires du Mirail and Presses de l'Université du Québec, 17-34.

Lefebvre, A. and Tremblay, G. (eds) (1998) *Autoroutes de l'Information et Dynamiques Territoriales*, Toulouse and Québec: Presses Universitaires du Mirail and Presses de l'Université du Québec.

Lefèvre, C. (1998) 'Metropolitan government and governance in western countries: a critical review', *International Journal Of Urban And Regional Research* 22: 9-25.

Le Galès, P. (1995) 'Du gouvernement des villes à la gouvernance urbaine', *Revue Française de Science Politique* 45(1).

Le Galès, P. (1998) 'Regulations and governance in European cities', *International Journal of Urban and Regional Research* 22: 482-506.

Lever, W. (1993) 'Competition within the European urban system', *Urban Studies* 30: 935-948.

Lipietz, A. (1994) 'The national and the regional: their autonomy vis-à-vis the capitalist world crisis'. In Palan, R. and Gills, B. (eds) *Transcending the State – Global Divide: A Neostructuralist Agenda in International Relations*, London: Lynne Rienner, 23-43.

Lipietz, A. (1995) 'Avoiding megapolisation: the battle of Ile-de-France', *European Planning Studies* 3(2): 143-154.

Llewelyn-Davies, UCL Bartlett School of Planning and Comedia (1996) *Four World Cities: A Comparative Study of London, Paris, New York and Tokyo*, London: Llewelyn-Davies.

Local Futures Group (1999a) *The Role of the City in London's Knowledge-Driven Information Economy*, London: Corporation of London, May.

Local Futures Group (1999b) *Telecommunications and Regional Development: Focus on London*, London: BT / Local Futures Group.

Local Loop Report (1999) 'France: the regulator wants to unbundle, but not the government', September: 9.

London Borough of Lewisham (2000) *Where it's @: Lewisham's Strategy for Getting Connected*, London: London Borough of Lewisham.

London Business School (1995) *City Research Project – Final Report: The Competitive Position Of London's Financial Services*, London: Corporation of London.

London Connects (2000) *Proposal for a Structure for Taking Forward the London Connects Initiative*, London: London Connects.

London Connects (2001) *Towards an E-Strategy for London*, London: London Connects.

London Development Agency (LDA) (2000) *Draft Economic Development Strategy*, London: LDA, November.

London First Centre (LFC) (1998) *London Business Briefing: Telecommunications*, London: London First Centre.

London MAN website, Available from: http://www.lonman.net.uk [Accessed 3 Feb 2000].

The London Pride Partnership (1994) *London Pride Prospectus*, London: The London Pride Partnership.

London Research Centre (LRC) (1999) *LondonNet ISB Proposal* [online], London: London Research Centre. Available from: http://www.london-research.gov.uk [Accessed 13 Feb 2001].

Longcore, T. and Rees, P. (1996) 'Information technology and downtown restructuring: the case of New York City's financial district', *Urban Geography* 17(4): 354-372.

Lorrain, D. (1991) 'Public goods and private operators in France'. In Batley, R. and Stoker, G. (eds) *Local Government in Europe*, London: Macmillan.

Luke, T. and Ó Tuathail, G. (2000) 'Thinking geopolitical space: the spatiality of war, speed and vision in the work of Paul Virilio'. In Crang, M. and Thrift, N. (eds) *Thinking Space*, London: Routledge, 360-379.

Lynch, G. (2000) 'Hubs: New York City rated as world's no. 1 telecom center', *America's Network Weekly* [online], 8 September, Available from: http://www.americasnetwork.com/enews/2000/Sept/20000908.htm [Accessed 14 Oct 2000].

MacLeod, G. and Goodwin, M. (1999) 'Space, scale and state strategy: rethinking urban and regional governance', *Progress In Human Geography* 23(4): 503-527.

Madon, S. (1997) 'Information-based global economy and socioeconomic development: the case of Bangalore', *The Information Society* 13: 227-243.

Malecki, E. (2003) 'Fiber tracks: explaining investment in fiber optic backbones', *Entrepreneurship & Regional Development* (forthcoming).

Martin, R. (1994) 'Stateless monies, global financial integration and national economic autonomy: the end of geography?'. In Corbridge, S., Thrift, N. and Martin, R. (eds) *Money, Power and Space*, Oxford: Blackwell, 253-278.

Massey, D. (1993) 'Power-geometry and a progressive sense of place'. In Bird, J., Curtis, B., Putnam, T., Robertson, G. and Tickner, L. (eds) *Mapping the Futures: Local Cultures, Global Change*, London: Routledge, 59-69.

Massey, D. (1994) *Space, Place and Gender*, Cambridge: Polity Press.

Matheron, P. (2000) *L'Urbanisme et les Technologies de l'Information et de la Communication à Paris: Économie, géographie des activités et réflexions sur le Plan d'occupation des sols*, Mémoire de DESS 'Aménagement' de l'Institut Français d'Urbanisme, Universités de Paris I-Sorbonne et de Paris VIII-Vincennes, Saint Denis.

Mattelart, A. (1994) *Mapping World Communication: War, Progress, Culture*, Minneapolis: University of Minnesota Press.

Mattelart, A. (1999) 'Mapping modernity: utopia and communications networks'. In Cosgrove, D. (ed) *Mappings*, London: Reaktion Books, 169-192.

Mattelart, A. (2000) *Networking the World, 1794-2000*, Minneapolis: University of Minnesota Press.

Maxwell, W. (1999) 'French licensing and interconnection'. In Eliassen, K. and Sjøvaag, M. (eds) *European Telecommunications Liberalisation*, London: Routledge, 128-151.

Mayer, M. (1994) 'Post-Fordist city politics'. In Amin, A. (ed) *Post-Fordism: A Reader*, Oxford: Blackwell, 316-337.

McDowell, L. (1998) 'Elites in the City of London: some methodological considerations', *Environment And Planning A* 30: 2133-2146.

MCI website, Available from: http://www.mci.com.

MCI WorldCom website, Available from: http://www.wcom.com [Accessed 8 Aug 2000].

McLuhan, M. (1964) *Understanding Media: The Extensions of Man*, London: Sphere Books.

Mercier, P-A. (1988) 'La maille et le réseau', *Quaderni* 3: 41-49.

Metromedia Fiber Networks (MMFN) website, Available from: http://www.mmfn.com [Accessed 29 Nov 2001].

Ministère de l'Aménagement du Territoire et de l'Environnement (2000) *Schéma de services collectifs de l'information et de la communication: document soumis à la consultation*, Paris: DATAR, Automne.

Mitchell, W. (1995) *City of Bits: Space, Place, and the Infobahn*, London: MIT Press.

Moreau, F. (2001) 'Interview de Jacques Douffiagues, membre du Collège de l'ART', *Stratégies Télécoms & Multimédia*, 30 May.

Morgan, K. (1989) 'Telecom strategies in Britain and France: the scope and limits of neo-liberalism and dirigisme'. In Sharp, M. and Holmes, P. (eds) *Strategies for New Technologies: Case Studies from Britain and France*, London: Phillip Allan, 19-55.

Moss, M. (1987) 'Telecommunications, world cities, and urban policy', *Urban Studies* 24: 534-546.

Mouline, A. (1996) 'Les stratégies internationales des opérateurs de télécommunications', *Communications & Stratégies* 21: 77-93.

Musso, P. (1997) *Télécommunications et Philosophie des Réseaux: La Postérité Paradoxale de Saint-Simon*, Paris: Presses Universitaires de France.

Musso, P. and Rallet, A. (eds) (1995) *Stratégies de Communication et Territoires*, Paris: L'Harmattan.

Négrier, E. (1990) 'The politics of territorial network policies: the example of videocommunications networks in France', *Flux* 1: 13-20.

Négrier, E. (1994) 'La ville saisie par les télécommunications', *Le Courrier du CNRS* 81 (La Ville): 99-101.

Négrier, E. (1995) 'Pouvoir régional et télécommunications'. In Musso, P. and Rallet, A. (eds) *Stratégies de Communication et Territoires*, Paris: L'Harmattan, 89-114.

Negroponte, N. (1995) *Being Digital*, London: Hodder & Stoughton.

Nelson, S. (2001) 'The nature of partnership in urban renewal in Paris and London', *European Planning Studies* 9(4): 483-502.

Newman, P. (2000) 'The new government of London', *Annales de Géographie* 613: 317-327.

Newman, P. and Thornley, A. (1994) 'Economic and Political Influences on Urban Planning: A Comparison of London, Paris and Berlin', *Working Papers in Land Management and Development* 20, Department of Land Management and Development, University of Reading.

Newman, P. and Thornley, A. (1996) *Urban Planning in Europe: International Competition, National Systems and Planning Projects*, London: Routledge.

Newman, P. and Thornley, A. (1997) 'Fragmentation and centralisation in the governance of London: influencing the urban policy and planning agenda', *Urban Studies* 34: 967-988.

Newman, P. and Verpraet, G. (1999) 'The impacts of partnership on urban governance: conclusions from recent European research', *Regional Studies* 33(5): 487-491.

Noam, E. (1992) *Telecommunications in Europe*, Oxford: Oxford University Press.

Noam, E. and Wolfson, A. (eds) (1997) *Globalism and Localism in Telecommunications*, Amsterdam: Elsevier.

Noin, D. and White, P. (1997) *Paris*, Chichester: John Wiley & Sons.

O'Brien, R. (1992) *Global Financial Integration: The End of Geography*, London: Pinter.

ODEP (Observatoire du Développement Economique Parisien) (2000) *Compte-Rendu de la Table Ronde NTIC du 20/09/00*, Paris: ODEP.

Office of the e-Envoy (2001) *UK online: the broadband future – An action plan to facilitate roll-out of higher bandwidth and broadband services* [online], London: Office of the e-Envoy. Available from: http://www.e-envoy.gov.uk [Accessed 4 May 2001].

Offner, J-M. (1999) 'Are there such things as small networks?'. In Coutard, O. (ed) *The Governance of Large Technical Systems*, London: Routledge, 217-238.

Offner, J-M. (2000a) '"Territorial deregulation': local authorities at risk from technical networks', *International Journal of Urban and Regional Research* 24(1): 165-182.

Offner, J-M. (2000b) 'L'action publique urbaine innovante'. In Wachter, S., Bourdin, A., Lévy, J., Offner, J-M., Padioleau, J-G., Scherrer, F. and Theys, J. *Repenser le Territoire: Un Dictionnaire Critique*, Paris: DATAR / Éditions de l'Aube, 139-155.

Offner, J-M. and Pumain, D. (eds) (1996) *Réseaux et Territoires: Significations Croisées*, Paris: Éditions de l'Aube.

Oftel (2000) *Competition Bulletin – Issue 18* [online], London: Oftel. Available from: http://www.oftel.gov.uk [Accessed 12 Jan 2001].

OTV (Observatoire des Télécommunications dans la Ville) (2000) *Dossier: Municipales 2001 – Villes moyennes et grandes villes* [online], Paris: OTV. Available from: http://www.telecomville.org/obs/inf723d.html [Accessed 3 Dec 2000].

OTV (Observatoire des Télécommunications dans la Ville) (2001a) *Dossier: TIC et collectivités locales en Ile-de-France* [online], Paris: OTV. Available from: http://www.telecomville.org/obs/inf726.html [Accessed 14 Aug 2001].

OTV (Observatoire des Télécommunications dans la Ville) (2001b) *Dossier: Boucle locale et réseaux hauts débits* [online], Paris: OTV. Available from: http://www.telecomville.org/obs/inf729.html [Accessed 14 Aug 2001].

OTV (Observatoire des Télécommunications dans la Ville) (2001c) *Dossier: La boucle locale radio* [online], Paris: OTV. Available from:
http://www.telecomville.org/obs/inf730.html [Accessed 14 Aug 2001].

OTV (Observatoire des Télécommunications dans la Ville) (2001d) *Les actes des 10èmes Rencontres de l'Observatoire: 'Vivre en réseau'* [online], Paris: OTV. Available from: http://www.telecomville.org [Accessed 19 Dec 2001].

Painter, J. (1995) 'Regulation theory, post-Fordism and urban politics'. In Judge, D., Stoker, G. and Wolman, H. (eds) *Theories of Urban Politics*, London: Sage, 276-295.

Painter, J. (1999) 'The aterritorial city: diversity, spatiality, democratisation', *Mimeo*.

Palmer, M. and Tunstall, J. (1990) *Liberating Communications: Policy-Making in France and Britain*, Oxford: Blackwell.

Parkinson, M. (1998) 'The United Kingdom'. In van den Berg, L., Braun, E. and van der Meer, J. (eds) *National Urban Policies in the European Union: Responses to Urban Issues in the Fifteen Member States*, Aldershot: Ashgate, 402-433.

Parkinson, M. (2001) 'Key challenges for European cities: achieving competitiveness, cohesion and sustainability', *Area* 33(1): 78-80.

Parkinson, M. and Le Galès, P. (1995) 'Urban policy in France and Britain: cross-Channel lessons', *Policy Studies* 16(2): 31-42.

Pascal, A. (1987) 'The vanishing city', *Urban Studies* 24: 597-603.

Pawley, M. (1998) *Terminal Architecture*, London: Reaktion Books.

Payen, F. (2000) 'Les collectivités tournées vers le haut débit', *Stratégies Télécoms & Multimédia* 192, 5 May: 10-11.

Peck, F. (1996) 'Regional development and the production of space: the role of infrastructure in the attraction of new inward investment', *Environment and Planning A* 28: 327-339.

Peck, J. and Tickell, A. (1994) 'Searching for a new institutional fix: the *after*-Fordist crisis and the global-local disorder'. In Amin, A. (ed) *Post-Fordism: A Reader*, Oxford: Blackwell, 280-315.

Pickvance, C. (1991) 'The difficulty of control and the ease of structural reform: British local government in the 1980s'. In Pickvance, C. and Preteceille, E. (eds) *State Restructuring and Local Power: A Comparative Perspective*, London: Pinter, 48-89.

Pickvance, C. and Preteceille, E. (eds) (1991a) *State Restructuring and Local Power: A Comparative Perspective*, London: Pinter.

Pickvance, C. and Preteceille, E. (1991b) 'Conclusion'. In Pickvance, C. and Preteceille, E. (eds) *State Restructuring and Local Power: A Comparative Perspective*, London: Pinter, 197-224.

Picon, A. (1998) *La Ville Territoire des Cyborgs*, Paris: Les Éditions de l'Imprimeur.

Powell, D., Page, A. and Bax, A. (1997) *London Local Government in the Information Society: A Progress Report on the London Boroughs' Electronic Information Project and Related Work* [online], London: London Research Centre. Available from: http://www.london-research.gov.uk [Accessed 10 Feb 2001].

Power, D. (2000a) 'Technology and structuring the financial district of London', *Mimeo*.

Power, D. (2000b) 'Exchanges, markets, cities and futures', *Mimeo*.

Preteceille, E. (1991) 'From centralisation to decentralisation: social restructuring and French local government'. In Pickvance, C. and Preteceille, E. (eds) *State Restructuring and Local Power: A Comparative Perspective*, London: Pinter, 123-149.

Pryke, M. (1991) 'An international city going 'global': spatial change in the City of London', *Environment and Planning D: Society and Space* 9: 197-222.

Pryke, M. (1994) 'Looking back on the space of a boom: (re)developing spatial matrices in the City of London', *Environment and Planning A* 26: 235-264.

Pryke, M. and Lee, R. (1995) 'Place your bets: towards an understanding of globalisation, socio-financial engineering and competition within a financial centre', *Urban Studies* 32(2): 329-344.

Pullen, M. (2001) *The Dotcom opportunity in Docklands & East London: The Location Case for East London* [online], Presentation, London: Insignia Richard Ellis. Available from: http://www.docklandseastlondon.com [Accessed 28 Oct 2001].

Pumain, D. (1993) 'Villes, métropoles, régions urbaines... un essai de clarification des concepts'. Paper presented at the *Métropoles et aménagement du territoire* conference, Université de Paris-Dauphine, 12-13 May.

Renaud, J-P. (1993) *Paris: Un État dans L'État?*, Paris: L'Harmattan.

Réseaux & Télécoms (1997) 'Boucle en fibres optiques à La Défense', 11 March.

Réseaux & Télécoms (2000) 'Boucle de La Défense: l'EPAD épinglé!', 24 January.

Richardson, R., Gillespie, A. and Cornford, J. (1994) 'Requiem for the teleport? The teleport as a metropolitan development and planning tool in western Europe', *Newcastle Programme on Information and Communications Technologies, Working Paper* 17.

Ricono, G. (1994) 'Marne-la-Vallée site d'aménagement téléportuaire', *Cahiers de l'IAURIF* 107: 26-30.

Rimmer, P. (1997) 'Global network firms in transport and communications: Japan's NYK, KDD and JAL?'. In Rimmer, P. (ed) *Pacific Rim Development: Integration and Globalisation in the Asia-Pacific Economy*, St Leonards, New South Wales: Allen & Unwin, 83-114.

Roberts, M., Lloyd-Jones, T., Erickson, B. and Nice, S. (1999) 'Place and space in the networked city: conceptualising the integrated metropolis', *Journal of Urban Design* 4(1): 51-66.

Robins, K. (1999) 'Foreclosing on the city? The bad idea of virtual urbanism'. In Downey, J. and McGuigan, J. (eds) *Technocities*, London: Sage, 34-59.

Rocco, A-M. (2001) 'Faut-il privatiser France Télécom?', *Le Monde Interactif* [online], 15 February. Available from: http://www.lemonde.fr [Accessed 19 Feb 2001].

Roubach, G. (1994) 'Saint-Quentin-en-Yvelines, partie prenante dans le Téléport Paris – Ile-de-France', *Cahiers de l'IAURIF* 107: 42-43.

Ryser, J. (1994) *The Future of European Capitals: Knowledge-Based Development Berlin – London – Paris*, London: Goethe-Institut.

Sallez, A. (1998) 'France'. In van den Berg, L., Braun, E. and van der Meer, J. (eds) *National Urban Policies in the European Union: Responses to Urban Issues in the Fifteen Member States*, Aldershot: Ashgate, 97-131.

Sandercock, L. (1998) *Towards Cosmopolis*, Chichester: John Wiley & Sons.

Sassen, S. (1991) *The Global City: New York, London, Tokyo*, Princeton: Princeton University Press.

Sassen, S. (2000) *Cities in a World Economy: Second Edition*, Thousand Oaks: Pine Forge Press.

Sassen, S. (ed) (2002) *Global Networks, Linked Cities*, New York: Routledge.

Savitch, H. (1988) *Post-Industrial Cities: Politics and Planning in New York, Paris, and London*, Princeton: Princeton University Press.

Schiller, D. (1999) *Digital Capitalism: Networking the Global Market System*, London: MIT Press.

Schmandt, J., Williams, F., Wilson, R. and Strover, S. (eds) (1990) *The New Urban Infrastructure: Cities and Telecommunications*, London: Praeger.

Short, J., Kim, Y., Kuss, M. and Wells, H. (1996) 'The dirty little secret of world cities research', *International Journal of Urban and Regional Research* 20: 697-717.

Silva, R.T. (2000) 'The connectivity of infrastructure networks and the urban space of São Paulo in the 1990s', *International Journal of Urban and Regional Research* 24(1): 139-164.

Sipperec (Syndicat Intercommunal de la Périphérie de Paris pour l'Électricité et les Réseaux de Communication) (2000) *Procédure de Publicité – Rapport de la Commission des Experts*, Paris: Sipperec, 18 April.

Sipperec (Syndicat Intercommunal de la Péripherie de Paris pour l'Électricité et les Réseaux de Communication) website, Available from: http://www.sipperec.fr [Accessed 12 Feb 2000].

Sipperec (Syndicat Intercommunal de la Péripherie de Paris pour l'Électricité et les Réseaux de Communication) promotional brochure, Paris: Sipperec.

Sipperec (Syndicat Intercommunal de la Péripherie de Paris pour l'Électricité et les Réseaux de Communication), AVICAM (Association des Villes pour le Câble et le Multimedia), Communauté Urbaine du Grand Nancy and District du Grand Toulouse (2000) *Les Hauts Débits au Service de Tous les Territoires: 7 Propositions pour Développer la Fibre Noire*, Paris: Sipperec.

Smith, N. (1993) 'Homeless / global: scaling places'. In Bird, J., Curtis, B., Putnam, T., Robertson, G. and Tickner, L. (eds) *Mapping the Futures: Local Cultures, Global Change*, London: Routledge, 87-119.

Smith, N. (1995) 'Remaking scale: competition and cooperation in prenational and postnational Europe'. In Eskelinen, H. and Snickars, F. (eds) *Competitive European Peripheries*, Berlin: Springer Verlag, 59-74.

Southern, A. (1997) 'Re-booting the local economy: information and communication technologies in local economic strategy', *Local Economy* 12(1): 8-25.

Southern, A. (2000) 'The political salience of the space of flows: information and communication technologies and the restructuring city'. In Wheeler, J., Aoyama, Y. and Warf, B. (eds) *Cities in the Telecommunications Age: The Fracturing of Geographies*, London: Routledge, 249-266.

Standage, T. (1998) *The Victorian Internet: The Remarkable Story of the Telegraph and the Nineteenth Century's Online Pioneers*, London: Weidenfeld & Nicolson.

Stein, J. (1999) 'The telephone: its social shaping and public negotiation in late nineteenth- and early twentieth-century London'. In Crang, M., Crang, P. and May, J. (eds) *Virtual Geographies*, London: Routledge, 44-62.

Stoker, G. (1995) 'Regime theory and urban politics'. In Judge, D., Stoker, G. and Wolman, H. (eds) *Theories of Urban Politics*, London: Sage, 54-71.

Storper, M. (1997) *The Regional World: Territorial Development in a Global Economy*, London: The Guilford Press.

Sussman, G. (1997) *Communication, Technology, and Politics in the Information Age*, London: Sage.

Sussman, G. and Lent, J. (eds) (1998) *Global Productions: Labour in the Making of the 'Information Society'*, Cresskill, New Jersey: Hampton Press.

Swyngedouw, E. (1993) 'Communication, mobility and the struggle for power over space'. In Giannopoulos, G. and Gillespie, A. (eds) *Transport and Communications Innovation in Europe*, London: Belhaven Press, 305-325.

Swyngedouw, E. (1997) 'Neither global nor local: 'glocalization' and the politics of scale'. In Cox, K. (ed) *Spaces of Globalization: Reasserting the Power of the Local*, London: Guilford Press, 137-166.

Swyngedouw, E. (2000) 'Authoritarian governance, power, and the politics of rescaling', *Environment And Planning D: Society And Space* 18: 63-76.

Swyngedouw, E., Lemattre, M. and Wells, P. (1992) 'The regional patterns of computing and communications industries in the UK and France'. In Cooke, P., Moulaert, F., Swyngedouw, E., Weinstein, O. and Wells, P. (eds) *Towards Global Localisation*, London: UCL Press, 79-128.

Taaffe, J. (1999) 'Entrants irked by slow unbundling', *Communications Week International*, 13 December: 5.

Taaffe, J. (2000) 'France Telecom charges steep price for DSL map', *Communications Week International*, 23 October: 3.

Taaffe, J. (2001) 'French operators attack new tariffs', *Communications Week International*, 19 February.

Tactis (2001) *Dans un contexte concurrentiel la diffusion du 'haut débit' sur les territoires se traduit par une France à 3 vitesses* [online], Paris: Tactis. Available from: http://www.tactis.fr [Accessed 26 Oct 2001].

Taylor, P. (1997) 'Is the United Kingdom big enough for both London and England?', *Environment and Planning A* 29: 766-770.

Taylor, P. (2000) 'World cities and territorial states under conditions of contemporary globalisation', *Political Geography* 19: 5-32.

Teather, D. (2001) 'The last mile is the longest', *Guardian Unlimited* [online], 5 February. Available from: http://www.guardian.co.uk/Archive/Article/0,4273,4130623,00.html [Accessed 12 Jun 2001].

Telcité website, Available from: http://www.telcite.fr [Accessed 23 Sep 1999].

Thatcher, M. (1992) 'Telecommunications in Britain and France: the impact of national institutions', *Communications & Strategies* 6: 35-61.

Thatcher, M. (1999) *The Politics of Telecommunications: National Institutions, Convergence, and Change in Britain and France*, Oxford: Oxford University Press.

Théry, G. (1994) *Les Autoroutes de l'Information: Rapport au Premier Ministre*, Paris: La Documentation Française.

Thrift, N. (1994) 'On the social and cultural determinants of international financial centres: the case of the City of London'. In Corbridge, S., Thrift, N. and Martin, R. (eds) *Money, Power and Space*, Oxford: Blackwell, 327-355.

Thrift, N. (1996a) 'Inhuman geographies: landscapes of speed, light and power'. In Thrift, N. *Spatial Formations*, London: Sage, 256-310.

Thrift, N. (1996b) 'New urban eras and old technological fears: reconfiguring the goodwill of electronic things', *Urban Studies* 33: 1463-1493.

Thrift, N. (2000) 'Less mystery, more imagination: the future of the City of London', *Environment and Planning A* 32: 381-384.

Thrift, N. and Leyshon, A. (1994) 'A phantom state? The de-traditionalisation of money, the international financial system and international financial centres', *Political Geography* 13(4): 299-327.

Toffler, A. (1981) *The Third Wave*, New York: Morrow.

Tomaney, J. (2001) 'The new governance of London: a case of post-democracy?', *City* 5(2): 225-248.

Trench, R. and Hillman, E. (1993) *London Under London: A Subterranean Guide 2nd Edition*, London: John Murray.

La Tribune (2000a) 'La justice casse le monopole de Cegetel à La Défense', 24 January.

La Tribune (2000b) 'La boucle locale radio suscite des vocations dans les télécoms', 31 January: 2.

Urban Geography (1999) Review symposium: 'Postmodern Urbanism', 20(5): 393-416.

Veltz, P. (1996) *Mondialisation Villes et Territoires: L'Économie d'Archipel*, Paris: Presses Universitaires de France.

Veyret, A. (2000) 'Quelle politique publique locale mettre en oeuvre dans la net-économie?', *IDEE Télécom Débat* [online] 8, 12 May. Available from: http://www.telecom.gouv.fr/francais/activ/telecom/cridee8.htm [Accessed 18 Oct 2000].

Ville d'Issy-les-Moulineaux (1999) 'Issy l'aud@cieuse', promotional brochure.

Vinchon, M-C. (1998a) 'Communication and French local governments: towards a new form of local public policy?', *Mimeo*.

Vinchon, M-C. (1998b) *Politiques Locales de Télécommunications: Un Panorama des Pratiques Territoriales et Institutionnelles*, Paris: Groupement de Recherche Réseaux, Ecole Nationale des Ponts et Chaussées.

Virilio, P. (1986) *Speed and Politics*, New York: Semiotext(e).

Virilio, P. (1987) 'The overexposed city', *Zone* 1(2): 14-31.

Virilio, P. (1997) *Open Sky*, London: Verso.

Wachter, S. (2000) 'L'agenda de l'aménagement du territoire entre prospective et rétrospective (1980, 1990, 2000)'. In Wachter, S., Bourdin, A., Lévy, J., Offner, J-M., Padioleau, J-G., Scherrer, F. and Theys, J. *Repenser le Territoire: Un Dictionnaire Critique*, Paris: DATAR / Éditions de l'Aube, 81-111.

Wachter, S., Bourdin, A., Lévy, J., Offner, J-M., Padioleau, J-G., Scherrer, F. and Theys, J. (2000) *Repenser le Territoire: Un Dictionnaire Critique*, Paris: DATAR / Éditions de l'Aube.

Ward, A. (1987) 'The roles of both London teleports on the developments of the Docklands'. In Noothoven Van Goor, J. and Lefcoe, G. (eds) *Teleports in the Information Age*, Amsterdam: Elsevier, 291-293.

Warf, B. (1995) 'Telecommunications and the changing geographies of knowledge transmission in the late 20th century', *Urban Studies* 32(2): 361-378.

Warf, B. (1998) 'Reach out and touch someone: AT&T's global operations in the 1990s', *Professional Geographer* 50(2): 255-267.

Webber, M. (1964) 'The urban place and the non place urban realm'. In Webber, M., Dyckman, J., Foley, D., Guttenberg, A., Wheaton, W. and Whurster, C. (eds) *Explorations Into Urban Structure*, Philadelphia: University of Pennsylvania Press, 79-153.

Webber, M. (1968) 'The post-city age', *Daedalus* 97(4): 1091-1110.

Wells, P. and Cooke, P. (1991) 'The geography of international strategic alliances in the telecommunications industry: the cases of Cable and Wireless, Ericsson, and Fujitsu', *Environment And Planning A* 23: 87-106.

Wheeler, J., Aoyama, Y. and Warf, B. (eds) (2000) *Cities in the Telecommunications Age: The Fracturing of Geographies*, London: Routledge.

White, E. (2001) *The Flâneur: A Stroll Through the Paradoxes of Paris*, London: Bloomsbury.

Wilson, G. (1976) *The Old Telegraphs*, London: Phillimore.

Wollmann, H. (2000) 'Local government systems: from historic divergence towards convergence? Great Britain, France, and Germany as comparative cases in point', *Environment and Planning C: Government and Policy* 18: 33-55.

Wood, A. (1998) 'Questions of scale in the entrepreneurial city'. In Hall, T. and Hubbard, P. (eds) *The Entrepreneurial City: Geographies of Politics, Regime and Representation*, Chichester: John Wiley and Sons, 275-284.

The Yankee Group Europe (2001) *European Telecom in 2001: Reversal of Fortune*, London: The Yankee Group Europe, January.

Yeomans, K. (1991) *Electronic Communications in London: The Local Authority Role*, London: London Research Centre and BT, December.

Yin, R. (1994) *Case Study Research: Design and Methods: Second Edition*, London: Sage.

01 Informatique (1999) 'Une concentration très forte', 2 July: 7.

Zukin, S. (1992) 'The city as a landscape of power: London and New York as global financial capitals'. In Budd, L. and Whimster, S. (eds) *Global Finance and Urban Living*, London: Routledge, 195-223.

Index